FUNDAMENTALS
— OF —
HAZARDOUS
WASTE SITE
REMEDIATION

— Kathleen Sellers —

LEWIS PUBLISHERS
Boca Raton London New York Washington, D.C.

Library of Congress Cataloging-in-Publication Data

Sellers, Kathleen.
    Fundamentals of hazardous waste site remediation / Kathleen
Sellers.
       p. cm.
    Includes bibliographical references and index.
    ISBN 1-56670-281-X (alk. paper)
    1. Hazardous waste site remediation.   I. Title.
TD1030.S45   1998
628.5--dc21
                                                     98-4419
                                                        CIP

© 1999 by CRC Press LLC
Lewis Publishers is an imprint of CRC Press LLC

No claim to original U.S. Government works
International Standard Book Number 1-56670-281-X
Library of Congress Card Number 98-4419
Printed in the United States of America 1 2 3 4 5 6 7 8 9 0
Printed on acid-free paper

The Author

During over twelve years of experience, Ms. Sellers has focused on developing and designing solutions to hazardous waste problems. Ms. Sellers received a B.S. in Chemistry from Indiana University and a M.S. in Environmental Engineering from the University of Massachusetts. She began her career with the U.S. Environmental Protection Agency and currently works for Ogden Environmental and Energy Services in Westford, MA.

Ms. Sellers has worked on a variety of sites, including former chemical manufacturing facilities, coal-gasification waste sites, and landfills. Her work, under both state and federal regulatory programs, has included development of site investigation plans, management of focused field investigations, feasibility studies, remedial design, negotiations with regulatory agencies, and public involvement programs. Ms. Sellers has also helped clients with their industrial wastewater treatment plant operations. She is a registered professional engineer in Massachusetts and Vermont.

Acknowledgments

This book is based on the syllabus and handouts that I prepared for a course that I taught at Tufts University in 1992. Kim M. Henry helped me to prepare the original profile of the hypothetical Ace Solvent Recovery Facility, discussed in Chapter 1, for that class. It is used herein with her permission. Doug Fannon drafted the figures in Chapter 1 based on Ms. Henry's original drawings.

I am grateful to the engineers and scientists who reviewed portions of this book in draft and provided helpful comments: Paul Anderson, Ph.D.; Ralph Baker, Ph.D.; Kristine Carbonneau, P.E.; Jay Clausen, P.G.; Diane Heineman; Kim M. Henry, LSP; David R. Palmer; Steven M. Pause, P.E., LSP; Kaela Sotnik; Douglas Simmons; and Allison Nightingale. The editorial team at CRC Press provided me encouragement and advice throughout this project. Most of all, thanks to my husband Dave, for his support.

Finally, I dedicate this book to my mothers and my daughters, and my other teachers.

Contents

CHAPTER 1

Introduction

1.1 ACE SOLVENT RECOVERY FACILITY

Consider the Ace Solvent Recovery (ASR) facility.* Between 1974 and 1990 the facility accepted waste solvents for reclamation. The facility had interim status under the Resource Conservation and Recovery Act (RCRA), but never received a final Part B permit. In 1990, the facility lost interim status due to the failure to obtain required insurance policies or alternative forms of liability coverage. Shortly thereafter, the owner/operator declared bankruptcy and abandoned the facility. The owner/operator cannot be located.

In 1990, the state environmental agency removed drums and tanks of waste solvents from the site. The removal action also included decontamination of the building and removal of trash and debris from the site. The agency began a site investigation in the late 1990s. The results of that investigation are summarized below.

1.1.1 Former Operations

ASR accepted waste solvents, primarily tetrachloroethylene (also known as perchloroethylene, or PCE) and trichloroethylene (TCE). The facility stored waste solvents in tanks and drums on concrete pads in an area west of the office/warehouse building (see Figure 1.1). These solvents were pumped through above-ground lines to stills located in the same area. Distilled solvents were collected in tanks and drums for off-site shipment. Still bottoms were often incinerated; they were occasionally disposed of in a shallow lagoon near the western edge of the property. According to a former employee, the lagoon was backfilled in the early 1980s.

For a short period in the early 1970s, ASR also accepted used transformers. According to a former employee, these transformers were stored on open ground to the south of the building (see Figure 1.1). When the transformers were stripped to recover the metals, the transformer oils were simply drained onto the ground.

---

* Note: The Ace Solvent Recovery facility is a hypothetical site, created to illustrate many of the common problems found at hazardous waste sites. It was created with the help of Kim M. Henry, LSP, and is used with her permission.

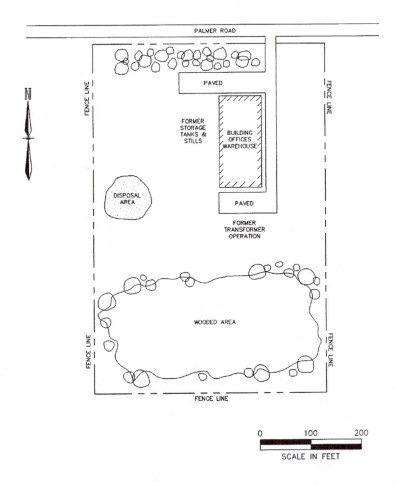

Figure 1.1   Site plan, Ace Solvent Recovery facility.

1.1.2   Scope of the Site Investigation

As shown on Figure 1.2, 35 borings were installed on the site, 7 of which were completed as monitoring wells. Two additional wells were installed off-site.

Soil samples were collected from the unsaturated zone and analyzed for volatile organic compounds (VOCs), semivolatile organic compounds (SVOCs), metals, and polychlorinated biphenyls (PCBs). Soil samples were also tested using the Toxicity Characteristic Leaching Procedure (TCLP).

Groundwater samples collected from the monitoring wells were analyzed for VOCs, SVOCs, metals, and PCBs. Samples were also analyzed for conventional wastewater parameters used to evaluate and design groundwater treatment systems. Pumping tests were performed in selected wells.

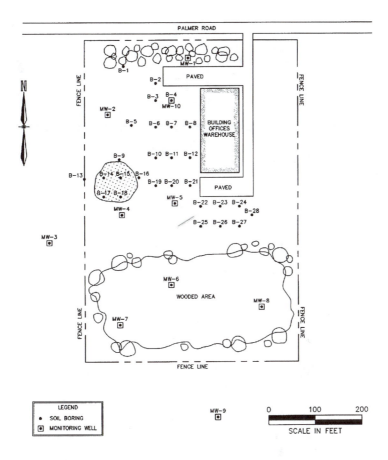

Figure 1.2 Locations of borings and monitoring wells, Ace Solvent Recovery facility.

1.1.3 Results of the Site Investigation — Soils

Boring logs were used to develop cross sections showing the site stratigraphy, as shown on Figures 1.3 and 1.4. The surface of the site is covered with a thin layer of fill. A layer of till, composed of fine sand and silt, underlies the fill in the northern portion of the site in a layer approximately 20 feet thick; in the southern portion of the site, the fill is underlain by glacial outwash composed of sand and gravel. Fractured bedrock is present beneath the site and slopes to the south. The bedrock is highly fractured to a depth approximately 20 feet below the top of rock. Investigations to a greater depth have not been conducted at this site.

Screening tests of the soil samples collected from the till indicate that approximately 15% of the soil volume exceeded 0.5 in. in size. The material used as backfill in the former lagoon contained some debris. An estimated 20% of the fill in the lagoon exceeded 0.5 in. in size.

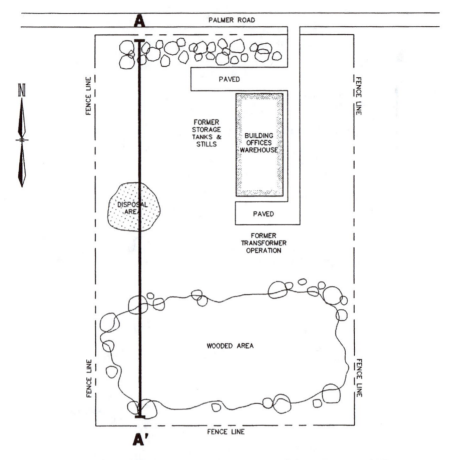

Figure 1.3 Location of cross-section A-A′, Ace Solvent Recovery facility.

The soil near the location of the former tanks and stills is contaminated with PCE and TCE, as is the soil in the former disposal lagoon. Concentrations of total VOCs ranged up to 730 mg/kg. Tables 1.1 and 1.2 summarize the pertinent data. As indicated in Table 1.2, several of the samples failed the TCLP test for VOCs. The soil near the former transformer operation is contaminated with PCBs at concentrations of up to 158 mg/kg. These data are also provided in Table 1.1.

1.1.4   Results of the Site Investigation — Groundwater

Water levels were measured in all monitoring wells during May and August to evaluate the seasonal fluctuation. Table 1.3 indicates the water table elevations in the shallow unconsolidated aquifer during the August monitoring period, and Figure 1.5 shows the groundwater contours. Groundwater elevations were approximately 1 foot higher throughout the site during May. Although groundwater elevations are not provided for the deeper bedrock aquifer, groundwater flow in the bedrock occurs in the same direction as the unconsolidated aquifer, to the south.

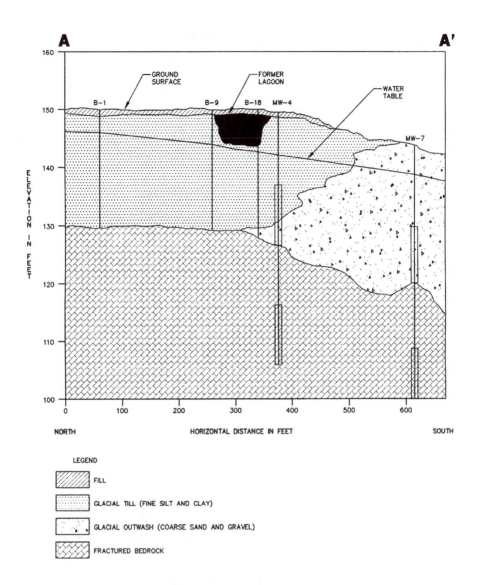

Figure 1.4 Cross-section A-A', Ace Solvent Recovery facility.

The glacial till layer in the northern portion of the site has a hydraulic conductivity of 1×10^{-4} cm/sec and an effective porosity of 10%. Pumping test data indicate that a well fully screened in this material has a radius of influence of approximately 25 feet and can yield a maximum flow of approximately 0.2 gallons per minute (gpm).

The glacial outwash (sand and gravel layer) in the southern portion of the site has a hydraulic conductivity of 1×10^{-1} cm/sec and an effective porosity of 25%. Pumping test data indicate that a well fully screened in this material has a radius of influence of approximately 500 feet and can yield a maximum flow of approximately 200 gpm.

**Table 1.1   Summary of Soil Data
                 ASR Site**

Location	Depth (ft)	PCE	TCE	PCB
B-1	1 to 2	ND	ND	2
B-2	0 to 2	ND	ND	ND
	4 to 6	2	1	NA
B-3	0 to 2	ND	ND	NA
	4 to 6	4	ND	NA
B-4	0 to 2	1	ND	NA
	2 to 4	10	5	NA
B-5	2 to 4	3	2	NA
B-6	0 to 2	31	32	NA
	4 to 6	112	87	NA
B-7	0 to 2	50	29	NA
	2 to 4	171	150	NA
	4 to 6	208	170	NA
B-8	0 to 2	39	17	NA
	4 to 6	128	83	NA
B-9	0 to 2	2	4	NA
	4 to 6	15	23	NA
B-10	0 to 2	7	6	NA
	2 to 4	49	23	NA
	4 to 6	107	87	NA
B-11	0 to 2	43	22	NA
	2 to 4	199	104	NA
	4 to 6	218	143	NA
B-12	0 to 2	12	16	NA
	2 to 4	72	87	NA
B-13	0 to 2	ND	ND	NA
	4 to 6	2	1	NA
B-14	0 to 2	2	3	NA
	4 to 6	67	76	NA
B-15	0 to 2	19	9	ND
	2 to 4	270	324	3
	4 to 6	295	367	NA
B-16	0 to 2	3	4	NA
	4 to 6	176	198	NA
B-17	0 to 2	11	13	NA
	2 to 4	263	427	NA
B-18	2 to 4	198	429	8
	4 to 6	227	503	14
B-19	0 to 2	4	2	NA
	4 to 6	23	16	NA
B-20	0 to 2	14	8	ND
	4 to 6	57	63	ND
B-21	0 to 2	5	4	2
	4 to 6	38	32	ND
B-22	0 to 2	1	ND	10
	2 to 4	3	1	2
B-23	0 to 2	NA	NA	158
	2 to 4	NA	NA	43
	4 to 6	NA	NA	1
B-24	0 to 2	NA	NA	67
	4 to 6	NA	NA	12

Table 1.1   Summary of Soil Data
            ASR Site (continued)

Location	Depth (ft)	PCE	TCE	PCB
B-25	0 to 2	NA	NA	5
	2 to 4	NA	NA	1
B-26	0 to 2	NA	NA	23
	2 to 4	NA	NA	7
B-27	0 to 2	NA	NA	12
	4 to 6	NA	NA	3
B-28	0 to 2	NA	NA	4
MW-1	0 to 2	ND	ND	1
	2 to 4	ND	ND	ND
MW-3	0 to 2	ND	ND	3
	2 to 4	ND	ND	ND
MW-4	2 to 4	2	2	NA
	4 to 6	12	17	NA

Notes: All data in mg/kg. ND, not detected; NA,
not analyzed.

Table 1.2   Summary of TCLP Data
            ASR Site

Location	Sample depth (ft)	PCE	TCE
B-11	2 to 4	4.1	2.3
B-15	2 to 4	4.3	5.6
B-18	2 to 4	3.5	7.4
B-20	2 to 4	ND	ND
Regulatory standard		0.7	0.5

Notes: All data in mg/L. ND, not detected.

Table 1.3   Groundwater Elevation
            Data — Overburden Wells
            August 28, 1997
            ASR Site

Well	Water level (ft MSL)
MW-1S	145.4
MW-2S	144.4
MW-3S	141.5
MW-4S	142.2
MW-5S	142.5
MW-6S	139.9
MW-7S	138.7
MW-8S	138.8
MW-9S	135.5
MW-10S	144.6

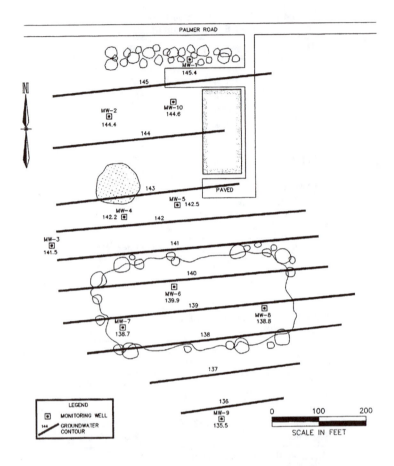

Figure 1.5 Groundwater contour map, Ace Solvent Recovery facility.

Analytical data from shallow monitoring wells screened at the bottom of the unconsolidated aquifer (indicated with an "S" in the well identification number) and from wells screened approximately 20 feet below the top of the bedrock aquifer (indicated with an "R" in the well identification number) are provided in Tables 1.4 and 1.5. Based on these data, chlorinated organic compounds have migrated off-site in the bedrock aquifer, but are still confined to the site property in the shallow aquifer.

Does the ASR site need to be cleaned up? How?

1.2 WHAT DO I DO NOW?

Every site presents a unique challenge. Nonetheless, every site can be analyzed in terms of a limited set of fundamental features, and a cost-effective solution based

Table 1.4 Groundwater Quality Data
Organic Compounds
ASR Site

Well	TCE	PCE	1,2-DCE	VC	PCB
MW-1S	ND	ND	ND	ND	ND
MW-1R	ND	ND	ND	ND	ND
MW-2S	0.002	0.002	ND	ND	NA
MW-2R	0.003	0.001	ND	ND	NA
MW-3S	ND	ND	ND	ND	NA
MW-3R	0.001	ND	ND	ND	NA
MW-4S	97.080	82.140	37.450	4.730	NA
MW-4R	120.000	98.000	42.691	3.781	NA
MW-5S	52.430	41.870	29.800	1.370	0.001
MW-5R	61.587	37.894	12.798	0.119	0.001
MW-6S	0.015	0.008	ND	0.003	ND
MW-6R	12.489	7.623	1.847	0.241	ND
MW-7S	0.005	0.001	ND	ND	NA
MW-7R	3.417	1.245	1.007	0.125	NA
MW-8S	ND	ND	ND	ND	ND
MW-8R	ND	0.001	ND	ND	ND
MW-9S	ND	ND	ND	ND	ND
MW-9R	0.090	0.025	0.045	ND	NA
MW-10S	0.950	0.035	0.009	ND	ND
MW-10R	1.145	1.497	ND	ND	NA

Notes: All data in mg/L. Abbreviations: DCE, dichloroethylene; VC, vinyl chloride; ND, not detected; NA, not analyzed.

Table 1.5 Groundwater Quality Data
Inorganic Parameters
ASR Site

Well	Iron	Manganese	Calcium	Magnesium	TDS	TSS	pH
MW-2S	10	4	23	10	250	32	7
MW-2R	5	1	15	6	187	1	6.8
MW-4S	24	10	61	26	230	152	6.7
MW-4R	7	ND	12	5	163	ND	6.9
MW-5S	19	7	17	9	212	56	7.1
MW-10S	15	4	26	15	243	41	6.6

Notes: All data except pH in mg/L. Abbreviations: TDS, total dissolved solids; TSS, total suspended solids.

on those features. Remedial solutions for hazardous waste sites depend on four broad considerations:

- The types of contaminants and their chemical characteristics determine whether and how a site requires remediation. The physical properties of a contaminant determine how it moves through or persists in the environment. These properties also determine how a contaminant may be treated. The chemical structure of a contaminant determines its toxicity to people and to the ecosystem, and thus the extent to which a site must be cleaned up.

- Environmental regulations specify the procedures to be used to study a waste site and select a remedy; requirements for handling and disposing of certain types of waste and wastewater (groundwater); and many other features of waste site remediation.
- The location, use, and physical features of a site fundamentally affect the clean-up goal and the methods used to reach that goal. Consider the following examples. Far more stringent clean-up requirements are typically imposed on a site in a residential area than in an industrial area. A community group may oppose on-site incineration. Contamination in a wetland cannot be managed simply by containment beneath an impermeable cap. Man-made features such as process piping or underground utilities can hinder excavation and increase remediation costs if they must be relocated. These examples show how fundamentally the location, use, and physical features of a site can affect the extent and type of remediation required.
- The natural characteristics of soil and groundwater often determine the systems needed to treat soil or groundwater. While the chemical characteristics of a waste indicate how it *may* be treated, the physical characteristics of a waste determine how it *can* be treated. For soils or sludges, materials handling — that is, converting the waste into a form that can be treated, and conveying it from the source through the treatment process — is the critical step in most treatment processes. Characteristics such as viscosity, for sludges, and particle size, for soil or other solid waste, can complicate or preclude treatment by certain methods. Certain natural components of groundwater can also complicate site remediation. Many groundwaters contain naturally high levels of hardness, iron, or manganese. These metals can precipitate, forming solid particles that can clog or foul common types of treatment equipment. Pretreatment to modify a natural characteristic of soil, sludge, or groundwater can be a large, expensive component of remediation.
- The capabilities of remedial technologies under site-specific conditions can vary widely. Remediation technologies act to contain contamination, separate it from the soil or groundwater, or destroy the contaminants. A limited set of physical mechanisms forms the basis for most of the hundreds of technologies that have been proposed or used to clean up hazardous waste sites. Understanding these mechanisms and the properties of contaminants enables an engineer to analyze and ultimately to apply an unfamiliar technology. The use of a particular technology depends, in addition to the site-specific factors described above, on its availability, proven or projected reliability, status (bench-scale, pilot-scale, or full-scale), and cost.

Chapter 2 of this book provides a framework for the reader to begin this fundamental analysis. Chapter 3 describes groundwater remediation technologies, and Chapter 4, soil remediation technologies. The objective of this book is to help the reader to develop a multidisciplinary approach to solving hazardous waste problems.

Rules of thumb are presented throughout this book. Rules of thumb should always be applied with a strong dose of common sense; they are guidelines, based on practical experience, and not absolute rules.

Problems

1.1 Reread the description of the ASR site. Write down all of the technical terms that you do not understand.

1.2 This exercise is intended to help you to explore and define your personal views on environmental protection, as personal opinions almost invariably affect the selection and design of remediation options for hazardous waste sites. Consider a hypothetical site contaminated with tetrachlorogoop (TCG), a potential human carcinogen which has caused cancer in laboratory animals at relatively low doses. TCG is leaching from the soil into the groundwater, which is migrating toward a wetland. Four remediation options are under consideration for the soil:

- Option 1: Cap the contaminated soil, to prevent people from contacting the soil and to minimize the leaching of TCG to groundwater. Fence the site. Maintain the cap and fence indefinitely. Estimated cost: $120,000.
- Option 2: Excavate contaminated soil, including some soil beneath the groundwater table, so that residual contamination is at a level which should not pose a human health threat or leach to groundwater above trace levels. Dispose of the soil in a hazardous waste landfill in another state. Backfill the excavation. Estimated cost: $800,000.
- Option 3: This option is similar to option 2, but the contaminated soil would be incinerated at an off-site facility to destroy the TCG. Estimated cost: $3.5 million.
- Option 4: This option is similar to option 3, except that the site would be restored to background conditions rather than to risk-based clean-up levels. (Since TCG is not a naturally occurring compound, "background conditions" means that no TCG could be detected in soil that would remain at the site.) Estimated cost: $25 million.

What should be done to clean up the site? Consider two alternative site settings; determine what should be done in each case, and present your rationale.

[a] The site is on the grounds of a large chemical-manufacturing facility. TCG was accidentally spilled from a storage tank. The multinational corporation which owns the facility made marginal profits in the last year.

[b] The site is located behind an elementary school. Contamination resulted from activities in the maintenance shed. Because of a large influx of new students, the school department is low on funds and may need to lay off art and music teachers, close the computer lab, and limit after-school sports. Funding for the school comes from property taxes in this affluent town; the townspeople have voted down property tax increases for ten years in a row.

[c] Did your answers for (a) and (b) differ? If so, why?

CHAPTER **2**

Basic Principles

Developing and designing solutions to hazardous waste problems requires knowledge of many disciplines: chemistry, toxicology, chemical engineering, civil engineering, hydrogeology, and environmental regulations. Most problems are solved by teams of people in different disciplines. However, each team member needs to have a basic understanding of the fundamental principles of each discipline to work effectively. This chapter summarizes those principles. For more detailed information, see the references cited in this chapter.

2.1 THE SIGNIFICANCE OF NUMBERS

The concepts of *significant figures* and *accuracy* are central to understanding analytical data, risk-assessment conclusions, groundwater-transport models, or cost estimates. Their importance is best explained by example:

- A risk-assessment specialist has calculated that the average concentration of contamination in the soil on a site should not exceed 6.2389 mg/kg tetrachlorogoop (TCG). He has also calculated that the average of TCG concentrations measured on the site is 6.24 mg/kg. Does the site need to be cleaned up?
- For another project, an engineer has prepared two preliminary cost estimates for site remediation. She estimated that option 1 would cost $1,326,279.86 and option 2 would cost $1,343,623.67. Based on these estimates, which option would be the most cost-effective choice?

The TCG-contaminated site may not need to be cleaned up, and neither option 1 nor option 2 may be more cost effective based on the estimates presented. The answers depend upon the accuracy of the estimates and the number of significant figures.

Accuracy is a simple concept: the more accurate an estimate, the closer it is to the true value. Increasing accuracy generally requires increasing cost, whether for collecting more data, running more sophisticated chemical analyses, or spending more labor hours working on the project.

13

The concept of significant figures is less intuitive. Calculated values are often reported simply as they appear on a calculator or spreadsheet which has been set to show an arbitrary number of figures after the decimal point. Those figures may not all be meaningful (i.e., significant figures). Depending on the assumptions used in the calculation, the clean-up level described above could be 6 mg/kg, 6.2 mg/kg, or 6.2389 mg/kg. Thus, the number of significant figures in that calculation and in the estimate of the average concentration on site may determine whether or not the site must be remediated.

How are significant figures determined? The number of significant figures is at least the number of nonzero digits. Zeros may be significant figures, unless they are used only to locate the decimal point.[1] For example, the following numbers all have three significant figures: 6.24, 0.000233, 1.00×10^{-6}, and 12,700.

The number of significant figures in a calculated value — such as the clean-up level or cost estimates in the examples above — depends on the uncertainty in the numbers used to calculate that value. Two types of uncertainty are considered: *absolute uncertainty* and *relative uncertainty*. The absolute uncertainty in a number is a single unit of measurement. The relative uncertainty is the ratio of the absolute uncertainty to the total quantity.[1] For example, a pump priced at $1499 may be a line item in a cost estimate. The absolute uncertainty in that cost is $1, and the relative uncertainty is 1/1499.

When numbers are added or subtracted, the absolute uncertainty in the result equals the largest absolute uncertainty among the components.[1] In less exact terms, the answer should contain digits only as far as the first column containing significant figures.[2,3] For example, consider the sum of 1200 + 1399.4 + 0.43. Without considering significant figures, the answer is 2599.83. However, only two figures are significant and thus the answer is 2600.

When numbers are multiplied or divided, the relative uncertainty in the result equals the largest relative uncertainty among the components.[1] Less precisely, the number of significant figures in the answer should be the same as the number of significant figures in the least exact factor.[2,3] Using this guideline, the product of 0.4375, 1.0, and 6 is 3, not 2.625.

When a series of calculations are performed, the final result should be rounded off to the proper number of significant figures, and not the intermediate results. Exact numbers, such as the number of tanks in a groundwater treatment system or the number of soil samples used to calculate an average concentration, are not considered in determining the number of significant figures.[2,3]

Examples 2.1 and 2.2 later in this chapter illustrate the use of significant figures.

2.2 CHEMISTRY OF HAZARDOUS MATERIALS

Remediation technologies attempt to

- separate contaminants from soil, water, or other media;
- isolate or immobilize contaminants; or
- destroy contaminants.

The behavior of chemical compounds determines how they may be remediated. Common contaminants fall into one of a relatively small number of categories. The information that follows describes the general behavior of common contaminants. The discussion begins with a review of basic terminology.

2.2.1 Terminology

Chemical compounds are classified as *organic* or *inorganic*. Organic compounds are based on carbon atoms (with the exception of cyanide, which is inorganic). They can be *anthropogenic* (man-made) or naturally occurring compounds. Inorganic compounds are non-carbon based, and commonly include metals. Whether organic or inorganic, the chemical structure of a compound determines its polarity or charge, solubility, volatility, and ability to react with other substances.

2.2.1.1 Units

Concentration is commonly reported in a variety of units. A *mole* is the mass of a chemical compound which contains 6.022×10^6 molecules. This unit is commonly used to balance chemical equations and to characterize concentrations in air and water. Chemists work with units of *gram moles*, which are calculated by dividing the mass in grams by the gram molecular weight. Chemical engineers often use *pound moles*, calculated by dividing the mass of a compound in pounds by the molecular weight. Gram moles are used in this book rather than pound moles. The *molecular weight*, sometimes called the *formula weight*, is determined by summing the *atomic weights* of the constituent atoms. Atomic weights are summarized in the Periodic Table or tabulated in chemistry texts and references.

Contaminant concentrations in water are commonly reported in units of *milligrams per liter (mg/L)* or *parts per million (ppm)*, and *micrograms per liter (μg/L)* or *parts per billion (ppb)*. Data for highly contaminated water samples are sometimes reported as a percentage. The concentration may be converted to *molarity*, or moles per liter (abbreviated *M*), for certain calculations. Units of *mole fraction (x_i)*, which is the ratio of the moles of substance *I* to the total moles of all components, are also used in some calculations.

Example 2.1: An analytical laboratory has reported that a groundwater sample contains 273 μg/L benzene (chemical formula C_6H_6). Convert the units to mg/L, ppm, ppb, %, *M*, and x_i.

Convert to mg/L:

1000 μg = 1 mg

(273 μg/L) · (1 mg/1000 μg) = 0.273 mg/L

Convert to ppm:

$$1\ ppm\ (by\ weight) = \frac{1\ mg\ contaminant}{10^6\ mg\ water} = 0.0001\ percent\ by\ weight \quad (2.1)$$

Density of water (STP) = 998 g/L = 998,000 mg/L

In general, 1 mg contaminant/1 L water = 1 mg/998,000 mg = 1.002004008 ppm;

$(0.273 \text{ mg/L}) \cdot (1 \text{ L}/998,000 \text{ mg}) = 0.2735470942$ ppm

Since the original concentration had only three significant figures, the concentration in ppm should only have three significant figures: 273 µg/L = 0.274 ppm

Convert to ppb:

1 µg/L = 1 ppb, analogous to the derivation above for ppm

so 273 µg/L = 274 ppb

Conversion to percentage:

From Equation 2.1, 1 ppm = 0.0001%

$(0.274 \text{ ppm}) \cdot (0.0001\%/1 \text{ ppm}) = 0.0000274\%$

Conversion to M:

The chemical formula of benzene is C_6H_6.

The atomic weight of carbon is 12.011; the atomic weight of hydrogen is 1.0079.

The molecular weight of benzene is $6 \cdot (12.011 + 1.0079) = 78.1134$ g/mole.

$(0.273 \text{ mg/L}) \cdot (1 \text{ g}/1000 \text{ mg})/(78.1134 \text{ g/mole}) = 3.494913926 \times 10^{-6}$ M.

Since the least number of significant figures in the numbers used in this calculation is 3, the molarity should be rounded off to 3.49×10^{-6} M.

Conversion to x_i:

Assume that the groundwater contains primarily water, which has a chemical formula of H_2O and a molecular weight of $(2 \cdot 1.0079) + 15.9994 = 18.0152$ g/mole. Using the density of water and the molarity of the solution (see above), $(3.494913926 \times 10^{-6}$ moles benzene/L water) $\cdot$ (1 L/998 g water) $\cdot$ (18.0152 g/mole water) = 6.3087748 $\times$ 10^{-8}, or, considering the number of significant figures, 6.31×10^{-8}. (Note that the mole fraction is unitless.)

Contaminant concentrations in solids such as soil, sediment, and sludge are presented in units of *milligrams per kilogram (mg/kg), micrograms per kilogram (µg/kg),* and percent. Concentrations may be expressed as *wet weight,* which includes the weight of the water in the sample as it was collected from the environment, or as *dry weight.* Common analytical methods for environmental analysis often report the data on a dry-weight basis, but may report it on a wet-weight basis (e.g., Reference 4).

Contaminant concentrations in air are commonly expressed in *parts per million by volume (ppmv), parts per billion by volume (ppbv),* mole fraction (y_i), and *partial pressure (p_i).* Highly concentrated contaminants may be reported as percentages.

Example 2.2: An air sample at 25°C and 1 atm contains 210 milligrams of benzene per cubic meter (mg/m³). Convert this concentration to ppmv, y_i, and p_i.

From Example 2.1, the molecular weight of benzene is 78.1134 g/mole. Air is approximately 78.1% nitrogen (N_2, molecular weight 28.0134 g/mole) by volume and 20.9%

oxygen (O_2, molecular weight 31.9988 g/mole). Therefore, the molecular weight of air is approximately $(0.781 \cdot 28.0134) + (0.209 \cdot 31.9988) = 28.566215$ g/mole.

Convert to ppmv:

At 25°C and 1 atm, 1 gram mole of an ideal gas has a volume of 22.41 liters. In general,

$$1\ ppmv = \frac{1\ volume\ of\ contaminant}{10^6\ volumes\ air,\ including\ contaminant}$$

$$= 0.0001\ percent\ by\ volume$$

(2.2)

Then,

$$\left(\frac{210\ mg\ benzene}{1\ m^3\ air}\right)\left(\frac{10^{-3}\ g}{mg}\right)\left(\frac{1\ mol\ benzene}{78.1134\ g}\right)\left(\frac{22.41\ L}{mol}\right)\left(\frac{10^{-3}\ m^3}{10^6}\right)$$

$$= 60.24702549\ \frac{volumes\ benzene}{volumes\ air}$$

or, considering significant figures, 60.2 ppmv.

Convert to y_i:

$$\left(\frac{210\ mg\ benzene}{1\ m^3\ air}\right)\left(\frac{10^{-3}\ g}{mg}\right)\left(\frac{1\ mol\ benzene}{78.1134\ g}\right)\left(\frac{22.41\ L}{mol}\right)\left(\frac{10^{-3}\ m^3}{L}\right)$$

$$= 0.0000602\ \frac{mol\ benzene}{mol\ air}$$

Convert to p_i:

In general,

$$p_i = p_{total} \cdot y_i$$

(2.3)

Since $p_{tot} = 1$ atm, $p_i = 0.0000602$ atm.

2.2.1.2 Charge and Polarity

Charge and polarity are similar properties which help to determine the solubility of a compound and its ability to adsorb to solids.

An *ion* is a charged atom or fragment of a molecule. Most inorganic chemicals dissolved in groundwater are ions. Certain organic compounds can also be ions under certain conditions. A *cation* is positively charged because it lacks one or more electrons from its structure. Examples include Ca^{+2}, Cr^{+3}, and Cr^{+6}. An *anion* is negatively charged because it carries excess electrons. Examples include Cl^- and

$Cr_2O_3^{-2}$. The *valence state* of an ion describes the degree of charge (e.g., +2, +3). It changes in redox reactions. (See the discussion of redox reactions in Section 2.2.1.6.)

Nonionic species can be characterized as *polar* or *nonpolar*. In a polar molecule, electrons are not shared equally between atoms. As a result, different parts of the molecule have slight positive and slight negative electronic charges. Certain atoms are more electronegative than other atoms; such atoms tend to attract electrons and thus carry a slight negative charge. Electronegativity increases as follows:[5]

$$F > O > Cl, \ N > Br > C, \ H$$

Polar compounds are more soluble in water than similar nonpolar compounds.

2.2.1.3 Solubility

Solubility, in common environmental applications, is the amount of a compound that can dissolve in water. In general, solubility decreases with increasing size for similar organic molecules, and polar compounds are more soluble than similar nonpolar compounds.

For ionic compounds, solubility is characterized by the *solubility product constant* (K_{sp}). For the equilibrium reaction

$$A_x B_y \rightleftharpoons A^{+y} + B^{-x}$$

(where A and B are ions and x and y are integers)

$$K_{sp} = [A]^x \cdot [B]^y \tag{2.4}$$

The solubility of a particular compound depends on temperature and pressure. It also depends on the types and concentrations of other compounds in the water. The presence of a relatively high concentration (i.e., percentage level) of one organic compound can increase the solubility of a second compound by *cosolvation*.[6,7] The pH of the water affects the solubility of ionic compounds. *Complexation* can increase the solubility of ionic compounds; for example, mercury can form soluble complexes with chloride in solution such as $HgCl_4^{-2}$. Complexation can increase the solubility of a compound above the concentration predicted by the solubility product constant for the hydroxide, for example.

Certain types of compounds with relatively low solubility may exist as a separate phase in an aquifer. *Light nonaqueous phase liquids (LNAPL)*, also known as *floaters*, are separate-phase liquids which are less dense than water. These substances have a *specific gravity* (i.e., density relative to that of water) of less than one. *Dense nonaqueous phase liquids (DNAPL)*, or *sinkers*, are more dense than water. They have a specific gravity greater than one. LNAPL and DNAPL are discussed further in Section 2.3.4.1.

2.2.1.4 Sorption

Sorption mechanisms include *adsorption*, which is the attraction of a chemical compound to a solid surface, and *absorption*, which is the penetration of a contaminant into a solid. Sorption is a reversible reaction (i.e., *desorption*). Sorption phenomena affect the fate and transport of groundwater contaminants, extraction of contaminants in groundwater or soil vapors, and certain forms of treatment for contaminated soil, groundwater, or air.

Sorption depends on the properties of both the contaminant and the solid. Critical contaminant properties include solubility and ionic or polar character. Important properties of the solid include the homogeneity, permeability and porosity, surface area, surface charge, and organic carbon content.[8]

Charged and polar species tend to sorb to charged and polar surfaces. Nonpolar compounds tend to sorb to nonpolar solids, typically those high in organic carbon content.[8] Large nonpolar compounds are *hydrophobic*. Such compounds have low solubilities in water and tend to sorb to solids. The *octanol–water partition coefficient* (K_{ow}) characterizes the hydrophobicity of a molecule. This coefficient is determined by measuring the concentrations of the compound which partition into an octanol phase and a water phase after mixing:

$$K_{ow} = \frac{Concentration_{Octanol}}{Concentration_{Water}} \tag{2.5}$$

The higher the valuate of K_{ow}, the more hydrophobic the compound. For compounds in the same family (e.g., chlorobenzene, dichlorobenzene, etc.), K_{ow} increases with the size of the molecule.

The physical and chemical characteristics of the solid also affect sorption:

- Variations in the properties of the solid, such as the different organic carbon content in layers of sand and silt in an aquifer, affect the distribution of sorbed materials.
- Increased permeability and porosity enhance fluid flow through the solid, and thus the opportunities for sorption to and desorption from the solid.
- Increased surface area on the solid increases the area available for sorption.
- Surface charge is often represented by the *cation exchange capacity* (CEC). The CEC of a solid, measured experimentally, characterizes its ability to attract and hold positively charged ions to negatively charged functional groups on the solid.
- The organic carbon content of soil sorbs contaminants, primarily hydrophobic contaminants.

Sorption of a nonpolar organic contaminant to a particular solid is sometimes characterized by a *partition coefficient* (K_p) or a *carbon-normalized partition coefficient* (K_{oc}):

$$K_p = \frac{Concentration_{solid\,phase}}{Concentration_{solution}} \; [=] \; \frac{L}{kg} \tag{2.6}$$

$$K_{oc} = \frac{K_p}{Fraction\ Organic\ Carbon\ of\ Solid} \qquad (2.7)$$

These parameters are often used in groundwater modeling of the fate and transport of contaminants. K_p can be estimated from K_{ow} or measured in laboratory experiments.[8]

K_p can be related to estimates of the mass of a contaminant which adsorbs to a solid using the *Freundlich isotherm*:

$$S = K_p \cdot C^{\frac{1}{n}} \qquad (2.8)$$

or

$$\log S = \log K_p + \tfrac{1}{n} \log C \qquad (2.9)$$

where: S = mass of sorbed contaminant per mass of solid (sometimes abbreviated x/m rather than S), in units of mg/kg,

n = experimentally determined factor for a given contaminant and solid, and

C = contaminant concentration, in units of mg/L.

The Freundlich isotherm is applicable at a relatively low concentration (i.e., $\leq 10^{-5}$ M or half the solubility, whichever is lower).[7] Note also that the Freundlich isotherm represents an equilibrium relationship; however, contaminant sorption from flowing groundwater is typically not at equilibrium.

2.2.1.5 Volatility

Volatility refers to the tendency of a compound to move from a solid or liquid phase to a gaseous phase. In general, volatility decreases with increasing size for organic molecules. Metals, except for certain forms of mercury, lead, and arsenic, are not volatile under typical environmental conditions.

The *Henry's law constant (k_H)* characterizes the volatility of a compound in water *at a specified temperature and air pressure.* For common environmental applications, Henry's law states that *at equilibrium* the concentration of a volatile compound "i" in water (x_i) is proportional to the partial pressure of the compound in the air (p_i):[9]

$$k_{H,i} = \frac{p_i}{x_i} \qquad (2.10)$$

The higher the Henry's law constant, the more readily the compound will volatilize from water to air.

k_H may be obtained from literature tabulations or calculated from the solubility and vapor pressure of a compound at the specified temperature and pressure. (The *vapor pressure* is the pressure exerted by a vapor in equilibrium with a pure liquid.) Literature values of k_H are given in different units: commonly, unitless, in atm · m³/mole, and atm (actually, atm/mole fraction).

Example 2.3:   The solubility and vapor pressure of naphthalene ($C_{10}H_8$) are 31 mg/L and 0.2336 mmHg, respectively, at approximately 25°C. Calculate the Henry's law constant in three units: unitless, atm · m³/mol, and atm.

Unit conversion of concentrations (see Example 2.1):

Molecular weight of naphthalene = $(10 · 12.011) + (8 · 1.0079) = 128.1732$ g/mol.

x_i = (31 mg/L) · (1 g/1000 mg) · (1 mol/128.1732 g) = 0.0002418602329 mol/L, or

x_i = (0.0002418602329 mol/L) · (1 L water/998 g) · (18.0152 g/mol water)

   = $4.365892252 \times 10^{-6}$ mole fraction (unitless)

p_i = (0.2336 mmHg) · (1 atm/760 mmHg) = 0.0003073684211 atm

Calculate k_H unitless:

This calculation is based on the *ideal gas law*:

$$PV = nRT \qquad\qquad (2.11)$$

where: P = pressure,
        V = gas volume,
        n = number of moles of gas,
        R = universal gas constant (0.0821 L · atm/mol K), and
        T = *absolute temperature*, in degrees *Kelvin* (°C + 273) or *Rankine* (°F + 460).

Substituting the ideal gas law into Equation 2.10 gives:

$$k_H \cong \frac{p_i}{x_i RT} \qquad\qquad (2.12)$$

Then,

k_H = (0.0003073684211 atm)/[(0.0002418602329 mol/L) · (0.0821 L · atm/mol K) · (298 K)]
   = 0.051943996 [unitless]

Since the solubility used to calculate k_H had two significant figures,

k_H = 0.052

Calculate k_H units of atm · m³/mole:

k_H = (0.0003073684211 atm)/[(0.0002418602329 mol/L) · 1000 L/m³]
   = 0.001270851386 atm · m³/mol

Since the solubility used to calculate k_H had two significant figures,

k_H = 0.0013 atm · m³/mol

Calculate k_H units of atm:

k_H = (0.0003073684211 atm)/($4.365892252 \times 10^{-6}$) = 70.40220095

Since the solubility used to calculate k_H had two significant figures,

k_H = 70 atm.

While Henry's law describes the behavior of a contaminant dissolved in water at a relatively low concentration, it does not accurately describe the volatilization of, for example, benzene in a layer of gasoline which is floating on the water table.

Raoult's law describes the equilibrium between a gas mixture which obeys the ideal gas law and an ideal liquid solution. Ideal liquids do not actually exist, but certain solutions approach ideality. Ideality would require that the molecules of the different components of the solution be similar in size, structure, and chemical nature. Mixtures of similar organic compounds can be modeled as ideal solutions. Raoult's law states that *at equilibrium* the mole fraction of a volatile compound (x_i) in an ideal solution times the vapor pressure (p_i^*) equals the partial pressure of the compound in the air (p_i):[9]

$$p_i^* \cdot x_i = p_i \qquad (2.13)$$

2.2.1.6 Common Reactions

Common reactions in the environment and in treatment systems include *oxidation*, *reduction*, and *precipitation* reactions.

Oxidation and reduction are complimentary reactions, sometimes referred to as *redox* reactions. In order for one compound to be oxidized, another must be reduced: the compound which is oxidized loses electrons, and the compound which is reduced loses electrons. An *oxidizing agent* or electron acceptor is a chemical species which readily accepts electrons; a *reducing agent* readily donates electrons. The *redox potential (E_h)* measures the relative oxidation of an aqueous solution.

Common oxidation reactions used in hazardous waste treatment and described in this book include combustion, ozonation, and aerobic biodegradation. In the combustion of methane, for example, methane is oxidized and oxygen is reduced:

$$CH_4 + 2O_2 \rightarrow CO_2 + 2H_2O$$

Reductive dehalogenation and reduction of Cr^{+6} to Cr^{+3} are common reduction reactions used to treat contaminated groundwater.

Precipitation reactions produce a solid by mixing two dissolved substances which react to form an insoluble precipitate. Precipitation reactions are commonly used to remove metals from groundwater. For example, dissolved iron can be removed from groundwater by oxidation and precipitation as the hydroxide:

$$Fe^{2+} + 2\,H^+ + \tfrac{1}{2}O_2 \rightarrow Fe^{3+} + H_2O$$

$$Fe^{3+} + 3OH^- \rightarrow Fe(OH)_3\,(\downarrow)$$

(The symbol "($\downarrow$)" indicates that the compound forms a solid that can precipitate; it is also indicated by the designation "(ppt)".)

Table 2.1   Characteristics of Petroleum Hydrocarbons

Source	Carbon Range[12]	Boiling Point Range (°C)[13]	Vapor Pressure[13] (mmHg, at 60°F)[a]	Specific Gravity (at 15.6°C)[13]
Gasoline	C_4 to C_{10}	37 to 185	7	0.7330
Kerosene	C_8 to C_{16}	160 to 285	2	0.7800
#2 oil	C_8 to C_{21}	150 to 400	1	0.8490
#4 oil	C_8 to C_{30}	150 to 500	12	0.9020
#6 oil	C_{12} to C_{30}	300 to 500	2.8	0.9650

[a] Based on compounds at the lower end of the boiling point range, at the atomizing temperature or 60°F, whichever is lowest.

2.2.2   Common Contaminants

Common organic contaminants include alkanes; *volatile organic compounds (VOCs)* such as aromatic hydrocarbons, chlorinated ethanes and ethenes, and ketones; *semivolatile organic compounds (SVOCs)* such as polynuclear aromatic hydrocarbons (PAH); and polychlorinated biphenyls (PCB). Common inorganic contaminants include metals and cyanides. The U.S. Environmental Protection Agency (EPA) has estimated that VOCs contaminate 73% of all federal Superfund sites, SVOCs contaminate 73% of sites, and metals 72%; over 70% of the sites are contaminated with chemicals in two or more of these categories.[10] The properties of the most common contaminants[11] are listed below.

2.2.2.1   Alkanes

Alkanes consist of carbon and hydrogen atoms; the molecules are based on chains of carbon atoms joined by a single bond. Alkanes comprise the major components of petroleum products such as gasoline and fuel oil, and may be characterized as *petroleum hydrocarbons* or *total petroleum hydrocarbons* (Table 2.1). They may be found in the environment as a result of spills or leaks from storage tanks. Certain hydrocarbons, such as hexane (C_6H_{14}), are also used in pure form as solvents.

As the number of carbon atoms in the chain increases, the boiling point increases, the solubility decreases, and the Henry's Law Constant decreases. Density also increases with the size of the molecule. Shorter-chain hydrocarbons can be LNAPLs; heavier hydrocarbons can be DNAPLs. Hydrocarbons can be readily oxidized.

2.2.2.2   Aromatic Hydrocarbons

Aromatic hydrocarbons include benzene, toluene, ethyl benzene, xylenes (commonly abbreviated together as BTEX), phenol, and cresols. These compounds are all based on a ring of six carbon atoms. *Aromatic compounds* are distinguished by the nature of the bonds between the carbon atoms in the ring, which is between a single and a double bond. (The aromatic structure is designated by drawing a circle, which represents these bonds, inside a hexagon; see Figure 2.1).

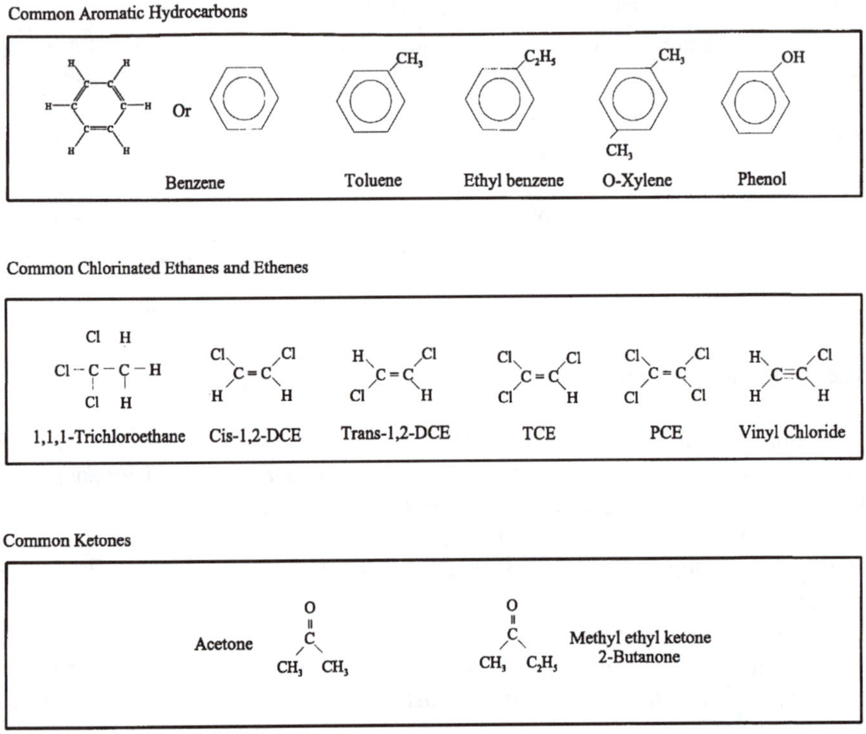

Common Aromatic Hydrocarbons

| Benzene | Toluene | Ethyl benzene | O-Xylene | Phenol |

Common Chlorinated Ethanes and Ethenes

| 1,1,1-Trichloroethane | Cis-1,2-DCE | Trans-1,2-DCE | TCE | PCE | Vinyl Chloride |

Common Ketones

Figure 2.1 Chemical structures of common volatile organic compounds.

BTEX compounds occur in petroleum products such as gasoline and in coal and wood tars, and are used as solvents. The individual compounds are also used as solvents. Phenol is used as a disinfectant, and in chemical manufacturing. Cresols are found in wood tar and have been used in disinfectants and other applications.

In general, aromatic hydrocarbons are characterized by low solubility in water and high volatility. BTEX compounds can form LNAPL on groundwater. The polarity of the hydroxyl functional group (–OH) on phenols and cresols makes those compounds more polar, more soluble, and less volatile than the analogous BTEX compounds. The nature of the bond in the aromatic ring makes aromatic hydrocarbons more stable than alkanes with a corresponding number of carbon atoms.

Table 2.2 summarizes the properties of some common aromatic compounds. Note the effect that various functional groups, such as –OH and –Cl, have on analogous compounds.

2.2.2.3 Chlorinated Methanes, Ethanes, and Ethenes

Chlorinated methanes consist of one carbon atom and one or more chlorine atoms. Chlorinated ethanes are based on a chain of two carbon atoms joined by a single bond. Chlorinated ethenes, or ethylenes, are based on a chain of two carbon atoms joined by a double bond. These compounds also commonly contain one to four chlorine atoms, as shown in Figure 2.1.

These compounds are (or were) commonly used as solvents and degreasing agents. Common users are electroplating facilities and dry cleaners.

Table 2.2 summarizes the properties of some common chlorinated methanes, ethanes, and ethenes. These compounds are characterized by their volatility and relatively low water solubility. Chlorinated ethanes and ethenes, in particular, are more dense than water and can form DNAPL when spilled in the environment. These chlorinated compounds are relatively stable, although they can be oxidized and reduced.

2.2.2.4 Ketones

Ketones contain an oxygen atom double-bonded to a carbon atom, as shown in Figure 2.1. Ketones such as acetone, methyl ethyl ketone, and methyl isobutyl ketone are used as solvents.

Because of their polarity, ketones are more soluble in water and less volatile than analogous alkanes. (See Table 2.2.) Ketones can be readily oxidized.

2.2.2.5 Polynuclear Aromatic Hydrocarbons (PAHs)

PAH compounds consist of fused aromatic rings (see Figure 2.2). PAHs are products of incomplete combustion of wood, gasoline, coal, and oil. As a result, PAHs are often found in soil in industrialized areas. PAHs are also found in creosote used for wood preserving, coal tar, and wood tar.

Table 2.3 summarizes the properties of some common PAHs. In general, PAHs have low volatility, water solubility, and reactivity; these characteristics decrease with the number of rings (i.e., size of the molecule).

2.2.2.6 Polychlorinated Biphenyls (PCBs)

The biphenyl molecule, which forms the basis for all PCBs, consists of two benzene rings joined by a single bond. PCBs include many different *congeners*, or biphenyl molecules with different numbers of chlorine atoms in different positions on the biphenyl molecule (see Figure 2.2 for an example). *Arochlors* were commercial mixtures of PCBs. Arochlor names indicate the amount of chlorine in a mixture. For example, Arochlor 1242 is 42% chlorine, and Arochlor 2154 is 54% chlorine by weight.

PCBs have been used as dielectric fluid in transformers and as flame retardants. The U.S. has banned production of PCBs since the late 1970s. Because of their persistence in the environment, however, PCBs are still found in soils and sediments.

Table 2.4 summarizes the characteristics of several Arochlors. PCBs are characterized by low water solubility. PCBs also have low volatility. Solubility and volatility decrease with the degree of chlorination. PCB molecules are highly stable, and do not react readily.

Table 2.2 Properties of Common Volatile Organic Compounds

Type	Compound	Boiling point (°C)[14]	Vapor pressure (mmHg) (at 25°C)[14]	Specific gravity (at 20°C)[15]	Solubility (mg/L) (at 25°C)[14]	Henry's Law constant atm · m³/mole (at 25°C)[14]	log K_{ow}[14]
Aromatic hydrocarbons	Benzene (C_6H_6)	80.09	95.2	0.8765	1,790	0.00555	2.13
	Toluene ($C_6H_5CH_3$)	110.63	28.4	0.8669	526	0.00664	2.73
	Ethyl benzene ($C_6H_5C_2H_5$)	136.193	9.60	0.867	169	0.00788	3.15
	o-Xylene ($C_6H_4(CH_3)_2$)	144.429	6.61	0.880	178	0.00518	3.12
	p-Xylene	138.359	8.84	0.8610	162	0.00753	3.15
	m-Xylene	139.12	8.29	0.8642	161	0.00718	3.20
	Chlorobenzene (C_6H_5Cl)	131.687	12.0	1.106	498	0.00377	2.84
	Phenol (C_6H_5OH)	181.839	0.350	1.0576 (at 41°C)	82,800	3.33×10^{-7}	1.46
	Pentachlorophenol (C_6Cl_5OH)[a]	309–310	0.000110	1.978	12,000	2.45×10^{-8}	0.94
Chlorinated methanes	Methylene chloride (CH_2Cl_2)	39.64	435	1.325	13,000	0.00325	1.25
	Chloroform ($CHCl_3$)	61.18	231	1.485	1,000,000	3.97×10^{-5}	-0.24
	Carbon tetrachloride (CCl_4)	76.7	115	1.5947	793	0.0276	2.83
Chlorinated ethanes and ethenes	Vinyl chloride (C_2H_3Cl)	-13.37	2,980	—	8,800	0.0378	1.62
	1,1-Dichloroethylene ($C_2H_2Cl_2$)	31.56	600	1.214	2,250	0.0261	2.13
	cis-1,2-Dichloroethylene ($C_2H_4Cl_2$)	60.2	201	1.285	3,500	0.00408	186
	trans-1,2-Dichloroethylene ($C_2H_4Cl_2$)	47.7	331	1.257	6,300	0.00938	2.09
	1,1-Dichloroethane ($C_2H_4Cl_2$)	57.30	227	1.175	5,060	0.00562	1.79
	1,1,1-Trichloroethane ($C_2H_3Cl_3$)	74.08	124	1.325	1,500	0.0172	2.49
	Trichloroethylene (C_2HCl_3)	86.7	69.0	1.462	1,100	0.00985	2.42
	Tetrachloroethylene (C_2Cl_4)	121.07	18.6	1.625	150	0.0177	3.4
Ketones	Acetone ($CO(CH_3)_2$)	56.07	231	0.7899	1,000,000	3.97×10^{-5}	-0.24
	Methyl ethyl ketone; 2-butanone ($COCH_3C_2H_5$)	79.6	95.3	0.805	223,000	5.69×10^{-5}	0.29

[a] Pentachlorophenol is not a volatile organic compound, but was included in this table to illustrate the properties of a series of compounds.

All data except specific gravity are reprinted with permission from Howard, P. H. and Meylar, W. M., Eds., Handbook of Physical Properties of Organic Chemicals, CRC Lewis Publishers, Boca Raton, FL, 1997.

Common PAHs

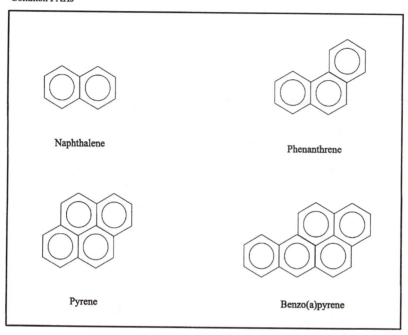

Naphthalene

Phenanthrene

Pyrene

Benzo(a)pyrene

Polychlorinated Biphenyls

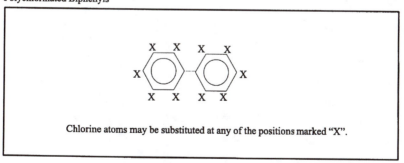

Chlorine atoms may be substituted at any of the positions marked "X".

Figure 2.2   Chemical structures of common PAH, PCB.

2.2.2.7 Metals and Other Inorganics

Metals and other inorganics may contaminate the environment from a variety of sources, as shown in Table 2.5. With the exception of mercury, and lead and arsenic under certain conditions, metals are not volatile. Their solubility depends upon the metal, its ionic charge, the pH of the solution, and the presence of other ions or compounds. As elements, metals cannot be destroyed.

Table 2.3　Properties of Common PAHs

Compound	Number of rings	Boiling point (°C)[14]	Vapor pressure (mmHg) (at 25°C)[14]	Specific gravity (20°C)[15]	Solubility (mg/L) (at 20°C)[14]	Henry's Law constant atm · m³/mole (at 25°C)[14]	log K_{ow}[14]
Naphthalene ($C_{10}H_8$)	2	217.94	8.50×10^{-2}	1.162	31.0	4.83×10^{-3}	3.30
Phenanthrene ($C_{14}H_{10}$)	3	340	1.12×10^{-4}	0.98	1.15	3.60×10^{-5}	4.46
Pyrene ($C_{16}H_{10}$)	4	393	2.46×10^{-6}	1.271	1.35×10^{-7}	1.10×10^{-5}	4.88
Benzo(a)pyrene ($C_{20}H_{12}$)	5	311	5.49×10^{-9}		0.00162	1.45×10^{-6}	5.97

All data except specific gravity are reprinted with permission from Howard, P. H. and Meylar, W. M., Eds., *Handbook of Physical Properties of Organic Chemicals*, CRC Lewis Publishers, Boca Raton, FL, 1997.

Table 2.4　Characteristics of Arochlors

Arochlor	Molecular weight[16]	Vapor pressure (mmHg) (at 25°C)[15]	Specific gravity[15]	Solubility (mg/L) (at 25°C)[15]	Henry's Law constant atm · m³/mole (at 25°C)[15]	Log K_{ow}[15]
Arochlor 1242	266.5	4.06×10^{-4}	1.385 at 20°C	0.45	3.4×10^{-4}	5.58
Arochlor 1254	328.4	7.71×10^{-5}	1.538 at 25°C	0.012	2.8×10^{-4}	6.03
Arochlor 1260	377.5	4.05×10^{-5}	1.44 at 30°C	0.0027	3.4×10^{-4}	7.15

Note:　As Arochlors are mixtures of compounds, these data are imprecise.

Table 2.5   Common Inorganics

Inorganics	Typical Sources	Notes
Arsenic (As)	Certain herbicides, pesticides, fungicides; certain dyestuffs; wood preservatives[17]	Arsenic is not a true metal, but is usually grouped with the metals for analytical purposes; arsenic can be found in four valence states (-3, 0, $+3$, $+5$); As^{+5} is typical in aerobic soil environments; arsenic can exist as either inorganic or organic compounds[18,19]
Cadmium (Cd)	Certain alloys; electroplating; pigments; stabilizers for polyvinyl plastics; Ni-Cd batteries; combustion of fossil fuels[20]	Cadmium is divalent ($+2$) in stable compounds[21]
Chromium (Cr)	Metallurgical industry; electroplating; certain pigments; tanning; wood preservatives; chromium also enters the environment from combustion of coal[22]	Chromium is found in the environment in two forms: Cr(III) (low solubility, low toxicity), and Cr(VI) (more soluble, higher toxicity)
Lead (Pb)	Paint; leaded gasoline (formerly); lead–acid batteries; various other uses; also, combustion of coal[23]	Lead can exist in three valence states (0, $+2$, $+4$), typically $+2$[19]
Mercury (Hg)	Chlor-alkali industry (manufacture of chlorine and caustic soda by hydrolysis); electrical and control instrument industry; laboratory products; dentistry; agriculture[24]	Mercury is commonly found in three forms: elemental, ionic ($+1$, more commonly $+2$), and methylmercury $(CH_3)_2Hg$[24]
Cyanide (CN)	Naturally occurs in some foods and plants; environmental sources include discharges from some metal mining processes, organic chemical manufacture, iron and steel works; cyanide is also used in electroplating, metallurgy, and making plastics, and is found in coal[25]	This nonmetallic inorganic functional group contains carbon and nitrogen atoms, joined by a triple bond

2.3 OVERVIEW OF HYDROGEOLOGY

Contaminants spilled or dumped on land often leach through the soil to contaminate groundwater. Their fate in the groundwater depends upon the chemical nature of the contaminant and its resulting tendency to degrade, dissolve, or sorb to soils, and on the hydrogeology of the aquifer. These factors also determine whether or not and how groundwater can be remediated. Following is a brief overview of the basic principles of hydrogeology and contaminant transport.

2.3.1   Hydrologic Cycle

The *hydrologic cycle* describes the circulation of water between the land, atmosphere, and oceans. When precipitation falls on the earth, some of the water runs overland to surface water bodies such as ponds and lakes, and streams and rivers, which lead to the ocean. Some of the water soaks into the earth, to be taken up in the roots of plants or to *recharge* the groundwater. Groundwater ultimately flows to surface-water *discharge* points. Water returns to the atmosphere when it evaporates from surface water, and by evapotranspiration from the leaves of plants. This cycle can be described in terms of a *hydrologic budget* or *water balance*:

$$P = Q + E + \Delta S_s + \Delta S_g \qquad (2.14)$$

where: P = precipitation,
        Q = runoff,
        E = evapotranspiration,
        ΔS_s = change in surface-water storage, and
        ΔS_g = change in groundwater storage.

While the hydrologic cycle primarily describes the movement of water, it also describes mechanisms for distributing or carrying contaminants through the environment.

2.3.2 Geologic Formations

The rate of groundwater flow depends primarily on the type of geologic formation. Geologic formations also affect groundwater chemistry and the transport of contaminants through an aquifer.

A layer of *fill* covers many sites in developed areas. Fill, which may comprise soil, ashes, and/or mixed debris, was placed at many sites to fill in a wetland and make usable land, to level natural ground contours, or to dispose of waste. Natural geologic formations include unconsolidated deposits and rocks. *Unconsolidated deposits* are soils originally deposited by glaciers, water (such as river sediments), or wind (such as sand dunes). Unconsolidated deposits include materials such as gravel, sand, silt, and clay. These materials are characterized by their particle size, as indicated in Table 2.6. Rock formations include consolidated sedimentary rocks such as sandstone, shale, and limestone, igneous rocks, and metamorphic rocks such as granite, basalt, slate, or gneiss.

Table 2.6 Common Soil Types and Characteristics

Soil type	Particle size (mm)[26]	Typical porosity (%)	Hydraulic conductivity (cm/sec)[27]
Clay	<0.002	50[27]	10^{-10} to 10^{-6}
Silt	0.002–0.05	35–50[28]	Silt, loess: 10^{-6} to 10^{-3} Silty sand: 10^{-4} to 10^{-2}
Sand	0.05–2.0	25[27]	10^{-3} to 10^{-1}
Gravel	>2.0	20[27]	10^{-1} to 1

Note: These parameters are quite site and location specific. Literature values vary widely. (See also common hydrogeology texts, e.g., References 29 and 30.)

Groundwater flows through pores and fractures in geologic formations. *Porosity*, abbreviated *n*, is the ratio of the volume of pore spaces in soil or rock to the volume of the solid portion. Porosity includes *primary porosity* and *secondary porosity*. Primary porosity is associated with the pores between grains in unconsolidated deposits and in sedimentary rocks. Secondary porosity is generally associated with fractures in rocks.[31] Fractures can also occur in soils such as clays.[32] Table 2.6 indicates the porosity of common types of soil and rock.

Geologic formations affect groundwater chemistry primarily from dissolution of rock materials. The natural constituents of groundwater can include iron, manganese, calcium and magnesium (collectively, *hardness*), sodium, carbonate and bicarbonate (collectively, *alkalinity*), sulfate, chloride, fluoride, and nitrate. High concentrations of these natural constituents can limit the usability of water for drinking water or industrial purposes, particularly due to the level of *total dissolved solids* or hardness. Those limitations often enter into the development of remediation goals. Relatively high concentrations of these natural constituents can also foul groundwater treatment equipment, or trigger limitations on the discharge of treated groundwater. As a result, the natural characteristics of groundwater may determine the type of treatment needed.

2.3.3 Groundwater Flow

Groundwater flows through zones of soil or rock called *aquifers*. Aquifers can be unconfined or confined, as described below. The *vadose zone*, or *unsaturated zone* at a site may be underlain by an unconfined aquifer. In some areas an unconfined aquifer may be underlain by a series of confined aquifers at progressively increasing depths. Groundwater may also be perched in lenses within the unsaturated zone.

An *unconfined aquifer* is sometimes known as a *water table aquifer*. As the name implies, the groundwater is not confined beneath a low-permeability layer (although it may be underlain by a low-permeability layer). Because it is not confined, the *water table*, or upper surface of the unconfined aquifer, can fluctuate in response to infiltration of precipitation or discharge of the water to a surface water body or well. Water rises from the water table into the overlying soil by capillary action to form a *capillary fringe*.

The term "water table" can be described intuitively. Imagine digging a hole in the ground. The soil at the top of the hole may be moist but is usually not wet. This soil is in the *unsaturated zone*, or *vadose zone*. Dig deeper, and groundwater starts to seep into the hole. At that point, the hole has intersected the water table, and is in the *saturated zone*. More formally, the water table is the surface on which the water pressure in the pores of the soil or rock is equal to atmospheric pressure.[31]

Perched groundwater is a limited form of an unconfined aquifer. It is sometimes encountered in the unsaturated zone above lenses of low-permeability material such as clay. A *perched water table* occurs when water seeping downward is blocked by the low-permeability layer and saturates the sand above it. The sand below the clay is not saturated, so the water perched above the clay lens is not connected to the aquifer beneath.[33]

Under natural conditions, the water table generally (but does not always) follows the contours of the surface topography. In humid and semiarid regions, the water table is generally 0 to 20 feet below ground surface. The elevation of the water table can be determined from water levels in excavations, surface water bodies such as ponds or streams, and in wells screened in unconfined aquifers.[34] Water-level measurements indicate the *hydraulic head* in the unconfined aquifer at the measurement point. Hydraulic head reflects both the hydraulic pressure (pressure head) and the elevation head (related to elevation above sea level or an arbitrary datum). For an unconfined aquifer which is not being pumped, the pressure head is zero and the hydraulic head is simply the elevation head.[31]

A *confined aquifer* is a water-bearing zone confined between two low-permeability layers such as clay or silt. Because the aquifer is confined, the groundwater is under pressure and water will rise above the top of the aquifer in a well drilled through the confining layer into the aquifer. If the water can rise to the ground surface, the well is known as a *flowing artesian* well. The height to which water can rise in a well indicates the hydraulic head. In a confined aquifer, the hydraulic head includes both the pressure head and the elevation head. In contrast to an unconfined aquifer, the pressure head in a confined aquifer is not zero.

Groundwater elevation data are usually plotted on a site plan to map the water table for an unconfined aquifer, or to define the *potentiometric surface* for a confined aquifer. Measured water levels are noted at the locations of monitoring wells, and *equipotential lines* (sometimes called *groundwater contour lines* for an unconfined aquifer) drawn through known or estimated points of the same elevation. Data from at least three wells are necessary to characterize the direction of groundwater flow by *triangulation*.[27] See Figure 1.5 for an example.

Groundwater flows primarily in response to differences in pressure or head. This change in pressure can result from the natural difference in hydraulic head at two points, or be imposed by a pump in a groundwater-extraction system. The *hydraulic gradient* describes the change in pressure in an aquifer which induces flow. It is the slope of the water table or potentiometric surface, or:[27]

$$I = \frac{dH}{dL} \qquad (2.15)$$

where *I* is the hydraulic gradient and *dH/dL* is the change in head (*H*) over a distance (*L*) perpendicular to the direction of flow.

The hydraulic gradient at a site can usually be determined from a water table map or potentiometric surface map as illustrated in Example 2.4. However, this method may not be accurate if there is a significant vertical component of groundwater flow in a confined aquifer, nonaqueous phase liquids are present, or in a fractured bedrock aquifer.[35] Groundwater flows from a point of high hydraulic head (*upgradient*) to low hydraulic head (*downgradient*) along a *flow line* perpendicular to the equipotential lines. A map which shows both the equipotential lines and flow lines is known as a *flow net*.

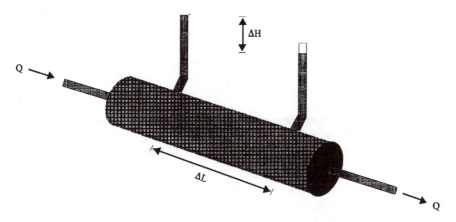

Figure 2.3 Experimental apparatus for demonstrating Darcy's Law.

Groundwater can also flow in response to temperature gradients and electrical gradients.[36] Those phenomena are beyond the scope of this book, except as discussed for soil remediation by electrokinetics in Chapter 4.

The flow of groundwater through an aquifer is described in most cases by an empirical relationship known as *Darcy's law*.[37] Equation 2.16 and Figure 2.3 illustrate Darcy's law.

$$Q = K \cdot I \cdot A \qquad\qquad (2.16)$$

where: Q = the volumetric flow rate of water [length^3/time],
   K = hydraulic conductivity [length/time], as described below,
   I = hydraulic gradient [unitless], and
   A = cross-sectional area in flow [length^2].

Darcy's law holds for laminar flow; it is not accurate for turbulent groundwater flow which occurs at high velocities, e.g., at the well screens of high-capacity wells and in fractured rock.[38,39]

Hydraulic conductivity, sometimes referred to as *permeability* (especially in older references), characterizes the ease with which groundwater can flow through a permeable medium. It is a measured property of an aquifer. Hydraulic conductivity also depends on the fluid, as discussed for nonaqueous phase liquids in Section 2.3.4.1. Table 2.6 indicates the range of hydraulic conductivities found for groundwater in various types of formations. Hydraulic conductivity is typically reported in units of centimeters per second (cm/sec), meters per second (m/sec), feet per day (ft/day), or gallons per day per square foot of cross-sectional area in flow (gal/day/ft^2).

Example 2.4: Estimate the flow of groundwater migrating off-site in the shallow aquifer at the Ace Solvent Recovery facility, using the data presented in Chapter 1.

The site investigation indicated that the glacial outwash in the southern portion of the site has a hydraulic conductivity of 1×10^{-1} cm/sec. The hydraulic gradient in the glacial outwash is

$$I = \frac{\Delta H}{\Delta L} = \frac{(140\ ft - 136\ ft)}{275\ ft} = 0.015$$

The saturated thickness (b) of the aquifer at the southern end of the site is approximately 25 ft. The cross-sectional width of the site is 460 ft. The cross-sectional area in flow is

$$A = 25\ ft \cdot 460\ ft = 11500\ ft^2$$

Substituting those values into Equation 2.16 (and including unit conversion factors),

$$Q = \left(1 \times 10^{-1}\ cm/sec\right) \cdot \left(1.969 \frac{ft/min}{cm/sec}\right) \cdot (0.015) \cdot \left(11500\ ft^2\right) \cdot \left(7.48052 \frac{gal}{ft^3}\right)$$

$$= 250\ gal/min$$

In addition to hydraulic conductivity, two other properties are used to characterize an aquifer: transmissivity (T) and storativity (S). *Transmissivity* is the rate at which water is transmitted through a unit width of aquifer or aquitard under a unit hydraulic gradient, and is calculated from the hydraulic conductivity:

$$T = K \cdot b \qquad\qquad (2.17)$$

where b is the saturated thickness of the geologic layer. Transmissivity is typically reported in units of square meters per day (m²/day), ft²/day, or gallons per day per foot (gal/day/ft). *Storativity* describes the volume of water that an aquifer will release from or take into storage per unit surface area of the aquifer per unit change in head. For unconfined aquifers, it is referred to as *specific yield*, and ranges from 0.1 to 0.3. For confined aquifers, S ranges from 0.0001 to 0.00001; for a leaky confined aquifer (i.e., the aquitard is not completely impermeable), it is roughly 0.001. Transmissivity and storativity are used to evaluate the response of an aquifer to stress, such as pumping; they are also used as input to groundwater flow models.[40]

Slug tests, pumping tests, and pressure tests are used to measure aquifer properties in the field. A comprehensive discussion of aquifer tests is beyond the scope of this book. However, a brief description of common tests follows.

In a *slug test*, a known volume of water is added to or removed from a well. The resulting change in the water level is measured over time to estimate the transmissivity (and thus hydraulic conductivity) in the vicinity of the well. Because a slug test affects the aquifer only in the vicinity of the test well, it does not provide a highly reliable estimate of aquifer properties. Slug tests are typically used in shallow unconfined aquifers, although they can also be used in confined aquifers. Slug tests are often performed early in a site investigation, when highly reliable data are not necessary. They cost relatively little to perform.

In a *pumping test*, groundwater is pumped from a well for a sustained period (typically at least 8 hours). At a hazardous waste site, the water pumped out during the test must usually be treated. The water levels in nearby observation wells are measured over time to determine the change in the potentiometric surface of the aquifer resulting from pumping. Pumping tests yield estimates of storativity and transmissivity (and thus hydraulic conductivity). Pumping tests provide more reliable data than slug tests. Because a pumping test also costs significantly more than a slug test, a pumping test is performed only when the more reliable data are necessary. For example, a site investigation team might initially perform slug tests to provide data for a remedial investigation and feasibility study of remediation options. A pumping test would be performed to provide more reliable data to design a ground-water extraction and treatment system.

Pressure tests are used to determine hydraulic conductivity in low-yielding frac-tured bedrock aquifers. A zone of an aquifer is isolated by inserting inflatable packers into a borehole. Water or air is injected under pressure to perform a slug test on the isolated segment of the borehole.[41]

2.3.4 Contaminant Transport

The movement of contaminants through an aquifer depends on whether the contamination is present as a nonaqueous phase liquid or as dissolved contamination. The contaminant transport mechanisms for each case are briefly summarized below.

2.3.4.1 Nonaqueous Phase Liquids

Certain types of compounds with relatively low water solubility can exist as a separate phase in an aquifer. *LNAPLs*, also known as *floaters*, are less dense than water. Gasoline is a LNAPL. *DNAPLs*, or *sinkers*, are more dense than water. DNAPLs include chlorinated solvents such as trichloroethylene (TCE) and perchlo-roethylene (PCE), coal tar, creosote, and PCBs.

Terminology and behavior common to both LNAPL and DNAPL are described below. Sections 2.3.4.1.1 and 2.3.4.1.2 describe the behaviors particular to LNAPL and DNAPL, respectively.

A separate organic phase persists when the mass of contamination exceeds its solubility in groundwater. LNAPL can pool above the water table and DNAPL can pool on layers or lenses of low-permeability materials. Kueper et al.[42] note that "pooled NAPL represents a continuous fluid distribution and is potentially mobile in the subsurface." These pools can serve as a continuing source of dissolved contamination in the groundwater.

NAPL residuals left behind when the separate phase flows through soil or rock also serve as long-term sources of dissolved groundwater contamination. *Residual NAPL* consists of blobs and ganglia of NAPL left in a geologic medium by flowing NAPL.[43]

Three forces control the migration of NAPL in the subsurface: gravity, viscous forces, and capillary forces. These forces depend on three characteristics of the NAPL: density, viscosity, and the NAPL–water interfacial tension. Certain soil properties also affect NAPL migration, including the heterogeneity, permeability, pore geometry, and pore size.[44] Several of these terms have not been discussed previously. Briefly,

- *Viscosity* is a measure of the resistance to flow of a fluid. A NAPL such as coal tar, which is highly viscous, flows much less readily than water.
- Kueper et al.[42] define *interfacial tension* as "the tensile force that exists in an interface separating two immiscible fluids (e.g., phases). This force arises due to a mutual attraction between molecules in the vicinity of the interface and like molecules in bulk solution. This can also be thought of as a mutual dislike for molecules on the opposite side of the interface." Interfacial tension is a site-specific parameter.
- *Capillary forces* hold fluids in soil pores, and can resist the penetration of NAPL into a water-filled soil pore or of water into a NAPL-filled soil pore. *Capillary pressure* is the change in pressure across the interface between NAPL and water. It is proportional to the interfacial tension and inversely proportional to the radius of curvature of the interface. The latter depends upon the size of the opening to the soil pore spaces.[42]

NAPL must displace water and air as it flows through soil or rock.[45] In order to do so, the pressure head (weight) of the NAPL must overcome the capillary forces holding the water or air within the soil pores or rock fractures.

Typically, NAPL will move through the center of the pore spaces while water coats the soil grains. This phenomenon is sometimes described as follows: water is the wetting phase relative to NAPL and the NAPL is nonwetting. NAPL tends to flow through an aquifer more slowly than pure water because NAPLs are more viscous than water and because water occupies a portion of the pore spaces, blocking the flow of NAPL. As the fraction of NAPL in the pore spaces in the saturated zone increases, the permeability to water decreases and the permeability to NAPL increases. Conversely, as the fraction of NAPL in the pore spaces decreases, the permeability relative to water increases and the permeability to NAPL decreases. When the fraction of NAPL in the pore spaces reaches the level of residual NAPL saturation, the relative permeability for the NAPL is effectively zero and the NAPL is not mobile as a separate phase.[32,45] Put another way, *residual saturation* is the maximum saturation level at which a NAPL comprises discontinuous blobs and ganglia in soil pores and fractures. At residual saturation, NAPL is trapped in pore spaces by capillary forces and cannot flow.[46] At saturation levels above residual saturation, the NAPL becomes a continuous mass and can flow through the soil.

In the unsaturated zone, residual NAPL will dissolve in water which is infiltrating from the surface; in the saturated zone, residual NAPL will dissolve in flowing groundwater. By either mechanism, residual NAPL will contribute to dissolved groundwater contamination. Residual NAPL with a high vapor pressure can also form *vapor plumes* in the air in the soil pores in the vadose zone. These vapor plumes can also contaminate infiltrating water.[32,45]

In addition to providing a long-term source of dissolved contamination, NAPL presents special problems in characterizing and remediating hazardous waste sites. These problems, discussed further below and in Section 3.6, are generally more severe for DNAPLs than for LNAPLs. LNAPLs do not penetrate far below the groundwater table, so they do not migrate vertically as far as DNAPLs and are easier to detect and remove. Further, the most mobile LNAPL compounds are generally more biodegradable than DNAPL compounds.[47]

2.3.4.1.1 LNAPL — LNAPL spilled on soil will flow downward through the center of the soil pores, leaving residual NAPL trapped in the pore spaces by capillary forces. For most LNAPLs, residual saturation occurs in the unsaturated zone when the LNAPL occupies 10 to 20% of the total pore volume.[46] If the entire mass of LNAPL is deposited in the vadose zone as residual, the LNAPL will not reach the water table. The residual can, however, serve as a long-term source of groundwater contamination as water infiltrating from the surface dissolves the LNAPL residual or LNAPL compounds in a vapor plume.

If a sufficient mass of a LNAPL spills or leaks, LNAPL flows through the unsaturated zone to the top of the capillary fringe to form a pool of LNAPL which floats on the capillary fringe. The weight of the LNAPL depresses the water table. LNAPL accumulates in the depression and continues to spread laterally above the capillary fringe. As the LNAPL spreads, it leaves residual NAPL in the soil pores. If the water table fluctuates due to seasonal variations or in response to pumping, the LNAPL will rise and fall with the water table and leave a smear of NAPL at residual saturation on the soil.[45] At residual saturation below the water table, most LNAPLs occupy 15 to 20% of the total pore volume.[46]

2.3.4.1.2 DNAPL — DNAPLs move downward through soil and groundwater in response to gravity. The description of DNAPL migration which follows is based primarily upon the conceptual model described by Feenstra et al.[32]

DNAPLs spilled at the ground surface sink through the unsaturated zone, moving preferentially through coarser-grained porous material and through fractures and root holes in lower-permeability materials such as clay. As a result of this preferential migration the distribution of DNAPL in soils can be quite complex and widespread. As the DNAPL moves through unsaturated soils, it leaves residual DNAPL trapped in pore spaces by capillary forces. Experiments with DNAPL spills in dry sandy soils suggest that the residual DNAPL can occupy between 1% and 10% of the pore space in the soil; in experiments in moist sands, DNAPL occupied 2% to 18% of the pore space.[32]

No absolute rules can predict when a spill of a DNAPL compound will cause DNAPL to migrate through the unsaturated zone to reach the groundwater. Feenstra et al.[32] suggest that a solvent DNAPL spill of tens of liters can reach groundwater where the depth to the water table is a few tens of meters or less and there are no exceptionally impermeable geologic formations below the spill. If the water table is much deeper, penetration of the groundwater may require a release of several hundred liters (or several drums). However, Feenstra et al.[32] note that repeated small releases at the same location, as from a leaking tank or pipeline, can also reach the groundwater.

When DNAPL reaches the saturated zone, the DNAPL must displace the water already in the soil pores. DNAPL can penetrate the saturated zone when the pressure head exerted by the accumulated DNAPL can overcome the capillary (entry) pressure.[44]

DNAPL which penetrates to the groundwater will continue to sink through the aquifer until either (1) the DNAPL reaches a barrier to flow, or (2) the mass is entirely trapped in pore spaces or small fractures as residual DNAPL. As a result of the preferential migration of DNAPL through seams of higher permeability and through fractures, the ultimate distribution of DNAPL is quite unpredictable. DNAPL can even migrate in a direction that is hydraulically upgradient of the release point with respect to groundwater flow.

A barrier to the flow of DNAPL may comprise a lens of unfractured low-permeability material such as clay, or the bottom of the aquifer. DNAPL will spread laterally when it reaches such a barrier. If the barrier is a lens of clay, DNAPL may spread laterally to the edge of the lens and then continue to migrate downward. If the barrier is an aquitard, pools of DNAPL may form on depressions in the surface of an aquitard.

Residual DNAPL in saturated soils, DNAPL trapped in fractures, and pools of DNAPL serve as long-term sources of dissolved groundwater contamination. DNAPL residuals in saturated soils can be more highly concentrated than residual DNAPL in unsaturated soil, occupying 2 to 40% of the pore space.[32,43,48]

Site investigations rarely detect DNAPL pools on an aquitard or DNAPL in fractures. The existence of DNAPL is usually inferred from the site history and groundwater monitoring data:[48]

- Substances that can form DNAPL were used, spilled, or disposed of at the site. In order to be a DNAPL, a substance must have a density greater than 1.01 g/cm³, a water solubility less than 2%, and a vapor pressure less than 300 mmHg.
- The characteristics, quantity, and release pattern of the DNAPL, and the nature of the geologic formation, would enable DNAPL to penetrate to a significant depth.
- Residuals of certain DNAPL, such as coal tar, are visually evident during soil sampling.
- Other field tests, such as using an organic vapor analyzer to detect volatile organic compounds emitted from the soil, may indicate residual DNAPL.
- The concentrations of DNAPL compounds in soil samples may suggest residual DNAPL. However, since DNAPL moves preferentially through zones of higher permeability which can be quite small, it can be difficult to collect representative samples.
- As a general rule of thumb, if the concentration of a DNAPL compound in a groundwater sample is 1 to 10% of the solubility of that compound, then DNAPL may be present in the aquifer. Careful definition of contaminant concentrations both vertically and horizontally within a plume can also be used to define a DNAPL problem.

Finally, the persistence of a plume of a DNAPL compound can indicate a continuing source such as DNAPL. Because the presence of DNAPL is usually inferred rather than observed, and DNAPL can occur as pools or residuals, the extent and mass of DNAPL in an aquifer is very difficult to determine precisely.

Data from contaminated sites illustrate the potential effects of DNAPL spills. MacKay and Cherry[49] estimated the chemical mass dissolved in plumes in sand and gravel aquifers in terms of the equivalent number of 55-gallon drums of solvent. For example,

- An aquifer in Cape Cod, Massachusetts was contaminated with TCE and PCE from sewage infiltration beds. A plume approximately 0.4 mile long contained roughly 10,000,000,000 gallons of contaminated groundwater. This contamination was equivalent to the mass of seven 55-gallon drums of solvent.
- A 0.6-mile-long plume in San Jose, California contained approximately 1,000,000,000 gallons of groundwater contaminated with 1,1,1-trichloroethane, Freon®, and dichloroethylene from an electronics plant. The mass of contamination was equivalent to 0.6 drum of solvent.

2.3.4.2 Dissolved Contamination

The fate and transport of dissolved contamination depends on a complex combination of hydrogeologic and chemical factors, briefly described below.[50-52] As shown in Figure 2.4, these factors change the size and shape of a plume of dissolved contamination over time. Advection, dispersion, and sorption are particularly critical transport functions.

- *Advection* means the transport of contaminants due to the flow of groundwater. Because the hydraulic conductivities of the materials in an aquifer can vary over a small distance, groundwater may move at different rates within a plume. This velocity difference can account for the movement of contaminants at different rates and thus spreading of a plume.
- *Hydrodynamic dispersion* causes dissolved contaminants to spread out in the aquifer, essentially becoming diluted with uncontaminated groundwater. It results from mechanical mixing and molecular diffusion. Mechanical mixing occurs as groundwater flows around particles of soil and rock, as a result of the different flow paths and within a given flow path as a result of friction with the surface of the particles. On a molecular level, contaminants diffuse from areas of higher concentration to areas of lower concentration.
- *Adsorption* is the attraction and adhesion of contaminant molecules to soil surfaces such as clay particles, metal oxides and hydroxides, and organic matter. Adsorbed chemicals can be released back to the groundwater. The kinetics of sorption/desorption are slow relative to the flow of groundwater. Sorption varies with the contaminant and the aquifer material. As different contaminants move through different aquifer materials, different rates of sorption and desorption cause those contaminants to migrate through an aquifer at different rates.

Other physical/chemical mechanisms include the following:

- *Ion exchange* refers to the process by which cations are attracted to anionic surfaces on soil particles, displacing existing ions. Like adsorption, ion exchange is a reversible process.

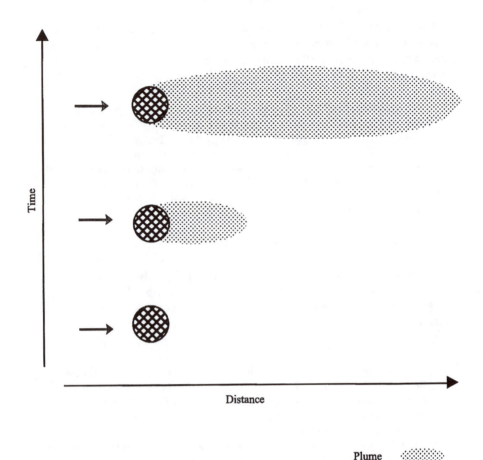

Distance

Plume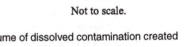
Source
Groundwater
flow
Not to scale.

Figure 2.4   Effects of advection and dispersion on a plume of dissolved contamination created by a point source in a uniform flow field.

- *Volatilization* may cause dissolved VOCs to migrate to the air in soil pores in the vadose zone.
- *Oxidation and reduction (redox) reactions* can degrade organic compounds, or convert metallic compounds into forms that are either more or less soluble than the original form of the contaminant. Redox reactions can include aerobic or anaerobic biodegradation. Of particular concern is the anaerobic biodegradation of chlorinated ethenes, which produces *daughter products* that are more toxic and mobile than the parent compounds.
- *Chemical precipitation* can result when dissolved metals form insoluble hydroxides as a result of the groundwater pH.
- *Filtration*, as the groundwater flows through the soil, can remove particles such as bacteria or precipitated metal hydroxides.

These phenomena are sometimes referred to collectively as *attenuation*. Over time, they will reduce the concentration of contaminants in an aquifer. *Natural attenuation*, according to a definition proposed by the U.S. EPA, is "the biodegradation, dispersion, dilution, sorption, volatilization, and/or chemical and biochemical stabilization of contaminants to effectively reduce contaminant toxicity, mobility or volume to levels that are protective of human health and the ecosystem."[53]

The *retardation factor* (R) describes the reduction in the rate of transport of a dissolved contaminant relative to groundwater flow, primarily due to sorption:[54]

$$R = \frac{v}{v_c} = 1 + \frac{\rho_b}{n} \cdot K_p \qquad (2.18)$$

where: v = average linear velocity of the groundwater (v = Q/nA),
$\quad$ v_c = velocity of contaminant at the point where the concentration is ½ the initial concentration;
$\quad$ ρ_b = bulk mass density of soil;
$\quad$ n = porosity of soil; and
$\quad$ K_p = soil–water partition coefficient (see Equation 2.6).

Equation 2.18 is deceptively simple. The fate and transport of groundwater contaminants are difficult to predict accurately due to the number of factors which can affect groundwater flow and contaminant retardation. Complex computer models are usually used to model contaminant fate and transport. Such models do not provide absolute predictions, but approximations which must be considered in light of the underlying assumptions.

2.4 HOW CLEAN IS CLEAN?

The objective of hazardous waste site remediation is to protect human health and the environment. In practice, this objective is deceptively simple. At most sites, it is translated into a series of *clean-up levels*, or concentrations in soil and water which should be safe for human health and the environment. Clean-up levels can vary widely depending on the basis used to develop those levels. Table 2.7 illustrates the range of clean-up levels that can be derived for a contaminant at similar sites: in a survey of former wood-preserving sites, the U.S. EPA found that the clean-up levels for PAHs in soils varied over three orders of magnitude. Different parties involved in the remediation of a particular site will often develop widely divergent clean-up levels for the site. Such variations dramatically affect the extent and cost of remediation.

Site-specific clean-up levels, or target goals for remediation, may be based on:

- Analytical detection limits;
- Background levels;
- Regulatory standards or criteria, generally derived using default risk-assessment assumptions;

Table 2.7   Soil Action Levels for Carcinogenic PAH at Superfund Wood Treating Sites[a]

Site	Action Level (mg/kg)	Risk	Notes
Residential Exposures			
United Creosoting, Conroe, TX	0.3	1×10^{-6}	PAH expressed as B(a)P equivalents
Mid South Wood Products, Mena, AR	3.0	1×10^{-5}	
North Cavalcade, Houston, TX	1.0	1×10^{-5}	
Bayou Bonfouca, Slidell, LA	2.0	3×10^{-5}	Carcinogenic PAH represent 2% of total PAH; action level is 100 mg/kg total PAH
Midland Products, Ola, AR	<0.2	$<10^{-7}$	Action level driven by ARAR is 1.0 mg/kg pentachlorophenol; imputed PAH level is far below detection limits; health risk is far less than 10^{-6}
Koppers, Orville, CA	0.19	1×10^{-6}	
Cape Fear Wood, NC	2.5	1×10^{-5}	
Southern Maryland Wood, MD	2.2	1×10^{-6}	
Commercial/industrial exposures			
United Creosoting, TX	40.0	3×10^{-6}	PAH expressed as B(a)P equivalents
South Cavalcade, TX	<700	$<1 \times 10^{-4}$	Action level is 700 mg/kg plus no leachate from soils; risk and action levels will be less than stated
Koppers, Texarkana, TX	100	1×10^{-5}	
Libby Groundwater, MT	88	1×10^{-5}	

[a] Remedies selected between 11/86 and 9/89.

From: U.S. Environmental Protection Agency, *Contaminants and Remedial Options at Wood Preserving Sites,* Office of Research and Development, Cincinnati, OH, 1994, Table 3.

- A site-specific assessment of the potential risks to human health or the ecosystem;
- Protection of groundwater from a continuing source of contaminants; and/or
- Mass removal.

Each of these bases is described below.

Clean-up levels may be applied to a site in different ways depending on their derivation and the philosophy of the decision maker. Clean-up levels may be *Not to Exceed* (NTE) values, i.e., concentrations that should not be exceeded at any single sampling point on site. Alternatively, clean-up levels may be concentrations which should not be exceeded *on average* across a site.

Regulatory agencies have historically applied clean-up levels on a NTE basis. More recently, in recognition that risk assessments often assume random contact with contaminanted materials across a site, some regulators have begun to accept proposals to remediate a site *as a whole* to a clean-up level. In effect, one area of a site can exceed a clean-up level if the average across the site is less than or equal to the clean-up level. The *residual risk*, or the risk associated with the whole site following remediation, will have been reduced to an acceptable level.

Table 2.8 Examples of Background Levels in Soils

Compound	Background concentration (mg/kg)		
	Massachusetts (suburban/rural)[57]	Typical levels in U.S. soils	Range in U.S. soils[58]
Arsenic	17	7.2[58]	<0.1 to 97
Cadmium	2	0.098[59]	
Chromium	29	34[58]	1 to 2000
Lead	99	19[58]	<10 to 700
Mercury	0.3	0.089[58]	<0.01 to 4.6

2.4.1 Detection Limits

For some compounds the risk-based clean-up level cannot be detected using available analytical methods. In such cases, the clean-up level is based on the analytical detection limit.

2.4.2 Background Levels

One objective for remediation is to return a site to the conditions present before the site was contaminated, or *background* conditions. Soil, sediments, and groundwater naturally contain low levels of metals, depending on the local geology. Certain organic compounds may also derive from natural phenomena. For example, forest fires can produce PAHs. Chlorinated organic compounds either do not occur naturally or occur at very low, generally undetectable levels. Some contaminants are also found throughout the environment as the result of the widespread use of certain materials. For example, lead is often found at elevated concentrations in urban soils as a result of the use of lead paint and from emissions from leaded gasoline, and PAHs are present from combustion of fossil fuels.[56]

Background levels may be obtained from literature reports, as shown in Table 2.8. A remedial investigation usually includes analysis of local upgradient or off-site samples to determine local background conditions.

2.4.3 Regulatory Standards, Criteria, and Screening Levels

Federal and state agencies have developed advisory levels and promulgated standards for contaminants in soil and water. (See Section 2.6 for a related discussion.) These levels are usually based on an evaluation of the potential human health or ecological risks. Those evaluations intentionally overestimate the potential risk to most people in order to protect the minority of people who are particularly susceptible to the health problems associated with the contaminant. (Risk-assessment assumptions and methods that overestimate risk are referred to as *conservative*.)

Health-based levels may be modified to account for technological limitations. For example, extremely low concentrations cannot be detected using state-of-the-art analytical methods, or, often, achieved at a reasonable cost using available treatment technology. Regulatory standards or criteria may be superseded by site-specific risk-based clean-up goals.

Table 2.9   Screening Levels for Soils and Sediments for Common Contaminants

Compound	Soil screening levels[60]			Sediment screening levels[61] (mg/kg)
	Soil ingestion, industrial (mg/kg)	Soil ingestion, residential (mg/kg)	Protection of groundwater (mg/kg)	
Benzene	200	22	0.5	0.057[a]
Toluene	410,000	16,000	5	0.67[a]
Phenol	1,000,000	47,000	49	—
Pentachlorophenol	48	5.3	0.2	—
Chloroform	940	100	0.3	—
Trichloroethylene	520	58	0.02	1.6[a]
Tetrachloroethylene	110	12	0.04	0.53[a]
PCB	0.74	0.083	—	0.023[b]
Benzo(a)pyrene	0.78	0.088	4	0.43[b]
Lead				47[b]
Chromium (III)	1,000,000	78,000	—	81[b]
Chromium (VI)	10,000	390	0.19	—
Arsenic	3.8	0.43	15	8.2[b]
Mercury (inorganic)	610	23	3	0.15[b]
Mercury (methyl)	200	7.8	—	—

Note: These values are periodically updated; values in this table are current as of 1996.

[a] EPA Sediment Quality Benchmark.
[b] Effects Range Low.

Some regulatory agencies have developed *screening levels*, in addition to or in lieu of promulgating standards or criteria. Screening levels are used to quickly evaluate site data. If the concentrations at a site exceed appropriate screening criteria, a site-specific risk assessment is usually used to evaluate potential risks more carefully or to develop site-specific clean-up levels. Screening levels are usually not enforceable standards.

Certain states have promulgated soil clean-up standards. These standards depend on the location of the soil and the use of the site. Regulators may consider the potential risks to human health from various exposure routes, including ingestion or inhalation of contaminants, or the potential for contaminants to leach to groundwater.

The U.S. EPA has developed *Soil Screening Levels* (SSLs) based on potential human health risks. SSLs are *not* intended to be final clean-up levels, but to determine whether or not conditions at a site deserve further evaluation.[62] Table 2.9 shows screening levels for common contaminants.

Standards for contaminated sediments are less common than standards or screening levels for soils. Several types of guidelines have been developed.[63] The U.S. EPA has proposed *Sediment Quality Criteria* (SQC) for certain PAHs. These criteria derive from a simple model of contaminant partitioning between sediment and water which incorporates the Ambient Water Quality Criterion (AWQC) of a contaminant (see below) as a gauge of potential toxicity. The National Oceanic and Atmospheric Administration (NOAA) has developed *Effects Range Low (ERL)* and *Effects Range Medium (ERM)* values for sediments. NOAA scientists derived these values by ranking, from lowest to highest concentration, the results of available studies showing the effects of sediment contamination on various organisms. An ERL represents the lower 10th-percentile concentration associated with biological effects. Concentrations below the ERL should not usually cause toxic effects. Similarly, the ERM represents the 50th-percentile concentration associated with biological effects. Finally, the U.S. EPA has developed *Ecotox Thresholds* (ETs) using a method similar to the model used to derive SQC. None of these levels — SQCs, ERLs, or ETs — are promulgated, enforceable clean-up standards. Rather, they are screening levels or guidelines.

Various standards and criteria have been developed for contaminants in water, depending on the use of the water. If an aquifer is or may be used as a drinking water source, groundwater remediation may be based on drinking water standards such as federal *Maximum Contaminant Levels* (MCLs), *Maximum Contaminant Level Goals* (MCLGs), or analogous state standards. Risk-based groundwater clean-up standards may also be based on the potential risks from inhalation of contaminants, or the risks to aquatic organisms in a surface-water body which receives a groundwater discharge.

Acute and chronic AWQC established under the Clean Water Act represent the contaminant concentrations in surface water that should not cause acute and chronic effects, respectively, to aquatic organisms in fresh water or salt water. Each criterion is based on at least eight acute toxicity tests from eight different families and three chronic tests. Criteria for metals depend on the hardness of the water (i.e., concentration of calcium and magnesium).[63]

Table 2.10 provides some federal and state standards for common contaminants for illustration. See Section 2.6.4 for further discussion of MCLs, MCLGs, and AWQC.

2.4.4 Site-Specific Risk Assessment

Clean-up levels may be based on a site-specific evaluation of the potential risks to exposed people or to the ecosystem, as discussed below.

2.4.4.1 *Clean-Up Levels Based on Potential Human Health Risks*

A remedial investigation usually includes a *baseline risk assessment*, or an assessment of the potential risks to human health resulting from exposure to the contaminants at the site should no action be taken to remediate the site. If the estimated risks exceed an acceptable level, the assumptions and equations used to calculate risks are used to calculate acceptable concentrations or clean-up levels. This two-step procedure was developed early in the Superfund program. As people gained more experience with evaluating and remediating hazardous waste sites, an alternative one-step procedure developed. For many contaminated sites, professional judgment or comparison to screening levels indicates that remediation is necessary, without performing a formal site-specific risk assessment. The project team can then develop or execute a remediation option, and then demonstrate that the residual risks meet acceptable levels.

Table 2.10 Standards and Criteria for Common Contaminants in Water

| Compound | Drinking water[64] | | Surface water[65] | |
	MCL (mg/L)	MCLG (mg/L)	Fresh water (mg/L)	Marine (mg/L)
Benzene	0.005	0	0.046[b]	—
Toluene	1	1	0.130[b]	—
Phenol	—	—	—	—
Pentachlorophenol	0.001	0	0.013[c]	0.0079[c]
Chloroform	0.1	0	—	—
Trichloroethylene	0.005	0	0.350[b]	—
Tetrachloroethylene	0.005	0	0.120[b]	—
PCB	0.0005	0	0.00019[b]	—
Benzo(a)pyrene	0.0002	0	0.000014[b]	—
Lead	0.015[a]	0	0.0025[c,d]	0.0081[c]
Chromium (total)	0.1	0.1	—	—
Chromium (III)	—	—	0.180[c,d]	—
Chromium (IV)	—	—	0.010[c]	0.050[c]
Arsenic	0.05	—	—	—
Arsenic (III)	—	—	0.190[c]	0.036[c]
Arsenic (V)	—	—	0.0081[b]	
Mercury (inorganic)	0.002	0.002	0.0013[c]	0.0011[c]
Mercury (methyl)	—	—	0.000003[b]	—

Note: These values are periodically updated; values in this table are current as of 1996.
[a] Action level, at tap.
[b] Tier II.
[c] Ambient water quality criterion or EPA-derived final chronic values (FCV).
[d] Criterion depends on hardness; value listed for 100 mg/L as $CaCO_3$.

Developing a risk-based remediation program requires both *risk management* and *risk assessment*. Decision makers must answer fundamental risk *management* questions: What level of residual risk is acceptable to society? How much money is a reduction in risk worth? How will the risks be controlled, by limiting exposure, or by removing or destroying the contaminants? Risk *assessment* is more scientific than policy oriented: for example, how toxic are the compounds at a site? What dose of a contaminant will people be exposed to? What kind of habitats are affected by site conditions? Do the contaminants bioaccumulate? Ideally, the distinction between risk management and risk assessment would be clear. In reality, the distinction blurs. Risk management philosophy colors risk assessment choices. Risk assessors must make many assumptions regarding the use of land, human behaviors, and human physiology. The level of conservatism in those assumptions often reflects risk management policies.

The discussion which follows describes how to calculate risk-based clean-up levels. These calculations begin with the selection of a target risk level, which is a risk-management decision. The same methodologies can be adapted to calculate pre- or post-remediation risks.

2.4.4.1.1 Types of Health Risks — Risk assessors consider two types of potential risks to human health: *carcinogenic* and *noncarcinogenic*. Carcinogenic compounds may cause cancer. Such compounds fall into four classes based on the weight of evidence of carcinogenicity:[66]

- Class A carcinogens are *known* human carcinogens. Few environmental contaminants fall into this category, because laboratory experiments are usually not performed on people, and environmental or industrial exposures can be difficult to isolate.
- Class B carcinogens are *probable* human carcinogens, based on limited human data (class B1) or laboratory data collected from testing animals (class B2).
- Class C carcinogens are *possible* human carcinogens, based on laboratory data collected from animal studies.
- Class D compounds are not classifiable.
- For Class E compounds, evidence suggests noncarcinogenicity for humans.

The *cancer slope factor* (CSF) characterizes the potential for a chemical to cause cancer when it is inhaled or ingested. The higher the CSF, the greater that potential. The CSF is derived from experimental data. It is the upper 95th percent confidence limit of the slope of the plot of the chemical dose vs. the response (incidence of cancer) and has units of $(mg/kg/d)^{-1}$. The U.S. EPA assumes that the dose–response plot passes through the origin, or, in other words, that no dose of a carcinogen is risk-free.[67] However, some regulatory agencies in the U.S. and in other countries assume that a threshold exists, and that exposures below the threshold do not increase the risk of cancer. The U.S. EPA revises CSFs periodically and maintains a library of slope factors in the *Integrated Risk Information System* (IRIS). (IRIS can be accessed through the web page for the U.S. EPA Office of Research and Development, at http://www.epa.gov.ord).

Contaminants may also have noncarcinogenic, or *systemic* health effects. These health effects can include, for example, damage to internal organs such as the liver, kidneys, or nervous system; reproductive impairment; or damage to a developing fetus. The U.S. EPA characterizes the dose-response relationship for a noncarcinogen with *reference doses* (RfDs).[67] A *chronic* RfD is an estimate of a daily exposure level that is likely to be without appreciable risk of adverse effects during a lifetime. Less commonly, subchronic (shorter-term) effects or developmental effects are considered.

RfDs are generally determined from animal studies, allowing for one or more uncertainty factors. An RfD is generally considered to have an uncertainty spanning an order of magnitude or more.[67] The U.S. EPA maintains a data base of verified RfDs in IRIS.

2.4.4.1.2 Acceptable Risk — Regulatory agencies have developed target risk levels for hazardous waste sites. These targets are levels at which society considers the risk from environmental contaminants to be acceptable.

For carcinogens, the target risk level is generally between 1×10^{-4} and 1×10^{-6}. A 1×10^{-4} risk is a 1 in ten thousand chance of getting cancer; a 1×10^{-6} risk is a one in a million chance. Some regulatory agencies specify a preference for the 10^{-6} level.

These risk levels represent the upper bound probability that an individual exposed to the contaminants at a site will develop cancer as a result of that exposure.[68] These probabilities apply only to people who are exposed to the contaminants under the conditions and to the extent that was assumed in estimating the risk. It is also important to recognize that these are upper-bound estimates of risk which depend on numerous assumptions. The actual risks are expected to be lower.[69]

The target level for noncarcinogens is based on the *Hazard Quotient* (HQ) or *Hazard Index* (HI). The HQ is the ratio of the estimated dose of a single contaminant to the RfD. For a site with multiple contaminants, the HI is the sum of the HQs for each contaminant, either in total or according to the toxic endpoint. A HI greater than or equal to 1 implies that unacceptable health effects may occur.[70]

The concentration corresponding to a specified theoretical cancer risk or HI depends on assumptions regarding the dose–response effect (i.e., CSF or RfD) and the frequency and amount of exposure.

2.4.4.1.3 *Exposure Assumptions* — The response to a toxic compound depends on the dose. How much of a dose does a child receive when she plays in contaminated soil? What about a construction worker digging in contaminated soil? In order to develop health-based clean-up levels for a site, one must hypothesize an exposure scenario and use scientific reports, regulatory conventions, and professional judgment to estimate the resulting dose.

The U.S. EPA offers the following definition:[71]

"Exposure is defined as the contact of an organism (humans in the case of health risk assessment) with a chemical or physical agent. The magnitude of exposure is determined by measuring or estimating the amount of an agent available at the exchange boundaries (i.e., the lungs, gut, skin) during a specified time period. Exposure assessment is the determination or estimation (qualitative or quantitative) of the magnitude, frequency, duration, and amount of exposure."

Exposure assessment occurs in three steps:[72] (1) characterize exposure setting, (2) identify exposure pathways, and (3) quantify exposure. Each step is described briefly below.

The first step is to characterize the exposure setting by considering the physical setting, land use, and the populations exposed as a result. This characterization depends on the current uses of a site and the likely future uses. Future site uses are often hotly debated for two reasons: first, human activities far into the future cannot be completely controlled by current actions; second, because assumptions regarding future use — for example, as a manufacturing site vs. a day-care center — can have a multi-million-dollar effect on the ultimate cost of remediation. Certain legal considerations affect assumptions regarding the future use of the land and water at a site. For example, a landowner can legally control the future uses of a site by putting a restriction on a deed. Land use may also be constrained by the regulations in many states which require that all groundwaters be maintained for use as a future drinking water supply.

The second step is to characterize exposure pathways. Each pathway describes a unique mechanism by which a population may be exposed to the chemicals at or originating from a site.[72] Exposure pathways can include, for example,

- residents drinking contaminated groundwater from wells downgradient from a site;
- residents inhaling contaminants volatilizing from groundwater during showering or other water use;
- incidental ingestion of soil on the hands of a resident gardening or playing in contaminated soil;
- incidental ingestion of soil on the hands of a construction worker who is working in contaminated soil; or
- inhalation of contaminated dust by an outdoor worker at an industrial site.

The third step is to quantify the magnitude, frequency, duration, and amount of exposure for each exposure pathway. This requires knowledge or assumptions regarding the behavior and physiology of the people who are actually or potentially exposed. For a residential exposure scenario, these factors could include, for example, the length of time people live in a home, their body weights, respiration rate, number of days spent outside (based in part on weather), and amount of skin exposed based on the amount of clothing worn. For an industrial exposure scenario, these factors could include, for example, the number of work days in a particular area in a year, the number of years spent on a job, and the amount of exposed skin. The U.S. EPA has developed default assumptions for various exposure scenarios.[73]

These assumptions are combined to estimate the *reasonable maximum exposure* (RME) for each scenario at a site. The RME is the exposure that falls within the 90 to 95th percentile of actual exposures.[74] The goal of developing RME estimates is to combine upper-bound and mid-range exposure factors to represent exposure scenarios protectively and reasonably. The goal is not to model the worst possible case.[73] Equation 2.19 shows how the intake (I, [mg/kg body weight/day]) of a chemical is calculated from the exposure factors.[74]

$$I = \frac{C \cdot CR \cdot EF \cdot ED}{W \cdot AT} \tag{2.19}$$

where: C = chemical concentration contacted over the exposure period [mg/kg or mg/L],

CR = contact rate, or the amount of contaminated medium contacted per unit time or event (e.g., 2 L/day drinking water),

EF = exposure frequency [days/year],

ED = exposure duration (number of years in a lifetime),

W = average body weight over the exposure period [kg], and

AT = averaging time, period over which exposure is averaged [days], which depends on the type of risk under consideration[74] as shown in the examples below.

The effect of this intake depends on the amount which is actually absorbed by the body. The amount absorbed depends upon, among other factors, the environmental medium and the form of the chemical. The amount absorbed by humans in an environmental setting can differ from the amount absorbed by the animals in the study from which a RfD or CSF is derived. The difference is accounted for by an *Absorbtion Adjustment Factor* (AAF). The AAF is often assigned a default value of 1. This factor is used to adjust RfDs and CSFs which are based on an administered dose, rather than an absorbed dose.[75]

The intake estimate described above is based on a single value for each parameter (e.g., a body weight of 70 kg). Such values are typically either upper-bound or average values, depending on the parameter. The intent, as described above, is to combine these values in order to characterize the RME and not the worst possible exposure. However, using many maximum or upper-bound values can overestimate the RME and thus overestimate the potential risk.

Anderson et al.[76] offer the following illustrations. Assume that the remediation goal is to keep exposure of 95% of the public to a contaminant below a given level. In this illustration, the 95% upper bound is defined as the RME. (Considering the worst possible exposure, the goal would be to protect 100% of the public.) Further assume — for simplicity — that only two parameters determine the level of exposure or intake. If 95% upper-bound estimates for each parameter are combined, the resulting intake estimate will represent a 99.75% upper bound $(100 - (0.05)(0.05))$. In order to achieve the stated goal of protecting 95% of the population, the intake estimate should be based on the upper bound of only one of the exposure parameters and the average of the other parameter. If the intake estimate is based on four parameters, and the 95% upper-bound estimate is assumed for each, the resulting value will overestimate exposure for 99.9994% of the population, assuming the parameters are independent. In other words, for 1,000,000 people exposed to the contaminant, only 6 would be expected to have an exposure greater than estimated by the four combined upper bounds. As indicated by Equation 2.19, most intake estimates are based on more than four parameters, so the use of upper-bound estimates for all of the parameters will greatly overestimate the actual exposures.

In order to more accurately model the RME and to account for the uncertainty or variability in certain exposure assumptions, risk assessors have turned to *Monte Carlo analysis*.[76-78] A Monte Carlo analysis is based on a known or estimated probability distribution for each parameter. A Monte Carlo simulation estimates a distribution of exposures or risk by repeatedly solving the model equations (e.g., Equation 2.19) using a computer. Each time the equation is solved, the computer selects randomly from the values for each parameter at random from the probability distribution. The analysis produces a range of estimates of exposure or risk. The risk assessor can identify the value corresponding to the desired percentile (e.g., 95%, 50%).[77] Example 2.8 describes an example of a site where a Monte Carlo analysis was used to assess potential human health risks and develop clean-up levels.

2.4.4.1.4 Calculation of Clean-Up Levels — The concentration corresponding to a specified theoretical cancer risk depends on assumptions regarding the frequency and amount of exposure, as described above, and the dose–response effect. In general,

$$Risk\ level = CSF \cdot I \qquad (2.20)$$

where the intake I is calculated averaging the exposure over a lifetime ($AT = 70$ years). Combining Equations 2.19 and 2.20 and rearranging,

$$Clean\text{-}up\ Level = \frac{R \cdot W \cdot AT}{CSF \cdot CR \cdot AAF \cdot EF \cdot ED} \qquad (2.21)$$

where R is the target risk level (e.g., 1×10^{-5}). This equation may be used to account for the effects of exposure by different pathways, as illustrated below. Each input parameter must be assumed; even the slope factors are estimated based on experimental data. As a result, this equation can produce many different clean-up levels for the same medium at a single site.

Example 2.5: Determine a clean-up level for methylene chloride in soil based on a target cancer risk of 1×10^{-6}. Consider potential exposure by ingestion and inhalation at a residential site, adapting default exposure assumptions developed by U.S. EPA to calculate conservative soil screening levels[73,79] and the properties of methylene chloride provided by EPA.[80]

Ingestion of Carcinogenic Compounds in Residential Soil:

This example models the potential risk from 30 years of exposure to a carcinogen through incidental ingestion of soil. More sophisticated calculations might consider the lifetime exposure using a soil ingestion rate based on a time-weighted average of childhood and adult ingestion rates.

Substitute the following values into Equation 2.21:

W = 70 kg
AT = averaging time, default value 70 years
CSF_o = oral slope factor, 0.0075 (mg/kg/day)$^{-1}$ for methylene chloride
CR = 100 mg/day
EF = exposure frequency, default value 350 d/year (assuming residents present except during two-week vacation each year)
ED = 30 years

Then,

$$Clean\text{-}up\ Level\ (mg/kg)$$

$$= \frac{\left(1 \times 10^{-6}\right)(70\,kg)(70\,yr)(365\,days/yr)}{\left(0.0075\,(mg/kg-d)^{-1}\right)\left(10^{-6}\,kg/mg\right)(100\,mg/day)(350\,d/yr)(30\,yr)}$$

$$Clean\text{-}up\ Level\ (mg/kg) = 227.1111111\ mg/kg$$

Considering that the target risk level has one significant figure,[69] this clean-up level reduces to 200 mg/kg.

Inhalation of Carcinogenic Compounds in Fugitive Dusts in Residential Soil:

People can inhale contaminants adsorbed to dust suspended in the air. Clean-up levels are usually not developed for this route of exposure for organic compounds or for most metals because the estimated risks from soil ingestion exceed the estimated risks from inhalation of those compounds in fugitive dusts.[79] The calculation which follows incorporates an estimate of the average level of dust in the air in a residential yard as a result of wind erosion of surface soils. Because it represents average conditions, this model does not represent acute exposures. The calculation incorporates an exposure period of 30 years, which is the national upper-bound (90th percentile) length of time at one residence.[72] Because toxicity criteria are based on the lifetime daily dose, the 30-year exposure period is averaged over a 70-year lifetime.

$$Clean\text{-}up\ Level\ (mg/kg) = \frac{R \cdot AT \cdot 365\,days/yr}{URF \cdot 1,000\,\mu g/mg \cdot EF \cdot ED \cdot \dfrac{1}{PEF}} \qquad (2.22)$$

where: URF = Inhalation unit risk factor, 4.7×10^{-7} $(\mu g/m^3)^{-1}$ for methylene chloride

ED = Exposure Duration, default value 30 year,

PEF = Particulate Emission Factor [m³/kg], default value of 1.32×10^9 calculated from Equation 2.23, below,

and other assumptions as indicated above.

$$PEF = \frac{(Q/C)(3,600\,s/h)}{(0.036)(1-V)(U_m/U_t)^3(F(x))} \qquad (2.23)$$

Q/C = inverse of the mean concentration at the center of a 0.5-acre-square source (which is the assumed size of a suburban residential yard), default value 90.80 g/m² per kg/m³,

V = fraction of vegetative cover, default value 0.5 [unitless],

U_m = mean annual wind speed, default value 4.69 m/sec,

U_t = equivalent threshold value of wind speed at 7 m, default value 11.32 m/sec, and

$F(x)$ is a unitless function dependent on U_m/U_t derived using a model by Cowherd et al.,[81] default value 0.194

The risk-based clean-up level for surface soil is then

$$Clean\text{-}up\ Level\ (mg/kg) =$$

$$\frac{(1\times 10^{-6})(70\,yr)(365\,days/yr)}{\left(4.7\times 10^{-7}(\mu g/m^3)^{-1}\right)(10^3\,\mu g/mg)(350\,d/yr)(30\,yr)\dfrac{1}{1.32\times 10^9(m^3/kg)}}$$

$$Clean\text{-}up\ Level\ (mg/kg) = 6,834,042.55\ mg/kg$$

Since this concentration cannot exist, this calculation implies that methylene chloride in soil cannot present a significant risk from inhalation of dust under the conditions assumed in this calculation.

Inhalation of Volatile Carcinogens in Residential Soil:

Clean-up levels based on this route of exposure incorporate two models: (1) volatil- ization of contaminants from the soil to the air which people breathe, and (2) inhalation, absorption into the body, and carcinogenicity. The calculation which follows is based on the assumptions that the soil contains an infinite mass of contamination, contami- nants are below saturation limits in soil (i.e., not NAPL), and vapor-phase diffusion is the only transport mechanism.[79] (*Diffusivity* describes the movement of a molecule in a liquid or gas as a result of differences in concentration. The higher the diffusivity, the more likely a chemical is to move in response to concentration gradients.)[72] Soil moisture and porosity strongly affect the rate of diffusion.

$$Clean\text{-}up\ Level\ (mg/kg) = \frac{R \cdot AT \cdot 365\ days/yr}{URF \cdot 1,000\ \mu g/mg \cdot EF \cdot ED \cdot \frac{1}{VF}} \quad (2.24)$$

R, AT, EF, ED, and *URF* are as noted above. *VF* is a soil-to-air volatilization factor [m³/kg] based on Equation 2.25:

$$VF = \frac{Q/C \cdot (3.14 \cdot D_A \cdot T)^{\frac{1}{2}} \cdot 10^{-4}\ (m^2/cm^2)}{2 \cdot \rho_b \cdot D_A} \quad (2.25)$$

where: D_A = apparent diffusivity [cm/sec²], calculated from Equation 2.26,

$$D_A = \frac{\left[(\Theta_a^{10/3} D_j k_H + \Theta_w^{10/3} D_w)/n^2\right]}{\rho_b K_d + \Theta_w + \Theta_a k_H} \quad (2.26)$$

Q/C = inverse of the mean concentration at the center of a 0.5-acre-square source, default value 68.81 g/m² sec per kg/m³,
T = exposure interval, default value 9.5×10^8 sec,
ρ_b = dry bulk soil density, default value 1.5 g/cm³,
Θ_a = air-filled soil porosity = $n - \Theta_w$,
n = total soil porosity = $1 - (\rho_b/\rho_s)$,
Θ_w = water-filled soil porosity, default value 0.15,
ρ_s = soil particle density, default value 2.65 g/cm³,
D_j = diffusivity in air, 0.101 cm²/sec for methylene chloride,
k_H = dimensionless Henry's law constant, 0.0898 for methylene chloride,
D_w = diffusivity in water, 1.17×10^{-5} cm²/sec for methylene chloride,
K_d = soil-water partition coefficient = $K_{oc}f_{oc}$ (organics) [cm³/g],

K_{oc} = soil organic carbon partition coefficient, 11.7 cm³/g for methylene chloride, and

f_{oc} = fraction organic carbon in soil, default value 0.006.

Working backward through Equations 2.26, 2.25, and 2.24,

$$D_A = 2.581660627 \times 10^{-3} \, cm^2/s$$

$$V_F = 2.465475725 \times 10^3 \, m^3/kg$$

$$Clean\text{-}up\ Level = 12.76451971 \, mg/kg = 10 \, mg/kg$$

As noted above, this calculation depends on the assumption that contaminant concentrations are below the saturation concentration. (As discussed in Section 2.3.4.1, soil saturation occurs when the air and water in the soil pores are saturated with the chemical and the absorptive capacity of the soil has been reached.)[79] Equation 2.26 incorporates Henry's law, which is valid for dilute concentrations of dissolved contaminants in water, not for NAPL. The maximum contaminant flux from soil to air (and the maximum resulting health risk) occurs at the saturation concentration. Therefore, if the clean-up level calculated from Equation 2.24 exceeds the saturation concentration, the calculations have predicted a physically impossible rate of volatilization under the assumed site conditions.

The saturation concentration, C_{sat} (mg/kg), can be estimated from Equation 2.27:[79]

$$C_{sat} = \frac{S}{\rho_b} \cdot \left(K_d \rho_b + \Theta_w + k_H \Theta_a \right) \tag{2.27}$$

with S = water solubility, 13,000 mg/L for methylene chloride, and the other variables as described above.

Using the default values and the values above for methylene chloride, C_{sat} = 2400 mg/kg. Since the calculated clean-up level of 10 mg/kg is below the saturation concentration, the estimate is legitimate.

Summary:

This example shows how to calculate conservative clean-up levels for a single carcinogen in soil based on incidental ingestion, inhalation of fugitive dusts, and inhalation of vapor-phase compounds. For methylene chloride, the following clean-up levels were calculated by each route of exposure using conservative assumptions:

Incidental ingestion	200 mg/kg
Inhalation of fugitive dusts	>1,000,000 mg/kg
Inhalation of methylene chloride in air	10 mg/kg

These calculations suggest that a clean-up level of 10 mg/kg should protect human health from the carcinogenic effects of methylene chloride. (That concentration does not account for the additive effect from risks by different routes of exposure. However, the calculations suggest that the risk contribution from ingestion and inhalation of dusts is relatively minor compared to the potential risk from inhaling methylene chloride which is volatilizing from the soil.) This clean-up level would apply to the *average* concentration of soil on site. Because the clean-up level was based on average exposures over time, individual samples could exceed this level without causing significant carcinogenic risk as long as the site-wide average was less than or equal to the clean-up level.

Although this example is quite lengthy, it is based on simplified models and assumptions, and does not address all of the possible routes of exposure which might occur on a hazardous waste site. For instance, this example does not address the potential health risks from absorption of contaminants through the skin. This example also does not address the potential risks from ingesting crops, meat, or milk which contain contaminants from farm land. For a brief discussion of such food intakes, see the U.S. EPA's risk assessment guidance.[82]

Example 2.6: Calculate a clean-up level for methylene chloride in groundwater, assuming that the groundwater is or may be used as a drinking water supply. From Equation 2.21, for a target risk level of 1×10^{-6} and assuming that a person weighing 70 kg drinks 2 L of water per day for 70 years,

$$Clean\text{-}up\ Level = \frac{\left(1 \times 10^{-6}\right)\left(70\,kg\right)\left(70\,yr\right)}{\left(0.0075\left(mg/kg-d\right)^{-1}\right)\left(2\,L/d\right)\left(70\,yr\right)}$$

$$Clean\text{-}up\ Level = 4.66666667 \times 10^{-3}\ mg/L$$

or, considering significant figures and converting the units, 5 µg/L.

This calculation was based on the average weight of an adult and the 90th percentile adult ingestion rate.[72] Note that 5 µg/L is the MCL for methylene chloride in public drinking water supplies.[80]

Clean-up levels for groundwater may be based on other routes of exposure, such as inhalation of volatile organic compounds while showering with the water, inhalation of vapors which migrated from groundwater to the air in basements, contact with surface water which receives groundwater discharge, or ingestion of fish living in that surface water. For additional examples of calculations, see the *EPA Region III Risk-Based Concentration Table*[83] or U.S. EPA's Risk Assessment Guidance.[84]

For a single contaminant with noncarcinogenic health effects, the clean-up level can be calculated as follows:

$$Clean\text{-}up\ Level = \frac{RfD \cdot W \cdot HI \cdot AT}{EF \cdot ED \cdot CR \cdot A} \tag{2.28}$$

This equation can be adapted to account for different exposure pathways, as illustrated below. For a site where there are multiple contaminants, the acceptable risk (HI value) is sometimes apportioned between compounds with the same toxic effect.

Example 2.7: In addition to being a potential human carcinogen, methylene chloride can cause noncarcinogenic health problems in humans. At high levels, methylene chloride impairs the central nervous system, causing dizziness, nausea, and tingling and numbness in the fingers and toes. Exposure to lower levels can impair hearing and vision slightly. Direct skin contact causes irritation.[85] In experimental animals, methylene chloride has reportedly caused kidney and liver damage and convulsions.[86]

Determine a clean-up level for methylene chloride in soil based on the potential noncarcinogenic effects on human health. Consider the potential risks from chronic exposure by inhalation and ingestion at a residential site, using default assumptions and equations developed by U.S. EPA to calculate conservative soil screening levels.[79]

Ingestion of Noncarcinogenic Compounds in Residential Soil:

Equation 2.29 models the potential risk to a child who incidentally ingests soil:

$$Clean\text{-}up\ Level\ (mg/kg) = \frac{RfD_o \cdot W \cdot HI \cdot AT \cdot 365\,days/yr}{10^{-6}\,kg/mg \cdot EF \cdot ED \cdot IR} \qquad (2.29)$$

where: RfD_o = oral RfD, 0.060 mg/kg -d for methylene chloride,[80]
   HI = target hazard index = 1 [unitless],
   W = body weight, default value 15 kg,
   AT = averaging time, default value 6 years (for noncarcinogens, AT = ED),
   EF = exposure frequency, default value 350 d/year,
   ED = exposure duration, default value 6 years, and
   IR = soil ingestion rate, default value 200 mg/d (estimated upper bound for ingestion).[73]

The risk-based clean-up level is then

$$Clean\text{-}up\ Level\ (mg/kg) = \frac{(0.060\,mg/kg/d)(15\,kg)(1)(6\,yr)(365\,days/yr)}{(10^{-6}\,kg/mg)(350\,d/yr)(6\,yr)(200\,mg/d)}$$

$$Clean\text{-}up\ Level\ (mg/kg) = 4692.857143$$

or, considering significant figures, 5000 mg/kg methylene chloride.

Inhalation of Noncarcinogenic Compounds in Fugitive Dusts from Residential Soil:

This calculation is similar to the procedure presented in Example 2.5, Equations 2.22 and 2.23.

$$Clean\text{-}up\ Level\ (mg/kg) = \frac{HI \cdot AT \cdot 365\,days/yr}{EF \cdot ED \cdot \left[\dfrac{1}{RfC} \cdot \dfrac{1}{PEF}\right]} \qquad (2.30)$$

where: HI = target hazard index = 1.0 (unitless),
   AT = averaging time, default value 30 years,
   EF = exposure frequency, default value 350 d/year,

ED = exposure duration, default value 30 years,

RfC = inhalation reference concentration, 3.0 mg/m³ for methylene chloride,[80] and

PEF = particulate emission factor, from Equation 2.24, default value 1.32×10^9.

The resulting risk-based clean-up level for surface soil is

$$Clean\text{-}up\ Level\ (mg/kg) = \frac{(1.0)(30\ yr)(365\ d/yr)}{(350\ d/yr)(30\ yr)\left[\dfrac{1}{(3.0\ mg/m^3)} \cdot \dfrac{1}{(1.32 \times 10^9)}\right]}$$

$$Clean\text{-}up\ Level = 4.129714289 \times 10^9\ mg/kg$$

This calculated value exceeds 1,000,000 ppm, and thus is an impossible concentration. The calculation indicates that the potential risk from inhalation of methylene chloride adsorbed to dust is not significant and that this route of exposure cannot be used to calculate a clean-up level.

Inhalation of Volatile Noncarcinogens in Residential Soil:

This calculation is based on the contaminant transport model used in Equations 2.25 and 2.26.

$$Clean\text{-}up\ Level\ (mg/kg) = \frac{HI \cdot AT \cdot 365\ days/yr}{EF \cdot ED \cdot \left[\dfrac{1}{RfC} \cdot \dfrac{1}{VF}\right]} \qquad (2.31)$$

Substituting the parameters defined above into Equation 2.31 yields a clean-up level of 7700 mg/kg. Since this level is above the saturation concentration, the model is not valid.

Summary:

This example showed how to calculate conservative clean-up levels for a single compound with noncarcinogenic health effects, based on incidental ingestion of soil, inhalation of fugitive dusts, and inhalation of mass volatilizing from soil. The calculations yielded the following clean-up levels for methylene chloride, using conservative assumptions:

Incidental ingestion	800 mg/kg
Inhalation of fugitive dusts	>1,000,000 mg/kg
Inhalation of methylene chloride in air	> C_{sat}

These levels would apply to the average concentration on a site.

These levels are all higher than the clean-up levels calculated in Example 2.5 based on the potential carcinogenic risk from exposure to methylene chloride. As a result, the clean-up level for a site containing methylene chloride in soil would be based on the potential carcinogenic risk rather than the potential noncarcinogenic health effects.

Table 2.11 Estimated Risks from Exposure to Soil American Creosote Works, Inc. Site

Area	Median	90th Percentile estimates
On site	2.6×10^{-4}	1.2×10^{-3}
Residential	1.6×10^{-6}	6.6×10^{-6}
Ditch	2.7×10^{-6}	1.2×10^{-4}
Condominium block	7.4×10^{-5}	1.9×10^{-4}

Examples 2.5, 2.6, and 2.7 are each based on a single contaminant and relatively simple sets of assumptions. Calculations of health-based clean-up levels can become very complicated. They may account for the cumulative risks resulting from exposures by different routes, exposures under different circumstances over a lifetime, and/or exposures to multiple contaminants. As discussed above and illustrated below, sophisticated calculations may include a Monte Carlo analysis, which incorporates the probabilities of various values for different parameters.

Example 2.8: In 1989, the U.S. EPA signed a Record of Decision (ROD) for the American Creosote Works, Inc. site in Pensacola, FL.[87] This 18-acre site was located in a moderately commercial and residential area of Pensacola, near Pensacola Bay and Bayou Chico. The American Creosote Works, Inc. operated a wood-preserving facility on the site from 1902 to 1981. At various times, creosote and pentachlorophenol (penta) were used to treat wood. As a result of historic wood-preserving and waste disposal activities, soils on site were contaminated with PAH, penta, and dioxin (originally an impurity in penta).

Three exposure pathways were considered in the risk assessment:

- oral and dermal exposure to surface soil at the site itself, in residential areas, in a drainage ditch, and on a condominium block;
- inhalation of airborne particulates from surface soils on site by nearby residents; and
- ingestion of crops grown in contaminated soil in gardens in residential areas.

The risk assessment for the site, including the development of soil clean-up levels, used a Monte Carlo analysis which incorported ranges of values for skin surface area, body weight, exposure frequencies, and soil ingestion rates. The assessment of potential risks from dermal-ingestion exposure projected the cancer risks shown in Table 2.11. The potential risks from inhalation of particulates and ingestion of vegetables were much lower.

EPA developed a clean-up level for PAHs based on the quantitative risk assessment and consideration of three factors:

- The cancer potency factor for benzo(*a*)pyrene (B(a)P), used to calculate the risk from all carcinogenic PAHs for this site, is very conservative;
- B(a)P comprised only 5 to 10% of the total concentration of carcinogenic PAHs detected in soil samples; and
- PAHs degrade naturally in soil; assuming a half-life of 1.25 years, the soil concentration will decrease naturally by an order of magnitude about every 5 years.

Based on this evaluation, the EPA selected a clean-up goal of 50 mg/kg carcinogenic PAHs in the surface soil at the site.

EPA also evaluated the need to remediate soils containing dioxin at the site. The term "dioxin" refers to a family of related compounds, called congeners. Risk assessors typically evaluate the toxicity of the members of this family relative to the toxicity of 2,3,7,8-tetrachlorodibenzodioxin (TCDD). The toxicity of other dioxins is calculated relative to that of TCDD and expressed as TCDD toxicological equivalents (TCDD-TE). (Note that PAHs are often evaluated similarly, as B(a)P-TE.) The ROD notes that, applying the median exposure estimates, a soil level of 0.3 μg/kg 2,3,7,8-TCDD-TE would yield an upper risk of 1×10^{-5}. The ROD also notes that the CSF for 2,3,7,8-TCDD was very conservative and under consideration for a tenfold increase, and that the Agency for Toxic Substances and Disease Registry (ATSDR) had indicated that 1 μg/kg was a reasonable level at which to begin consideration of remedial action. Therefore, EPA selected a clean-up level of 2.5 μg/kg 2,3,7,8-TCDD-TE for the soils at the site.

2.4.4.2 Clean-Up Goals Based on Potential Risks to an Ecosystem

As complicated as human health risk assessment is, evaluating the effects of contaminants on an ecosystem is far more complicated. Contaminants may affect individual species or communities within an ecosystem. The effects on a single species can cascade through the system, leading to a complicated relationship between site conditions and ecological effects. For example, consider a wet meadow. The primary effect of contamination may be to kill the plants in the meadow. "Death of the wetland plants leads to secondary effects such as loss of feeding habitat for ducks, breeding habitat for red-winged blackbirds, alterations of wetland hydrology that changes spawning habitat for fish, and so forth."[88] In addition to the effects on the ecosystem resulting from chemical stressors, a site can affect an ecosystem by changing a habitat. This change may result from past uses of the site, or from planned remediation activities. Finally, the effects of a stressor can depend upon the season of the year and the life cycle of the organisms.[88]

Ecological risk assessment is a rapidly developing field. Where the data are available, a *stressor–response* relationship is determined for a target organism similar to the dose–response relationships for human health effects, with careful consideration of the underlying uncertainties and simplifying assumptions. Clean-up levels for aquatic organisms are often based on AWQC.

Example 2.9: Potential risks to the ecosystem determined, in part, the need for remediation at the former Koppers Site in Charleston, SC. The Feasibility Study (FS) for this Superfund site describes the site setting and the basis for remediation.[89] The site is located on the west side of the peninsula formed by the Ashley and Cooper Rivers and consists of several pieces of property. Koppers Company, Inc. and other parties formerly operated a wood-preserving facility on a 45-acre parcel. Pentachlorophenol (penta) and creosote were used to treat wood. The site also includes 57 acres of land used for various other industrial and commercial purposes; EPA included that area in the site definition because the area may have been affected by the constituents of wood preservatives.

The ecological risk assessment focused on four areas: the Ashley River, Northwest Marsh, North Marsh, and South Marsh. Background sediment samples were collected from the Ashley River so that data from the site could be compared to local background conditions. The initial ecological risk assessment evaluated the potential risks to the benthic invertebrate community based on laboratory bioassays. *Neanthes arenaceodentata* and *Mysidopsis bahia* were exposed to sediment samples from various locations on the site. The data were used to designate areas of potential ecological concern (APECs) which required further evaluation or remediation. Some of the samples were acutely toxic to the test organisms. The corresponding locations within the APECs were designated for remediation.

The second phase of the assessment examined the potential risks to other organisms higher on the food chain. The U.S. EPA Environmental Response Team performed one assessment; Beazer East, Inc., a potentially responsible party, commissioned ENSR and Ogden Environmental and Energy Services, Inc., to perform an alternative assessment (referred to as the Alternative Scenario). These two risk assessments were intended to bound the range of potential risks, based on the reasonable maximum risk, in order to provide information for a risk management decision. Neither was intended to represent the actual ecological risks, which were expected to be lower. The two assessments illustrate the range of approaches and assumptions that can be made for a single site.

The EPA's study focused on:

- protection of the benthic invertebrate community (based on the first phase of work), and
- protection of piscivorous birds, insectivorous birds, omnivorous birds, fish communities, and omnivorous mammals from adverse effects on growth, survival, and/or reproductive success resulting from exposure to arsenic, chromium, copper, dioxin, lead, penta, and PAHs through the food chain.

The EPA assumed that the receptors were exposed to either the maximum or arithmetic average concentration of the target compounds in the river and marshes. Some of the assessment endpoints could not be evaluated due to a lack of toxicity data. For the other endpoints, EPA calculated Toxicity Quotients (TQ), akin to the HQs calculated in a human health risk assessment. The "acceptable dose" used in that calculation was either the Lowest Adverse Effects Level or the No Adverse Effects Level. Locations where TQ >1 were delineated on site maps.

The results of the EPA's study were also used to calculate sediment clean-up levels. To reduce the TQ to less than one, the sediments would have to be remediated to concentrations of less than 0.5 mg/kg PAHs, 3 ng/kg dioxin, 0.3 mg/kg arsenic, and 0.1 mg/kg lead. These concentrations are all below the local background concentrations in the Ashley River.

The alternative scenario had a slightly different focus than the EPA's study:

- protection of the structure and function of the benthic invertebrate community, and
- protection of the long-term health and reproductive capacity of populations of piscivorous birds, insectivorous birds, omnivorous birds, forage fish, game fish, and omnivorous mammals utilizing the marshes, the Ashley River, and the Barge Canal.

The alternative scenario evaluated the potential effects from arsenic, dioxin, lead, and PAHs. Receptors with relatively large feeding ranges — great blue heron, red drum, and racoon — were assumed to be exposed to the area-weighted average of concentrations detected at locations in the river and marshes. Receptors with relatively small feeding ranges — marsh wren, clapper rail, and mummichog — were assumed to be exposed to the concentrations at each sampling point. Toxicity indexes (TIs) were calculated and areas where TI > 1 plotted on site maps.

The site maps resulting from both analyses were compared to the original delineation of APECs. The FS concluded that:

- due to other sources of contamination in the area, local background concentrations exceeded the clean-up levels developed from EPA's risk assessment; and
- remediation of the APECs would eliminate the acute toxicity to benthic macroinvertebrates and reduce or eliminate the potential risks to other receptors through the food chain.

As a result, the FS evaluated various removal, containment, and treatment options for sediments in the river and marshes at the site.

Based upon the FS, the EPA proposed the following remedial actions for comment by the public:[90]

- Enhanced sedimentation of sediments in the river. Engineering measures would be taken to enhance the natural deposition of sediment particles, to cover sediments containing PAHs;
- Subaqueous capping of 3.2 acres of sediments in the Barge Canal, which contained PAHs; and
- Dredging selected sediments containing PAHs from tidal marshes for off-site disposal. The top foot of sediments (the biologically active zone) would be removed from a total area of 1.75 acres, replaced with clean soil, and revegetated; and
- *In situ* bioremediation of PAHs in the sediments in tidal marshes.

This proposal was based on the removal or active remediation of areas which could demonstrate significant aquatic toxicity to the indicator species, and less intrusive measures (capping, enhanced sedimentation, bioremediation) for areas which did not demonstrate sediment toxicity, but did contain sediment concentrations above relevent benchmarks.

2.4.5 Clean-Up Goals Based on Protection of Groundwater

Contaminants in soil can leach to groundwater when rainwater infiltrates through soil. One goal for soil remediation may be to reduce the source of groundwater contamination. Clean-up levels may be based on laboratory leaching tests, simple to elaborate models, or a combination. For example:

- Simulated rainwater or groundwater is passed through columns of representative soil samples containing different concentrations of the target contaminant. The concentrations in the leachate are compared to groundwater clean-up levels to determine the soil concentration at which the leachate reaches an unaccepptable level.

• Fate-and-transport modeling is used to estimate the contaminant transport from soil to groundwater, allowing for dilution, degradation, and attenuation on soil. Such models can be quite complex or relatively simple. Example 2.10 presents a simple model developed by the U.S. EPA.

Example 2.10: Estimate a clean-up level for methylene chloride in soil based on the potential leaching to groundwater. Use the simplified model for migration to groundwater which U.S. EPA developed in order to calculate soil screening levels,[91] using default assumptions rather than site-specific assumptions.

The model incorporates simplifying assumptions:

• Infinite source (i.e., the concentration in soil does not decrease over time);
• Uniform distribution of contamination throughout the vadose zone;
• No attenuation, either within the vadose zone or the aquifer;
• Instantaneous and linear equilibrium soil/water partitioning;
• Unconfined, unconsolidated aquifer with homogenous and isotropic hydrologic properties;
• Receptor well located at the downgradient edge of the source and screened within the plume; and
• No NAPL present.

Assuming these conditions, Equation 2.32 can be used to develop a clean-up level:

$$Clean\text{-}up\ Level = C_w \cdot \left(K_d + \frac{\left(\Theta_w + \Theta_a k_H \right)}{\rho_b} \right) \qquad (2.32)$$

where C_w is the target soil leachate concentration (mg/L), and the other parameters are as defined in previous examples. (The EPA uses the following default values in this calculation: $f_{oc} = 0.002$; $\Theta_w = 0.3$; and $\rho_b = 1.5$ kg/L.)

The target soil leachate concentration may be the clean-up level for groundwater, or may be based on that level assuming that the leachate is diluted with clean water as it reaches the groundwater. A *dilution attenuation factor* (DAF) of 1 implies that the leachate is not attenuated or diluted before it reaches an exposure point, and as a result, C_w equals the groundwater clean-up level. A site-specific DAF can be calculated using a water balance based on the hydrogeology of a site and typical infiltration:[91]

$$DAF = 1 + \frac{KId}{iL} \qquad (2.33)$$

where: K = hydraulic conductivity (m/year),
         I = hydraulic gradient (unitless),
         i = infiltration rate (m/year),
         L = source length parallel to the direction of groundwater flow (m), and
         d = mixing zone depth (m), calculated from Equation 2.34:

$$d = \left(0.112\,L^2\right)^{0.5} + d_a \left(1 - e^{\frac{-Li}{KId_a}}\right) \tag{2.34}$$

where d_a is the aquifer thickness (m), and the other variables are as defined previously.

Assume that the clean-up goal for groundwater is the MCL for methylene chloride in public drinking water supplies, 0.005 mg/L. For this simplified calculation, apply a DAF of 20 rather than calculating a DAF using Equations 2.33 and 2.34. The U.S. EPA has selected this default value as a protective level for contaminated soil sources up to 0.5 acre in size, for use in calculations of soil screening levels. Then,

$$C_w = (20)(0.005\ mg/L) = 0.010\ mg/L$$

and from Equation 2.32,

$$Clean\text{-}up\ Level = (0.10\ mg/L) \cdot \left(\left(0.0702\ cm^3/g\right) + \frac{(0.15) + (0.283962264)(0.0898)}{1.5\,g/cm^3}\right)$$

$$Clean\text{-}up\ Level = 0.019\ mg/kg$$

Note that this concentration is significantly lower than the clean-up levels in Examples 2.5 and 2.7 which were based on the potential risks to human health as a result of exposures to methylene chloride in soils or air. That finding illustrates the importance of considering all reasonably possible routes of exposure. The goal of groundwater protection often drives remediation of VOCs in soil. For less-soluble compounds, remediation may be driven by the need to protect people from the risks resulting from direct contact or the need to protect the ecosystem from the effects of bioaccumulation in the food chain.

The preceding model incorporated the simplifying assumption that the source of groundwater contamination would not decrease over time. As the mass of contamination in the soil leaches to the groundwater, the mass of the source decreases. As a result, the model is especially inaccurate for small sources. To account for this effect, the U.S. EPA has developed a simple model to estimate a mass-limit soil screening level.[91]

2.4.6 Clean-Up Levels Based on Mass Removal

One goal of remediation may be to reduce the mass of contamination at a site in order to protect the groundwater or to fulfill other regulatory objectives. This goal may be part of a two-tiered approach to a site, which incorporates removal or treatment of the most concentrated mass of contaminants and containment or isolation of contaminants remaining above health-based levels.

Remediation may focus on *hot spots*, or areas containing particularly high levels of contamination. Hot spots may be defined using a concentration criterion (e.g., 100 times the level in surrounding samples),[92] visual observations of stained or NAPL-saturated soil, or estimates of the mass of contamination within an area.

An analysis of the mass of contamination in soil volumes can be used to evaluate the cost effectiveness of mass removal. A site may be divided into different areas, based on concentrations, access, or use. The mass of soil contamination in each area can be estimated from the volume and density of the soil and the contaminant concentration. These estimates can be used to target areas for remediation. The cost effectiveness of removing successive increments of mass can be evaluated using these estimates by plotting the estimated remediation cost vs. the mass removed. Often, the plot will show a clear breakpoint where continued remediation becomes less cost effective.

2.5 REACTION ENGINEERING

Matter is neither created nor destroyed (short of nuclear reaction, which is not typically part of a remediation system!). Contaminants may transfer from one medium to another, or react to change chemical form. This section describes the basic principles used to characterize the transfer and transformation of matter.

2.5.1 Types of Reactors

Three types of ideal reactors are used to represent actual reactors:[93,94] batch reactors, continuously stirred tank reactors, and plug-flow reactors. Figure 2.5 shows these reactors schematically.

A *batch reactor* is filled with reactants, then the conditions necessary for the reaction to occur are imposed (e.g., mixing, temperature). When the reaction is complete, the reactor is emptied.

As the name implies, a reaction occurs as liquids flow through a well-mixed tank in a *continuous stirred tank reactor (CSTR)*. This type of reactor is sometimes called a *back-mix reactor*. The water flowing into the reactor is called the feed, or *influent*, and the water exiting the reactor is called the *effluent*. The *hydraulic retention time* or *detention time* in a CSTR is

$$Hydraulic\ retention\ time = \frac{tank\ volume}{flow\ rate} \tag{2.35}$$

Models of CSTRs assume that mixing and reactions are complete and instantaneous, so that the concentration of a compound anywhere in a CSTR is the same as the concentration in the effluent.

In a *plug-flow reactor*, the reaction occurs as the feed stream moves continuously through a tube. Complete mixing in the radial direction is assumed, so that the concentration is constant in a cross section along the tube. The tube may be hollow or filled with a packing material.

Reactors are sometimes used in *series* or in *parallel*, as shown in Figure 2.5. Reactors may be used in series to provide a sufficient degree of treatment or to provide a "fail-safe" should the first unit in the series fail. Reactors are sometimes

Types of ideal reactors

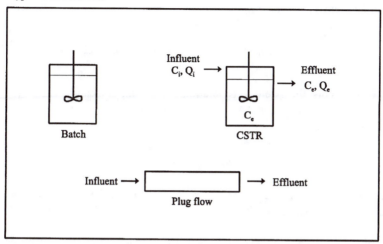

Reactor configurations

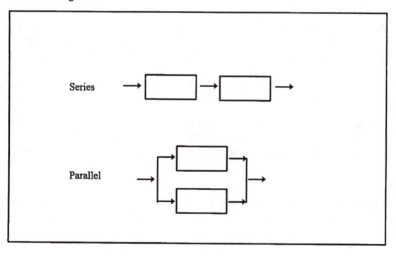

Figure 2.5 Schematics of ideal reactors.

used in parallel to provide flexible treatment capacity or to allow for the process to continue while a unit is taken off-line for maintenance.

2.5.2 Mass Balance

Designing remediation options frequently requires the use of a *mass balance*, sometimes called a *material balance*. (Systems which are not isothermal may also require an energy balance.) A mass balance accounts for the mass of each material entering and leaving a system:

$$Accumulation = Input - Output + Generation - Consumption \quad (2.36)$$

If a reaction occurs, the extent of generation or consumption can be calculated from the rate of reaction (Section 2.4.3). If no reaction occurs, Equation 2.36 can be simplified to:

$$Input = Output \quad (2.37)$$

The first step in performing a mass balance is to define the system and the time period over which the mass will be balanced, as illustrated by Example 2.11.

Example 2.11: An air stripper treats groundwater contaminated with organic compounds by transferring the contamination from the water to an air stream. The mass (or concentration) of organic compounds in the air stream exiting the air stripper may trigger permitting requirements and may require treatment to destroy or adsorb the contaminants before they are discharged to the atmosphere. This mass (or concentration) can be calculated using a mass balance.

Consider a groundwater treatment plant comprising an air stripper followed by two carbon adsorption units in series (see Figure 2.6). The untreated groundwater, pumped at a rate of 20 gallons per minute (gpm), contains 250 µg/L benzene. Air flows through the air stripper at 300 cubic feet per minute (cfm). The air stripper should remove 98% of the benzene in the groundwater. How many pounds of benzene are discharged in the air stream per day?

Step 1: Define the system and the time period over which the mass balance will be performed. The mass balance will be performed on the air stripper, not the entire treatment plant. (The system is bounded by a dashed line on Figure 2.6.) The calculation will estimate the mass per day.

Step 2: Since no reaction is occurring, use Equation 2.37 to perform the mass balance. Assume that the concentration of benzene in the influent air is negligible, and that the mass of water transferred to the air stream is negligible.

Let $M_{out, air}$ = mass of benzene in the air stream, kg/day.

Since the air stripper removes 98% of the benzene, $C_{out, water} = 0.02 C_{in, water} = 5$ µg/L

Unit conversion: $Q_{water} = 20$ gal/min = 109,020 L/day.

Input = $(Q_{water} \cdot C_{in, water}) + (Q_{air} \cdot C_{in, air})$
= $(250 \text{ µg/L}) \cdot (109,020 \text{ L/day}) \cdot (10^{-9} \text{ kg/µg}) = 0.027255$ kg/day

Output = $(Q_{water} \cdot C_{out, water}) + M_{out, air}$
= $[(109,020 \text{ L/day}) \cdot (5 \text{ µg/L}) \cdot (10^{-9} \text{ kg/µg})] + M_{out, air}$
= 0.0005451 kg/day + $M_{out, air}$

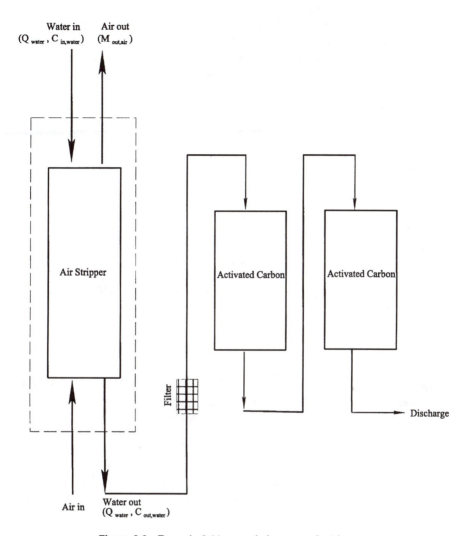

Figure 2.6   Example 2.11: mass balance on air stripper.

Since input = output,

$0.027255 = 0.0005451 + M_{out, air}$

$M_{out, air} = 0.0267099$ kg/day

Convert units:

$M_{out, air} = (0.0267099$ kg/day$) \cdot (2.2046$ lb/kg$) = 0.058884645$ lb/day

Considering significant figures, $M_{out, air} = 0.059$ lb/day benzene exiting the air stripper.

A mass balance can be used to determine a clean-up level for soil, as indicated in the discussion of "How Clean Is Clean?". It is also frequently used — and frequently forgotten — in evaluating remediation technologies. Remediation technologies often simply separate a contaminant from the soil or groundwater being treated. Many treatment technologies thus generate residuals of concentrated contamination. These byproducts of the treatment process can be costly to manage, and may trigger complicated environmental regulations, such as permit requirements.

2.5.3 Rate of Reaction

How long will it take to bioremediate the contaminants in soil at a site? What size tank is needed to treat groundwater? The answer to such questions depends in part upon the rate of the reaction used to treat a contaminant.

The rate at which a reaction occurs usually depends on the concentrations of compounds undergoing reaction. Many of the reactions which are involved in site remediation are *first-order reactions*. In an irreversible first-order reaction, the rate of reaction depends only on the concentration of a single compound:[93,94]

$$Reactant \Rightarrow Product \tag{2.38}$$

For a constant-volume batch system, the change in the concentration of the reactant (C) over time (t) is given by

$$\frac{\Delta C}{\Delta t} = k \cdot C \tag{2.39}$$

where k is an experimentally determined rate constant with the units of (time^{-1}). Rearranging Equation 2.39 and integrating the differential,

$$\frac{\Delta C}{C} = k \cdot \Delta t \tag{2.40}$$

$$\int_{C\,at\,t=0}^{C\,at\,t} \frac{1}{C} dC = k \int_{t=0}^{t} dt \tag{2.41}$$

$$C_t = C_0 e^{-kt} \tag{2.42}$$

$$\ln \frac{C_t}{C_0} = \ln C_t - \ln C_0 = -kt \tag{2.43}$$

Equation 2.42 or Equation 2.43 can be used to calculate the concentration (C_t) at any time from the initial concentration (C_0) and the rate constant.

Reactions are sometimes characterized by the *half-life* ($t_{1/2}$) of a compound, or the time required to reduce the concentration by half (i.e., $C_{1/2} = C_0/2$):

$$t_{1/2} = -\frac{\ln(0.5)}{k} \tag{2.44}$$

The reaction rate constant must be determined experimentally. Data may be collected in a batch reactor or continuous-flow reactor; often, reaction rates are determined in laboratory tests using beakers or flasks, which are essentially batch reactors. Beginning with different values of C_o, the experimentors measure C at various times t after the reaction begins. The data may be evaluated using various methods;[94,95] the *integral method* is described below as it applies to a first-order reaction.

If the reaction order is not known, the data must be tested to determine the reaction order. Equation 2.43 indicates that a plot of $(ln\ C)$ vs. t should be linear for a first-order reaction. The fit can be tested statistically. If this test indicates that the reaction is indeed first order, the reaction constant k is the slope of the line.

The reaction rate depends on the temperature, as represented by the *Arrhenius equation*:[94]

$$k = k_o e^{-\frac{E}{RT}} \tag{2.45}$$

where k_o is the preexponential factor (sometimes called a frequency factor), E is the activation energy, R is the Universal Gas Constant, and T is the absolute temperature (in K or R). If the reaction rate is known at two or more temperatures, the reaction rate at another temperature can be estimated from a plot of $\ln(k)$ vs. $1/T$.

The preceding discussion focused on first-order reactions to provide a basis for specific reactions discussed later in the book. Reaction kinetics can be much more complex. For information on the kinetics of other types of reactions, see the references cited in this section.

2.5.4 Treatability Testing

Engineers sometimes use a series of tests to determine whether or not a treatment technology can be effective and, if so, to design a full-scale remediation system. These tests are commonly referred to as bench-scale tests and pilot-scale tests.

Bench-scale tests are performed in a laboratory. In its simplest form, a bench-scale test is designed to provide a "yes/no" answer to the question: Will the technology work? More elaborate bench-scale tests are intended to provide information on the kinetics of a degradation reaction, to provide preliminary mass-transfer information, and/or to determine whether or not the treatment technology can meet the clean-up level. A bench-scale test may use a small batch reactor, soil column (essentially, a plug-flow reactor), or continuous-mix reactor. Depending upon the scope of the test, the U.S. EPA estimates that a bench-scale study can cost between $10,000 and $250,000.[96]

A *pilot test* is larger scale and more elaborate than a bench-scale test. Pilot tests are used to evaluate materials-handling limitations, mass-transfer limitations, and cost. A pilot-scale test provides more accurate information on the performance of a technology than a bench-scale test. A pilot-scale test may represent a portion of a full-scale system, e.g., one soil-vapor extraction well, rather than a network of wells.

Alternatively, a pilot-scale test may simply use a small version of full-scale equipment. The U.S. EPA indicates that a pilot test can cost over $250,000.[96]

2.6 ENVIRONMENTAL REGULATIONS

Understanding environmental regulations and their "alphabet soup" of acronyms can be daunting. Dozens of federal, state, and local regulations can apply to remedial actions at a hazardous waste site. However, many of these regulations contain similar requirements. Once these general points are understood, it is relatively easy to read and understand many regulations from different jurisdictions. The intent of this section is to provide the reader with a framework for understanding the environmental regulations which pertain to the remediation of hazardous waste sites. This section does not list every provision of the cited regulations. In professional practice, the reader should not rely solely on the regulatory information in this book, but should research, obtain, and use the pertinent laws, regulations, and guidance.

The examples in this section are based on federal regulatory requirements. State requirements often parallel these federal requirements. The federal requirements were used for illustration in this book because of their universal applicability in the U.S. and because they have served as models for many state requirements, in some cases being directly adopted by the states. Federal regulations are cited according to their place in the *Code of Federal Regulations* (CFR). For example, 40 CFR 268.43 refers to title 40, part 268, Section 43.

2.6.1 Terminology

A person working in hazardous waste site remediation must contend with environmental *laws*, *regulations*, *policy*, and *guidance*. Each of these regulatory mechanisms carries different weight and, often, different degrees of flexibility.

- *Laws* are passed by legislative bodies. They define restrictions and requirements for certain activities. For example, the Comprehensive Environmental Response, Compensation, and Liability Act (CERCLA, known as Superfund) requires that different remedial actions be evaluated for sites on the National Priorities List (NPL) and defines the criteria by which to evaluate options.[97] However, CERCLA does not define the procedures for evaluating different options.
- A *regulation* is a set of legal requirements which interprets and implements a law. Regulatory agencies *promulgate* regulations following a period of public comment on proposed regulations. For example, the National Contingency Plan implements CERCLA. It dictates, among other things, the steps that must be taken in a feasibility study (FS) to develop, evaluate, and select a remedial action for a hazardous waste site regulated under CERCLA [40 CFR 300.430(e)].
- A *policy* is a decision-making guideline developed by an agency to further a regulatory or environmental objective. Policies may be written or unwritten. For example, an agency might have the policy that all remedial actions must reach a certain set of unpromulgated clean-up levels, and that no other options should be evaluated in a FS.

- Regulatory agencies prepare *guidance* documents which describe the nuts-and-bolts procedures which they expect people to use in order to fulfill the requirements of regulations. For example, U.S. EPA guidance describes the procedures to be used to perform a FS, down to the format of the final report.[98]

The term *Applicable or Relevant and Appropriate Requirements,* abbreviated ARARs, is also used to describe regulatory requirements. The U.S. EPA created this term under the federal Superfund program to describe *promulgated regulations* that pertain to remedial actions. Other regulatory programs have adopted this concept. The U.S. EPA defines ARARs as follows:[99]

- *Applicable requirements* are "those cleanup standards, standards of control, and other substantive environmental protection requirements, criteria, or limitations promulated under Federal or State law that specifically address a hazardous substance, pollutant, contaminant, remedial action, location, or other circumstance at a CERCLA site."
- *Relevant and appropriate requirements* are "those cleanup standards, standards of control, and other substantive environmental protection requirements, criteria, or limitations promulated under Federal or State law that, while not 'applicable' to a hazardous substance, pollutant or contaminant, remedial action, location or other circumstance at a CERCLA site, address problems or situations sufficiently similar to those encountered at the CERCLA site that their use is well suited to the particular site."

The distinction between an applicable requirement and a relevant and appropriate requirement can be confusing. For example, consider regulations promulgated under the federal Resource Conservation and Recovery Act (RCRA) which describe the design standards for capping a landfill. These regulations would legally apply to a hazardous waste landfill that was permitted to operate under RCRA, but may be relevant and appropriate to an impermeable cap proposed to cover an area of contaminated soil. Guidance and policy are not ARARs, but are, in U.S. EPA jargon, *To-be-Considered Material* (TBC).[99] By definition, such material should contain more flexibility than promulgated regulations.

2.6.2 Waste Site Remediation

At the federal level, CERCLA, the RCRA Corrective Action provisions, and the corresponding regulations dictate how to investigate, evaluate, and clean up hazardous waste sites (40 CFR 300 and 40 CFR 264, Subpart S, respectively). Individual states have analogous programs. In general, the federal and state programs share the framework described below.

Setting priorities. Environmental regulations outline procedures for notifying environmental agencies of a contaminant spill or a suspected hazardous waste site. They also include criteria used to judge the conditions at a suspected hazardous waste site in order to determine whether or not it should be regulated. Not every contaminated site must be cleaned up in order to protect human health and the environment: our society has decided that certain relatively low levels of contamination are acceptable. For example, the U.S. EPA developed the Hazard Ranking System (HRS) to determine which sites should be placed on the NPL for action under CERCLA. The HRS is a numerical model used to calculate a score for a site. If the score is high enough, the site is placed on the NPL. The HRS considers (1) the potential for an uncontrolled release of a hazardous substance, (2) the characteristics of the substance such as its toxicity, quantity, and chemical behavior, and (3) the potential effects on exposed people and the environment [40 CFR 300, Appendix A]. The HRS is a screening model; consequently, the results may not reflect the ultimate need for or extent of remedial actions. While the HRS applies only to Superfund sites, other regulatory programs have other systems for setting priorities.

Removal actions control or clean up contaminants that pose a relatively urgent threat to public health or the environment. For example, a removal action would be taken upon discovery of a pile of abandoned drums containing chemicals that could explode. Depending on the situation, removal actions could include, for example, fencing an area to prevent exposure to contaminants, removing drums, or providing an alternative water supply to people drinking contaminated water [40 CFR 300.415(d)]. The requirements for selecting, carrying out, and documenting removal actions are necessarily less complicated than the requirements for final remedial actions. Removal actions may be taken at any time under CERCLA without, for example, going through the HRS or the extensive studies described below.

Remedial Investigation/Feasibility Study (RI/FS) or RCRA Facility Investigation/Corrective Measures Study (RFI/CMS). These studies characterize the extent of actual or potential environmental problems at a site, determine goals for clean-up of a site, and evaluate the alternatives for reaching those goals. The studies begin with a review of historical information to understand which contaminants may be at the site, how they got there, and where they may be located. This research forms the basis for a project work plan.

The RI or RFI may incorporate a variety of long-established methods or more innovative techniques to evaluate conditions in wastes, structures, soil, groundwater, sediments, and/or surface water. Site investigations commonly include several phases of investigation spanning several years. A sampling team may, for example, install groundwater-monitoring wells, excavate test pits with a backhoe to observe subsurface soils, install soil borings with a drilling rig to collect soil samples at specific depths and characterize the site geology, identify site features (such as subsurface piping or utilities) that could hinder remediation or present a hazard to site investigators, and perform hydrogeologic tests to characterize the underlying aquifer(s). Samples are analyzed for hazardous substances using standard analytical methods approved by regulatory agencies. Samples may also be subjected to bench-scale or pilot-scale testing of one of the treatment options under consideration. Finally, the project team assembles the data to create a *conceptual site model*, which is not a computer model but a three-dimensional "picture" of site conditions. This picture describes the distribution and migration of contaminants, exposure pathways, and potential receptors.[100] The last step is to write a remedial investigation report.

Risk assessment specialists evaluate the RI data to determine whether or not and to what extent a site must be cleaned up to protect public health and the environment. This may entail a human health risk assessment and an ecological risk assessment, which are documented along with the RI. If a site must be cleaned up, the risk assessment models may be used to develop numerical clean-up levels for the site, as described in Section 2.3.3.

The feasibility study or corrective measures study uses the results of the RI and the risk assessment to evaluate engineering options that may achieve the clean-up goals for a site. The project team begins the study, concurrent with the RI, by identifying all of the options which might possibly work. This list of options is narrowed down in an initial screening step. The project team develops a conceptual design for the remaining alternatives so that they may be evaluated in detail. Poor design work at this stage can result in wasted time (and cost) during detailed design, selection of ineffective remedial actions, and preparation of cost estimates which drastically underestimate final remediation costs.

The conceptual design for each alternative should include, as appropriate to the alternative, the area or volume of material to be remediated, a process flow diagram, the proposed layout on the site, ARARs and/or permit requirements, a mass balance to evaluate the ability to meet remediation goals or estimate treatment residuals, a cost estimate, and an estimate of the time required for remediation (see Section 2.7 for a more detailed discussion). The project team evaluates each conceptual design using criteria specified in the regulations. These criteria, couched in regulatory jargon, often seem arcane. However, most of the criteria translate to common-sense questions: Will this option effectively meet the remediation goals? Would it achieve those goals by permanently destroying the contaminants, or simply containing them? Is the cost — and the distribution of costs over time — reasonable? How long will this option take to become effective? How long will it require operation and maintenance? Can it meet regulatory requirements such as ARARs? This evaluation culminates in recommendations for remedial actions, subject to review by the regulating agency and public comment.

Remedial Design (RD) begins after the selection of remedial actions. The engineering team expands the conceptual design into complete plans and specifications that can be put out to bid. A complex design may be developed and submitted to the regulating agency for review in three phases: preliminary, draft final, and final. In addition to the plans and specifications, the design package usually includes monitoring plans, a health and safety plan which describes precautions to be taken during construction, a list of required permits which must be obtained before construction begins and/or a plan for complying with all regulatory requirements, a project schedule, and a quality assurance plan. In contrast to the level of detail in the requirements for an RI/FS, regulations and guidance for RD typically define the contents of the design package and submittal requirements, but not the specific engineering content or conventions to be used in the design.

Enforcement. Regulatory agencies compel the owner/operator of a RCRA-permitted facility or the *Potentially Responsible Parties* (PRPs) to study and remediate sites whenever possible, rather than doing the work themselves. Environmental laws and regulations provide mechanisms for agencies to compel PRPs to do this work; they also enable agencies to recover their own costs from a PRP. The requirements of an agency may be contained in an administrative document, such as a permit or *Administrative Order*, or in a *Consent Decree*, which is a document entered with a court.

Public involvement. Regulatory agencies must inform and involve the public in decision making. Public meetings are usually held to present the results of a site investigation, to discuss the content and recommendations of a feasibility study, before beginning construction, and at other critical points. Reports are available to the public for review.

Delegation of authority. Certain states are delegating the responsibility for environmental decisions to private individuals. For example, the Massachusetts Contingency Plan allows for work on certain sites to proceed under the direction of a Licensed Site Professional (LSP) rather than the oversight of the Department of Environmental Protection [310 CMR 40.0169]. An LSP must demonstrate suitable experience and pass a written examination in order to obtain a license. The Massachusetts Department of Environmental Protection periodically audits samples of LSPs' work in order to ensure compliance with the regulations.

Many regulatory programs also encourage *voluntary clean-ups*. By the end of 1995, 34 states had established voluntary clean-up programs by statute, regulation, or policy.[101] These programs provide guidelines for responsible parties to clean up sites without enforcement or detailed oversight by an agency.

Brownfields. Many states have promulgated *Brownfields* legislation or developed Brownfields programs within the context of existing legislation to encourage the redevelopment of contaminated industrial properties. Remediation and redevelopment of such properties can bring jobs to economically depressed areas and prevent the development of rural "green fields". The State of Rhode Island, for example, promulgated Chapter 19.14 of the General Laws, the *Industrial Property Remediation and Reuse Act*, in July 1995. This act offers a process for relief from liability for innocent parties such as people who lease or purchase a site, financial institutions who loan money on a Brownfields property, and parties who voluntarily clean up sites. If a site has been returned to conditions suitable for its intended use, or if a settling party has agreed to clean up a site according to a work plan approved by the Department of Environmental Management, the State will issue a Covenant Not to Sue. This covenant precludes further enforcement actions for certain environmental conditions.

The process of identifying, evaluating, and cleaning up a hazardous waste site can take many years and cost millions of dollars. Remediation of a federal Superfund site costs some $10 million, on average. RCRA Corrective Action costs, on average, $14.9 million per facility.[102]

2.6.3 Hazardous Waste Treatment, Storage, and Disposal

The treatment, storage, and disposal of hazardous wastes are regulated under the federal RCRA and corresponding regulations. State regulations can also apply. Following is a discussion of the RCRA regulations which apply most commonly to the remediation of hazardous waste sites.

2.6.3.1 Waste Classification

Perhaps counterintuitively, not all of the waste from a hazardous waste site is hazardous waste. Waste, soil, or sludge can be classified as hazardous waste or non-hazardous waste under federal and state regulations. Note that state regulations can be more stringent than federal regulations; in addition, certain states have "special waste" classifications.

At the federal level, the classification, handling, and disposal of hazardous waste is regulated under the RCRA. For a waste to be a hazardous waste under RCRA, it must (1) meet the definition of a *solid waste* and (2) be a *listed* waste or a *characteristic* waste.

In general, a "solid waste" is any material that is discarded by being abandoned (disposed of, burned, or accumulated prior to being disposed of or burned), recycled, or considered inherently waste-like. The regulations do not require that the waste actually be a solid material [40 CFR 261.2].

"Listed wastes" are literally listed in the regulations [40 CFR 261.30 through 261.33]. Wastes are listed in three ways:

- Hazardous wastes from nonspecific sources [40 CFR 261.31]; for example F001 waste is "the following spent halogenated solvents used in degreasing: tetrachloroethylene, trichloroethylene, methylene chloride, 1,1,1-trichloroethane, carbon tetrachloride, and chlorinated fluorocarbons; all spent solvent mixtures/blends used in degreasing containing, before use, a total of ten percent or more (by volume) of one or more of the above halogenated solvents or those solvents listed in F002, F004, or F005; and still bottoms from the recovery of these spent solvents and spent solvent mixtures."
- Hazardous wastes from specific sources [40 CFR 261.32]. For example, the column bottoms or heavy ends from the combined production of TCE and PCE are K030 waste.
- Discarded commercial chemical products, off-specification species, container residues, and spill residues thereof [40 CFR 261.33]. This category can include soil, water, or debris contaminated as a result of a spill of these materials. For example, discarded TCE or soil contaminated by a spill of discarded TCE would be classified as U228 waste.

The U.S. EPA recognizes that, due to variability in waste streams, a waste may be considered to be hazardous by definition, but may not meet the criteria that caused the waste to be listed. Therefore, RCRA regulations at 40 CFR 260.20 and 260.22 contain procedures which allow an individual to petition the EPA for a regulatory amendment to exclude a listed waste generated at a specific facility. In order to be *delisted*, the waste must not meet the criteria which caused the waste to be listed, and may not exhibit one or more of the characteristics of hazardous waste.

Characteristic hazardous wastes have one or more of four physical characteristics which would make them hazardous in the environment:

- *Ignitability.* This category includes: liquids with a flash point <60°C (140°F); nonliquids which can cause fire at standard temperature and pressure through friction, absorption of moisture, or spontaneous chemical changes; ignitable compressed gases; and oxidizers [40 CFR 261.21].
- *Corrosivity.* Corrosive materials are defined as aqueous materials with a pH ≤ 2 or pH ≥ 12.5, or those liquids which can corrode steel at a rate greater than 0.250 in./year at a test temperature of 130°F [40 CFR 261.22].

- *Reactivity.* A solid waste has this characteristic if it falls into one of eight categories of materials which are extraordinarily reactive, e.g., it reacts violently with water or forms potentially explosive mixtures with water [40 CFR 261.23].
- *Toxicity characteristic.* The *Toxicity Characteristic Leaching Procedure*, or *TCLP test*, measures the amount of specific hazardous substances that can leach from a solid waste under mildly acidic conditions; the test was designed to mimic the effects of disposal in a municipal landfill. The leachate is analyzed for 8 metals, 6 pesticides, and 25 organic compounds. Waste that leaches one or more contaminants in excess of specified levels is classified as a characteristic hazardous waste. For example, if a waste leaches TCE at a concentration greater than or equal to 0.5 mg/L, it is D040 waste [40 CFR 261.24]. The TCLP test replaced the *Extraction Procedure* (EP) *Toxicity Test* used previously.

As the examples listed above indicate, a single compound may be listed under many waste codes or may cause a waste to be a characteristic hazardous waste. In addition, a material may be classified as a hazardous waste if it has been mixed with, contains, or has been derived from a hazardous waste. Waste classification requires knowledge of the source of the waste and chemical analyses, and is not always clear cut.

2.6.3.2 Land Disposal Restrictions

Not all hazardous wastes can be put in a landfill. The Hazardous and Solid Waste Amendments (HSWA) to the RCRA, passed in 1984, include specific provisions restricting the land disposal of hazardous wastes. These land disposal restrictions, abbreviated *LDRs*, are sometimes called the *land ban*. LDRs prohibit the land disposal of certain wastes and specify treatment methods for other wastes.

Land disposal, as defined by the regulations, means "placement in or on the land, except in a corrective action management unit, and includes, but is not limited to, placement in a landfill, surface impoundment, waste pile, injection well, land treatment facility, salt dome formation, salt bed formation, underground mine or cave, or placement in a concrete vault or bunker intended for disposal purposes" [40 CFR 268.2(c)].

The LDRs apply to all characteristic and listed hazardous wastes with the exception of some wastes termed "newly listed" or "newly identified", which became subject to the regulations under RCRA after HSWA was passed in 1984, and for which EPA has not yet issued treatment standards. The general categories of wastes subject to the LDRs include: spent solvents [40 CFR 268.30]; dioxin-containing wastes [40 CFR 268.31]; the so-called California Wastes [40 CFR 268.32]; listed wastes; and certain newly identified/newly listed wastes for which treatment standards have been established. LDRs do not apply to wastes that do not meet the definition of a hazardous waste under RCRA.

LDRs include requirements for treating certain wastes before land disposal. The EPA established the treatment standards based on the *best demonstrated available technology (BDAT)* identified for the treatment of the waste being regulated. The types of treatment standards include:

- Treatment standards expressed as concentrations in a waste extract. For certain waste codes, EPA has determined the level in the TCLP extract which must be met

in order for the waste to be land disposed. This level may be met naturally or achieved by some form of treatment. With a few exceptions, these limits apply to nonwastewaters which contain metals [40 CFR 268.41].
- Treatment standards as waste concentrations. These regulations restrict the land disposal of waste which contains hazardous substances above allowed concentrations. For example, soil contaminated by a spill of discarded TCE (U228 waste) can contain no more than 5.6 mg/kg TCE if it is to be placed in a land dispoal unit [40 CFR 268.43].
- Treatment standard expressed as specified technologies. Certain wastes must be treated by a technology dictated by the regulations before the waste can be land disposed [40 CFR 268.42].

In addition, the underlying hazardous constituents in certain characteristic wastes must meet the *Universal Treatment Standards* [40 CFR 268.40(3)]. The *underlying hazardous constituents* are hazardous constituents (except fluoride, vanadium, and zinc), other than those that may cause a waste to exhibit a characteristic that can reasonably be expected to be present at the point of generation at a concentration above the Universal Treatment Standards. RCRA regulations specify these constituents as well as the corresponding Universal Treatment Standards, expressed as concentration limits [40 CFR 268.48].

RCRA regulations allow for exceptions to the LDRs under certain circumstances:

- Treatability variance. This option is available when the generator's waste is significantly different from the waste used to set the treatment standard, and, as a result, the treatment standard cannot be met or the BDAT technology is inappropriate [40 CFR 268.44].
- Equivalent treatment method petition. A generator may petition the EPA for approval to use a treatment method other than that specified in the regulations if it can be demonstrated that the alternative treatment method will achieve an equivalent measure of performance [40 CFR 268.42(b)].
- No Migration petition. The regulations include a procedure for requesting an exemption from LDR for a particular waste and disposal unit on the grounds that hazardous constituents will not migrate from the disposal unit (other than injection wells) for as long as the waste remains hazardous [40 CFR 268.6].
- Case-by-case extension. This option allows EPA to grant an extension of the effective date of an LDR requirement if an applicant submits a petition demonstrating a series of criteria as described in 40 CFR 268.5.
- Nationwide capacity variance. This provision of the regulations allowed for extensions of up to 2 years on the effective date of the treatment standards as they were phased in, recognizing that treatment capacity would have to be built in response to the regulations and waste treatment demand. Many of these extensions have expired.

The LDR include specific notification, certification, and recordkeeping requirements which a generator must follow.

2.6.3.3 Other RCRA Regulations

As indicated above, RCRA regulations govern the treatment, storage, and disposal (TSD) of hazardous wastes. The regulations include provisions that may apply to many aspects of hazardous waste site remediation, including, but not limited to:

- *Generator requirements* for storing and labeling wastes destined for off-site treatment or disposal, manifest requirements for shipment, limits on accumulation times, reporting requirements, and related provisions [40 CFR 262]. Generators must obtain an identification number from the U.S. EPA and fill out a manifest form (see Figure 2.7) which describes the waste, its origin, transporter, and its destination. Copies of the manifest are retained by the generator, transporter, TSD facility, and government agencies.
- Design, operation, and closure of specific TSD units: The RCRA regulations contain specific requirements for waste management units, including design criteria, operational requirements, and standards for *closure* when the unit will no longer be used. When waste materials are left in place at closure, *post-closure* requirements include inspection, maintenance, and groundwater monitoring, typically for a period of 30 years. Specific regulatory provisions apply to: containers [40 CFR 264.170 through 264.179]; tanks [40 CFR 264.190 through 264.200]; surface impoundments [40 CFR 264.220 through 264.232]; waste piles [40 CFR 264.250 through 264.259]; land-treatment units [40 CFR 264.270 through 264.283]; landfills [40 CFR 264.300 through 264.317]; and incinerators [40 CFR 264.340 through 264.351]. Additional requirements pertain to air emissions from tanks [40 CFR 264.200] and from impoundments [40 CFR 264.232].
- Air-emission standards for process equipment, including process vents [40 CFR 264.1030 through 264.1036] and equipment leaks [40 CFR 264.1065].
- Groundwater protection. RCRA-regulated TSD units may release contaminants to groundwater. For such occurrences, the regulations specify groundwater protection standards which are concentration limits for certain compounds [40 CFR 264.94], when and where those standards must be met [40 CFR 264.95, 96], and the monitoring program required to detect such releases [40 CFR 264.97].

2.6.4 Other Pertinent Regulations

Regulations pertaining to drinking water, wastewater treatment, air pollution control, wetlands and flood plains, transportation, and occupational safety and health, among others, may apply to remedial actions at hazardous waste sites. Some of the most commonly encountered federal regulations are described below. As noted previously, many of these federal regulations have state counterparts.

2.6.4.1 Clean Water Act

The objective of the Clean Water Act (CWA) is to restore and maintain the chemical, physical, and biological integrity of the nation's waters.[103] The CWA includes the following provisions:

- Effluent limitations on wastewater discharges to surface water must be based on the *best practicable control technology* currently available or *best conventional*

COMMONWEALTH OF MASSACHUSETTS
DEPARTMENT OF ENVIRONMENTAL QUALITY ENGINEERING
DIVISION OF HAZARDOUS WASTE
One Winter Street
Boston, Massachusetts 02108

Please print or type. (Form designed for use on elite (12-pitch) typewriter.)

UNIFORM HAZARDOUS WASTE MANIFEST

1. Generator US EPA ID No. | Manifest Document No. | 2. Page 1 of | Information in the shaded areas is not required by Federal law.

A. State Manifest Document Number MA C695275

3. Generator's Name and Mailing Address

B. State Gen. ID

4. Generator's Phone ( )

5. Transporter 1 Company Name | 6. US EPA ID Number | C. State Trans. ID

D. Transporter's Phone ( )

7. Transporter 2 Company Name | 8. US EPA ID Number | E. State Trans. ID

F. Transporter's Phone ( )

9. Designated Facility Name and Site Address | 10. US EPA ID Number | G. State Facility's ID Not Required

H. Facility's Phone ( )

11. US DOT Description *(Including Proper Shipping Name, Hazard Class, and ID Number)*	12. Containers No.	Type	13. Total Quantity	14. Unit Wt/Vol	I. Waste No.
a.					
b.					
c.					
d.					

J. Additional Descriptions for Materials Listed Above *(include physical state and hazard code.)* a. c. b. d.

K. Handling Codes for Wastes Listed Above a. c. b. d.

15. Special Handling Instructions and Additional Information

16. GENERATOR'S CERTIFICATION: I hereby declare that the contents of this consignment are fully and accurately described above by proper shipping name and are classified, packed, marked, and labeled, and are in all respects in proper condition for transport by highway according to applicable international and national government regulations.

If I am a large quantity generator, I certify that I have a program in place to reduce the volume and toxicity of waste generated to the degree I have determined to be economically practicable and that I have selected the practicable method of treatment, storage, or disposal currently available to me which minimizes the present and future threat to human health and the environment; OR, if I am a small quantity generator, I have made a good faith effort to minimize my waste generation and select the best waste management method that is available to me and that I can afford.

Printed/Typed Name | *Signature* | Date *Month Day Year*

17. Transporter 1 Acknowledgement of Receipt of Materials

Printed/Typed Name | *Signature* | Date *Month Day Year*

18. Transporter 2 Acknowledgement of Receipt of Materials

Printed/Typed Name | *Signature* | Date *Month Day Year*

19. Discrepancy Indication Space

20. Facility Owner or Operator: Certification of receipt of hazardous materials covered by this manifest except as noted in Item 19.

Printed/Typed Name | *Signature* | Date *Month Day Year*

GENERATOR / TRANSPORTER / FACILITY

In case of emergency or spill, immediately call the National Response Center (800) 424-8802.

MA C695275 COPY>1: FACILITY MAILS TO DESTINATION STATE

Form Approved OMB No. 2050-0039, Expires 9-30-91
EPA Form 8700-22 (Rev. 9-88) Previous editions are obsolete.

COPY>1: FACILITY MAILS TO DESTINATION STATE

Figure 2.7 Hazardous waste manifest.

pollutant control technology at a minimum. The *best available technology economically achievable* must be applied to toxic and nonconventional pollutants [40 CFR 125.3]. More stringent limits may be set on a wastewater discharge if necessary to protect the receiving water [40 CFR 122.44]. Such limits are based on *AWQC* for individual compounds and/or toxicity testing. Those criteria are set by states based on guidelines provided by the U.S. EPA. The limitations on the quantity and quality of a discharge to surface water are embodied in a *National Pollutant Discharge Elimination System* permit or the state equivalent. Sections 301, 302,

303, 304, 306, 307, 402, and 403 of the CWA contain the provisions which apply to discharges to surface water bodies.

- A wastewater discharge to a publicly owned treatment work (POTW) (i.e., sewage treatment plant) must meet pretreatment standards set by the POTW and approved by the regulatory agency. Sections 301 and 307 of the CWA pertain to such discharges.

- Discharge of dredge and fill material into the waters of the U.S. must be permitted under Section 404 of the CWA [33 CFR 323]. The *waters of the U.S.* include navigable waters, their tributaries, adjacent wetlands, and other waters or wetlands where degradation or destruction could affect interstate or foreign commerce. Thus, excavation and replacement of contaminated soil from a wetland, for example, requires a permit under the CWA. In addition, the state must certify that the discharge will not violate water quality standards; this is known as *Water Quality Certification*, and is required under Section 401 of the CWA.

2.6.4.2   Safe Drinking Water Act

The Safe Drinking Water Act (SDWA) regulates public water supplies, defined as systems having at least 15 service connections or serving an average of 25 year-round residents.[104] It specifies primary and secondary drinking water standards for drinking water:

- Maximum Contaminant Level Goals (MCLGs) are unenforceable health-based goals for public water systems. The U.S. EPA sets MCLGs at levels that would result in no known or anticipated adverse health effects with an adequate margin of safety;[105] accordingly, the MCLG for a probable carcinogen is zero. Regulations at 40 CFR 141 list promulgated MCLGs.

- Maximum Contaminant Levels (MCLs) are primary drinking water standards. The U.S. EPA has promulgated MCLs for common organic and inorganic contaminants [40 CFR 141]. These levels are enforceable standards. MCLs are set as close to MCLGs as possible, considering the technical and economic feasibility of removing the contaminant from the water supply.[105]

- Secondary Maximum Contaminant Levels (SMCLs) pertain to compounds that could cause an offensive taste, color, or odor in drinking water. They are not health based, nor are they enforceable standards. See 40 CFR 143 for promulgated SMCLs.

The SDWA also includes provisions to protect underground water supplies:

- The Underground Injection Control Program regulates the injection of hazardous waste or other waste, fluids associated with the production of oil or natural gas, or fluids associated with mineral extraction into injection wells [40 CFR 144 through 148]. These regulations protect underground water supplies.

- The Sole Source Aquifer Program [40 CFR 149] allows the U.S. EPA to designate an aquifer as a *sole source aquifer* if it is the sole or principal drinking water source for an area and, if contaminated, would present a significant hazard to human health. Such designation prevents federal activities that could contaminate a sole source aquifer and provides for demonstration programs designed to protect critical aquifer protection areas.

2.6.4.3 Occupational Safety and Health

Occupational Safety and Health Standards include regulations intended to protect the health and safety of workers engaged in hazardous waste operations and emergency response (HAZWOPER) [29 CFR 1910.120]. These regulations pertain to clean-up operations required by any government body at an uncontrolled hazardous waste site. They require, among other things:

- health and safety plans and precautions (personnel protective equipment, monitoring, decontamination) for dealing with hazardous materials;
- medical surveillance of workers; and
- employee training.

The HAZWOPER regulations specify four levels of personal protective equipment to be worn while working on a hazardous waste site. These levels refer primarily to the degree of respiratory protection.

- Level A is the highest level of personnel protection, used in highly hazardous conditions. Workers wear fully enclosed positive-pressure suits ("moon suits") and breathe supplied air.
- At Level B, workers breathe supplied air (e.g., compressed clean air in cylinders is supplied to workers through air lines attached to respirators, or workers wear self-contained breathing apparatus). Workers also wear protective clothing such as Tyvek®* coveralls, gloves, hard hats, and steel-toed boots.
- Level C is similar to level B, except that workers wear air-purifying respirators rather than breathing supplied air.
- Level D is the lowest level of protection. Workers do not use respiratory protection, but simply wear the protective clothing described for level B.

Occupational Safety and Health Standards also include requirements for routine construction work. Some of these requirements pertain to certain types of work at hazardous waste sites. Safety and Health Considerations for Construction [29 CFR 1926], for example, includes requirements for sloping or shoring excavations for personnel protection.

2.6.4.4 Transportation

The Department of Transportation (DOT) regulates the shipment of hazardous waste and hazardous materials [49 CFR Chapter I, Subchapter C, Hazardous Materials Regulations]. The DOT regulations identify hazardous materials for the purpose of transportation and dictate requirements for labeling, packaging, and quantity limits aboard aircraft and stowage of hazardous materials aboard vessels [49 CFR 172]. DOT regulations also include requirements for vehicle marking and manifests [49 CFR 171].

---

* Tyvek® is a registered trademark of the DuPont Company for its brand of spunbonded olefin.

2.7 COST ESTIMATION

Decisions which will determine the future of a piece of land or contaminated aquifer and commit millions of dollars of resources often hinge on cost estimates prepared during the feasibility study of a site. Such estimates depend on many assumptions, and, depending on the information used to develop the estimates, have varying degrees of accuracy.

2.7.1 Level of Accuracy

Cost estimates for site remediation may be prepared at different levels of accuracy, depending on the purpose for preparing the estimate and the level of design that has been completed. A cost estimate of $10 million may mean that the estimator expects the final project to cost between $5 and $20 million, or it may mean that she expects the final cost to reach $9 to $11.5 million. Common types of estimates include:

- *Screening-level cost estimate* — these estimates are very preliminary, and are intended only to be accurate to within +100/–50%. Such estimates are used only to identify extraordinarily expensive options during a preliminary evaluation of options.[106]
- *Order-of-magnitude cost estimate* — these estimates are intended to be accurate to within +50/–30% of the final cost. Order-of-magnitude estimates are based on a well-developed conceptual design that includes the volume of solid waste or flow rates of groundwater to be removed or treated; major equipment items; and estimates of residuals from treatment that will require disposal. These estimates are usually prepared during the detailed evaluation of alternatives in a FS, and form the basis for selection of a remedy.[106]
- *Final project cost estimate* — these estimates are based on the final design plans and specifications. Such estimates should be within +15/–10% of the actual cost.

The discussion of cost estimating which follows focuses on the preparation of order-of-magnitude estimates. Decisions based on such estimates may commit millions of dollars to remediation and determine the fate of contaminated land; nonetheless, order-of-magnitude estimates are often prepared with limited effort.

2.7.2 Basis for Estimates

A cost estimate can only be as good as the design used to prepare the estimate. Depending on the site and the remediation option, the conceptual design used to prepare an order-of-magnitude estimate should include:

- Remediation goals or clean-up levels;
- Area/volume of soil to be remediated, or extraction or treatment rate for groundwater;
- Site preparation requirements, such as clearing brush, surveying, grading the site, installing power lines, setting up temporary facilities such as an office trailer, decontamination facilities, and sanitary facilities;

- Depth of excavation, type of soil, need for shoring, depth to groundwater, and dewatering plans, if any;
- Site layout, showing major site features such as topography, buildings, utilities, and wetlands, and the proposed layout of the remediation system, temporary facilities, and staging areas;
- Site restoration plans, particularly if wetlands must be restored;
- If a treatment system is to be designed and built for the site: process flow diagram and/or preliminary piping and instrumentation diagram, preliminary component sizing and materials of construction, and preliminary indication of process control and instrumentation requirements;
- If a mobile treatment system is to be provided by a vendor, a list of potential vendors and their equipment capacities and limitations;
- Preliminary mass balance, indicating the ability to meet remediation goals, and/or estimating treatment residuals;
- Bench- and/or pilot-scale studies needed to complete design, if any;
- Waste classification and disposal plans for waste and/or treatment residuals;
- Monitoring requirements;
- Operation and maintenance requirements, including labor and materials;
- Utility requirements;
- Preliminary schedule, including assumed season of construction;
- Depending on the location and type of project, travel and per diem for workers;
- Permits required; and
- Health and safety precautions, including decontamination of personnel and equipment.

Uncertainties and assumptions should be noted.

The conceptual design provides the estimator with a basis for listing the major components to include in a cost estimate. The total project cost is calculated by summing the costs of these components. The estimator should focus on the highest cost items which will significantly affect the total estimate. Fine-tuning the estimates for lower-cost components can absorb a significant amount of the time and budget allotted to a project without making the total estimate significantly more accurate. For example, consider a groundwater extraction and treatment system which will cost approximately $100,000. For an order-of-magnitude estimate, it is not worth the effort to precisely determine whether a pump will cost $2000 or $2500, as the $500 difference is not significant in light of the magnitude and projected accuracy of the total estimate.

2.7.3 Components of Cost Estimates

Cost estimates include *capital costs* and *operation and maintenance costs*. Capital costs are those expenditures required to initiate and install a remedial action. Capital costs consist of *direct* (construction) and *indirect* (nonconstruction and overhead) costs.

Direct costs may include capital costs such as site preparation, excavation and staging of waste or soil, treatment or off-site disposal of wastes, materials and equipment, and labor for construction or installation. Indirect costs may include: engineering and supervision of field work, bench- and/or pilot-scale testing, contractor fees, permitting, and/or startup costs. Indirect costs usually include a contingency, which is added to the capital cost estimate to account for unforeseen circumstances that could add to the cost. Indirect costs are often estimated as a fraction of direct capital costs (see Section 2.6.5).

The costs for certain regulatory and procedural tasks are typically not included in engineering cost estimates, but can be quite significant for a hazardous waste project. These include: negotiating the scope and type of remedy with regulatory agencies, legal costs, agency oversight costs, and additional site investigation to provide information needed to complete the design.

Operation and maintenance (O & M) costs are post-construction costs necessary to ensure the continued effectiveness of the remedial action. O & M costs may include, for example, operators' labor, chemicals, power, monitoring, equipment replacement, disposal of treatment residuals, sewer use fees, and reporting.

O & M costs are usually calculated on two bases: *annual* and *present worth*. *Annual* costs are simply the yearly costs of operating and maintaining a remedial action. The *present-worth cost* (or *present value*) of O & M represents the amount of money which, if invested in the initial year of the remedial action and paid out for operation and maintenance as needed, would suffice to cover all of the operation and maintenance costs during the life of the project. These O & M costs may be annual costs or periodic costs. The cost estimate may also account for the present value of salvaging equipment at the end of a project. The present worth cost of annual costs and of periodic costs are calculated separately and then summed to determine the total present worth cost of O & M.

To calculate the present-worth cost, one must assume an interest rate and an inflation rate. Alternatively, one can assume a *discount rate*, which represents the rate of return on investments after inflation (therefore, inflation need not be factored into the calculation separately). The federal Office of Management and Budget (OMB) recommendations are usually used for federal Superfund projects. During the life of the Superfund program, the discount rate has generally fluctuated between 5 and 10%, depending on the state of the economy. As of 1996, OMB and the U.S. EPA recommend using a real discount rate (before taxes and after inflation) of approximately 3 to 7%, typically 7%.[107,108]

Present-worth calculations are also based on the project duration. For some projects, the project life can be estimated from design data. The default value for long-term projects is typically 30 years. While some projects might last longer than 30 years (e.g., for groundwater extraction/treatment; maintenance of a landfill cap), the present value of those future costs is usually negligible after 30 years.

The present-worth of *annual* O & M costs can be calculated as follows, assuming that the annual cost is constant:[109]

$$P = \{P/A, \, i\%, \, n\} \cdot A \qquad (2.46)$$

where: P = present-worth cost,
        A = annual cost,
        i = assumed interest (discount) rate,
        n = number of years (project life), and

Table 2.12   Factors for Calculating Present-Worth Costs

Year	i = 3% P/A	i = 3% P/F	i = 5% P/A	i = 5% P/F	i = 7% P/A	i = 7% P/F	i = 10% P/A	i = 10% P/F
1	0.9709	0.9709	0.9524	0.9524	0.9346	0.9346	0.9091	0.9091
2	1.9135	0.9426	1.8594	0.9070	1.8080	0.8734	1.7355	0.8264
3	2.8286	0.9151	2.7232	0.8638	2.6243	0.8163	2.4869	0.7513
4	3.7171	0.8885	3.5460	0.8227	3.3872	0.7629	3.1699	0.6830
5	4.5797	0.8626	4.3295	0.7835	4.1002	0.7130	3.7908	0.6209
6	5.4172	0.8375	5.0757	0.7462	4.7665	0.6663	4.3553	0.5645
7	6.2303	0.8131	5.7864	0.7107	5.3893	0.6228	4.8684	0.5132
8	7.0197	0.7894	6.4632	0.6768	5.9713	0.5820	5.3349	0.4665
9	7.7861	0.7664	7.1078	0.6446	6.5152	0.5439	5.7590	0.4241
10	8.5302	0.7441	7.7217	0.6139	7.0236	0.5083	6.1446	0.3855
11	9.2526	0.7224	8.3064	0.5847	7.4987	0.4751	6.4951	0.3505
12	9.9540	0.7014	8.8633	0.5568	7.9427	0.4440	6.8137	0.3186
13	10.6350	0.6810	9.3936	0.5303	8.3577	0.4150	7.1034	0.2897
14	11.2961	0.6611	9.8986	0.5051	8.7455	0.3878	7.3667	0.2633
15	11.9379	0.6419	10.3797	0.4810	9.1079	0.3624	7.6061	0.2394
16	12.5611	0.6232	10.8378	0.4581	9.4466	0.3387	7.8237	0.2176
17	13.1661	0.6050	11.2741	0.4363	9.7632	0.3166	8.0216	0.1978
18	13.7535	0.5874	11.6896	0.4155	10.0591	0.2959	8.2014	0.1799
19	14.3238	0.5703	12.0853	0.3957	10.3356	0.2765	8.3649	0.1635
20	14.8775	0.5537	12.4622	0.3769	10.5940	0.2584	8.5136	0.1486
21	15.4150	0.5375	12.8212	0.3589	10.8355	0.2415	8.6487	0.1351
22	15.9369	0.5219	13.1630	0.3419	11.0612	0.2257	8.7715	0.1228
23	16.4436	0.5067	13.4886	0.3256	11.2722	0.2109	8.8832	0.1117
24	16.9355	0.4919	13.7986	0.3101	11.4693	0.1971	8.9847	0.1015
25	17.4131	0.4776	14.0939	0.2953	11.6536	0.1842	9.0770	0.0923
26	17.8768	0.4637	14.3752	0.2812	11.8258	0.1722	9.1609	0.0839
27	18.3270	0.4502	14.6430	0.2678	11.9867	0.1609	9.2372	0.0763
28	18.7641	0.4371	14.8981	0.2551	12.1371	0.1504	9.3066	0.0693
29	19.1885	0.4243	15.1411	0.2429	12.2777	0.1406	9.3696	0.0630
30	19.6004	0.4120	15.3725	0.2314	12.4090	0.1314	9.4269	0.0573
50	25.7298	0.2281	18.2559	0.0872	13.8007	0.0339	9.9148	0.0085
100	31.5989	0.0520	19.8479	0.0076	14.2693	0.0012	9.9993	0.0001

{P/A, i%, n} is a factor calculated using Equation 2.47 or taken from reference tables (see Table 2.12).

$$\{P/A,\ i\%,\ n\} = \frac{(1+i)^n - 1}{i \cdot (1+i)^n} \qquad (2.47)$$

Note that in Equation 2.47 the interest rate should be expressed as a fraction, not as a percent (e.g., 0.10 rather than 10%).

Occasionally, remedial actions will entail periodic costs. For example, equipment may be replaced periodically during a long-term remedial action (e.g., pumps replaced every 5 years). The present worth of periodic costs can be calculated as follows:[109]

$$P = \{P/F, i, n\} \cdot F \qquad (2.48)$$

where F is the future cost and {P/F, i%, n} is a factor calculated from Equation 2.49 or taken from reference tables (see Table 2.12).

$$\{P/F, i\%, n\} = (1+i)^{-n} \tag{2.49}$$

When periodic costs will be incurred several times over the course of a project, a separate cost factor {P/F, i%, n} must be calculated for each time that the cost is incurred.

2.7.4　Sources of Information

Unit cost information can be obtained from experience at similar sites, vendor quotes, or literature reports. Use costs reported in the literature cautiously, for two reasons. First, these values become outdated quickly; by the time cost information is published, it is probably already 1 or 2 years old. Second, literature reports often provide very little of the information that is needed to extrapolate costs from one site to another. Unit costs can also be obtained from cost-estimating references. For examples, see References 110, 111, and 112. The U.S. EPA also compiles unit costs from vendors in the Vendor Information System for Treatment Technologies.[113] Use information from cost-estimating guides carefully. References prepared for routine construction work usually do not account for the increased cost resulting from conditions on a hazardous waste site. Personnel protective equipment (PPE), such as respirators and protective clothing, add to the cost of site remediation. Workers are also less productive wearing PPE.

Unit costs can be adjusted for productivity limitations based on the level of personnel protection used. Levels A through D, described previously, refer primarily to the level of respiratory protection. In addition to those precautions, workers on hazardous waste sites typically monitor the air to determine levels of dust and contaminants, and decontaminate clothing and equipment. All of these precautions decrease the productivity of workers and their equipment:[114]

- Working in level A reduces labor productivity to 10%–40% (typically 37%) of the productivity on a nonhazardous construction job, and equipment productivity to roughly 50%.
- Working in level B can reduce labor productivity to 25%–60% (typically 48%) and equipment productivity to roughly 60% of typical levels.
- Level C precautions can decrease labor productivity to 25%–60% (typically 55%) and equipment productivity to roughly 75% of levels for routine construction work.
- At level D, labor productivity can drop to 50%–90% (typically 82%) of the level typical for work at nonhazardous sites, but equipment productivity is not significantly affected.

Costs also depend on the region of the country in which a site is located. The estimating cost references provide indices that can be used to extrapolate costs from one region to another.

2.7.5   Shortcuts for Preliminary Estimates

Following are some rules of thumb for streamlining preliminary cost estimates. Use these rules of thumb with caution. They were developed for relatively large construction projects in traditional civil engineering applications or in the chemical process industry. These rules of thumb should not be applied blindly to small projects; for example, estimating engineering costs as a percentage of direct capital costs on a $50,000 job with heavy regulatory requirements will probably be very inaccurate. Cost estimates developed using general rules of thumb should always be given a good reality check.

2.7.5.1   Updating Capital Costs

Costs from past projects are frequently used to develop cost estimates. Prices can change considerably over time, so costs from a previous year may not yield accurate estimates for a current project. Prices can be updated to the current year by using one of several cost indices:

$$C_p = C_i \cdot \frac{I_p}{I_i} \qquad\qquad (2.50)$$

where: C_p = present cost,
        C_i = cost in the year i,
        I_p = index value at the present time, and
        I_i = index value at the time the original cost was obtained.

This method is reasonably accurate as long as the time period involved is less than 10 years.[115] One of several indices may be used, depending on the application:[115]

- Engineering News-Record (ENR) Construction Cost Index. This index tracks the variation in labor rates and materials costs for industrial construction. It is based on a composite cost for specified amounts of structural steel, lumber, concrete, and common labor. The index may be reported on one of three bases: an index value of 100 in 1913, 1949, or 1967. Current values are published weekly in the *Engineering News-Record*.
- Chemical Engineering Cost Index. This index tracks the variation in labor rates and materials costs for chemical plant construction based on four primary components: equipment, machinery and supports; erection and installation labor; buildings, materials, and labor; and engineering and supervision. The index is based on a value of 100 in 1957 to 1959. Current values are published monthly in *Chemical Engineering*.
- Marshall and Swift Equipment Cost Indices (all-industry equipment, process industry equipment). The all-industry equipment index is the arithmetic average of individual indices for 47 different types of industrial, commercial, and housing

equipment. The process-industry equipment index is a weighted average for eight categories: cement, chemicals, clay products, glass, paint, paper, petroleum, and rubber. The Marshall and Swift indices are based on an index value of 100 for 1926. Current values are published monthly in *Chemical Engineering*.

2.7.5.2 Scaling Equipment Costs

The cost for a piece of treatment equipment can be estimated from the cost for a differently sized (but similar) piece of equipment using the *Six-Tenths-Factor Rule:*[115]

$$C_a = C_b \cdot \left[\frac{S_a}{S_b} \right]^{0.6} \tag{2.51}$$

where: C_a = cost of equipment A,
   C_b = cost of equipment B,
   S_a = capacity (size) of equipment A, and
   S_b = capacity (size) of equipment B.

The exponential factor varies slightly for different types of equipment.[116] In general, this scaling method should not be used to extrapolate beyond a tenfold range in capacity.[115]

2.7.5.3 Estimating Costs as a Percentage of Capital Costs

Certain costs can be estimated as a percentage of the direct costs, as shown in Table 2.13. Apply these factors only to the pertinent capital costs. For example, consider a site where capital costs included excavation of an extraction trench, off-site incineration of the contaminated soil removed from the trench, and construction of a groundwater treatment plant. Estimates of the cost of piping for the treatment plant should not be based on the total capital costs for the entire project, but on the capital costs of the treatment plant alone. On the other hand, contingencies should be estimated as a percentage of the total capital cost, since contingency costs may be incurred for every aspect of construction.

2.7.6 Sensitivity Analysis

Preparation of preliminary cost estimates requires that the estimator make many assumptions. A sensitivity analysis tests the effect of varying some of the more critical assumptions on the total estimated cost. It may be used to identify uncertainties which should be factored into decision making, alert management to the potential range of costs, and/or determine the need for more data collection during design.

Table 2.13 Cost-Estimating Factors

Category	Item	Typical cost as percentage of total direct capital cost	Range (percentage)
Direct capital costs	Mobilization/demobilization		0.45–7[a][120]
	Installation of process equipment		25–55[b][118]
	Instrumentation and controls	5[117]	6–30[118]
	Piping[c]	10[117]	8–15[117,118]
	Electrical	10[117,118]	10–15[118]
Indirect capital costs	Engineering and supervision	15	7–30[118,119]
	Performance and payment bonds		2.5–3[120]
	Contractors' insurance[d]		1.7–11[120]
	Contractors' overhead and profit		2–15[e 118,120]
	Contingencies	15[117,119]	10–50
	Startup		5–20[118,119]
Annual operation and maintenance	General maintenance	5[118,119]	2–20[118]

[a] Depends on size of job, e.g., 0.45% for project > $75 million; 7% for project < $0.1 million.
[b] Applied as a percentage of the purchased cost of the particular piece of equipment.
[c] Refers to piping within a treatment process, not from groundwater extraction wells to the treatment unit.
[d] Includes Contractor's Pollution Liability Insurance, General Liability Insurance, Builder's Risk Insurance.
[e] The contractor's profit depends strongly on factors such as the distribution of work between prime and subcontractors, the size of the job, the degree of risk, the location, and the local economy. For more information, see for example the discussion in *Environmental Remediation Estimating Methods*.[120]

A sensitivity analysis can be used to estimate a range of probable costs for a remedial alternative, rather than a single estimate. A sensitivity analysis is particularly important when comparing two alternatives that have very similar order-of-magnitude costs and appear to be similarly feasible. It can also be useful in identifying which aspects of the site remediation affect the cost the most, so that the engineer can focus on those aspects of the design and (ideally) optimize the cost.

An estimator may vary capital costs or O & M costs in a sensitivity analysis; she should focus on factors that could significantly change the overall costs of a remediation program with small changes in their values. Examples of such factors follow.

Capital costs typically depend primarily on the estimated volume of soil or sludge or the estimated rate of groundwater to be treated. Such estimates are frequently based on preliminary clean-up levels or limited data. Relatively small changes to a clean-up level, or field conditions different than those projected from limited data, can significantly increase the cost of remediation. The unit cost of treatment or disposal may also be a critical component of a cost estimate. Frequently the estimator will obtain a range of unit prices from vendors, as site-specific testing is often necessary to develop a solid price. A sensitivity analysis would evaluate the sensitivity of the total capital cost to variations in treatment prices. The total estimate of O & M costs may depend strongly on the project life, the discount rate, or the volume of treatment residuals generated for disposal.

Problems

Note: Many of the problems in this book are based on real waste sites. However, the characteristics of the sites have been altered in order to simplify the problems or to illustrate particular points. Unless a site in a problem is named and a reference is provided for the information, the site descriptions in the examples and problems in this book are not intended to accurately represent the actual conditions at any one site.

2.1   Define the terms that you listed for Problem 1.1.

2.2   This problem concerns different liquids containing styrene which are each pumped into a separate tank. In each case, the tank is covered, but is vented so that it is at atmospheric pressure. Styrene is an aromatic hydrocarbon with the formula $C_6H_5C_2H_3$. It has a solubility of 300 mg/L and a vapor pressure of 5 mmHg at 20°C.[15] What is the concentration of styrene in the air in the headspace of the tank (in mg/m³) under each of the following scenarios? Note any assumptions used to calculate the answers.

[a]   The liquid is groundwater which contains styrene at an average concentration of 42 mg/L.
[b]   The liquid is a LNAPL, which contains styrene at a mole fraction of 0.15.
[c]   The liquid is pure styrene.

2.3   Illustrate the following general trends with two pairs of compounds. Base your answer on the information provided in the figures and tables in this chapter.

[a]   Increasing molecular weight decreases solubility, all else being equal.
[b]   Increasing molecular weight decreases volatility, all else being equal.
[c]   Increasing polarity increases water solubility, all else being equal.

2.4   The potential applicability of a treatment technology to a contaminant can usually be gauged by considering the basic mechanisms by which the technology works and the chemical characteristics of the contaminant. Consider three forms of groundwater treatment:

- Air stripping transfers a contaminant from the water to an air stream.
- Carbon adsorption removes contaminants from water by adsorption to granular activated carbon.
- Aerobic biological treatment destroys contaminants by enhancing biological oxidation.

Based only on the information in this chapter, which technologies should effectively remove each of the following contaminants:

[a]   acetone?
[b]   benzene?
[c]   PCBs?

Briefly present your reasoning.

2.5   Is LNAPL likely to be present at the ASR site described in Chapter 1? DNAPL? In each case, why or why not?

2.6   Figure 2.8 shows a plan view of the Hypothetical Chemical Company (HypoChem) facility. A limited site investigation was performed as part of a property transfer

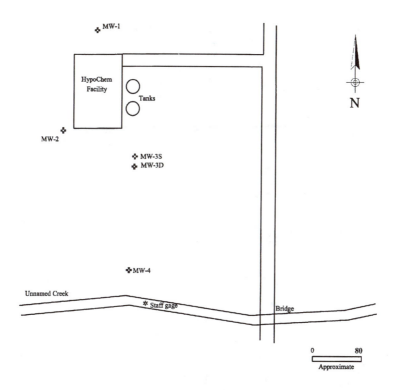

Figure 2.8 HypoChem Facility, Site Plan.

assessment by a prospective purchaser. Financial institutions commonly require a property transfer assessment when granting a real estate loan. The goal of a property transfer assessment is to determine whether or not contamination is present, whether or not that contamination requires remediation, and, if so, to estimate the financial liability.

The facility includes a building built in 1972 by the current occupant. Toluene is stored in two tanks adjacent to the building. The ground surface is relatively flat near the facility and slopes toward an unnamed creek to the south of the facility. Soil borings indicated that the soil at the site is a coarse sand with some silt to approximately 12 ft below ground surface, underlain by a layer of clay some 7 ft thick. Five monitoring wells (four shallow, one deep) were installed in the soil borings and a staff gage was placed in the creek to measure the water level. The elevations at the monitoring wells and the staff gage were surveyed, and water levels were measured and are reported in Table 2.14. Those data, with the field observations of the soils in the borings, were used to develop a north–south cross section (Figure 2.9).

The field team estimated that the creek is about 5 ft wide at the staff gage. Based on the flow of a leaf in the creek, one team member estimated the flow in the creek at 5 ft per minute. A person who had worked at the facility for over 20 years told the field team that the water in the creek flows even in dry weather.

Soil and groundwater samples were collected for analysis for VOCs. The laboratory has not yet completed the soil analysis, but has provided the data in Table 2.15 for groundwater samples.

Table 2.14   Groundwater and Surface Water Elevations, Hypothetical Chemical Company

Monitoring point	Ground surface elevation (ft)	Water level (ft)
MW-1	210.0	205.21
MW-2	208.8	204.10
MW-3S	207.4	202.73
MW-3D	—	204.92
MW-4	206.5	201.32
Staff gauge	198.3[a]	200.10

[a] Elevation at bottom of creek.

Table 2.15   Groundwater Data Hypothetical Chemical Company

Well	Toluene (mg/L)
MW-1	ND
MW-2	0.21
MW-3S	42.7
MW-3D	[a]
MW-4	18.2

[a] Sample bottle broke in shipment to the laboratory.

The hydrogeologist and ecologist in the office are both out sick, but the client wants an update on the project immediately, so you've been asked to help the project team evaluate the data by answering the following questions:

[a] Given the available data, is the lower aquifer likely to be contaminated? Why or why not?

[b] Are site conditions likely to cause a risk to aquatic organisms in the creek? A search of the ecologist's filing cabinet revealed that EPA does not have an AWQC for toluene. However, the freshwater criterion calculated using the Great Lakes Water Quality Initiative Tier II methodology is 130 µg/L. (This methodology is similar to that used to derive AWQC.)[121] (*Hints: This is a mass balance problem. Assume a hydraulic conductivity of* 1×10^{-3} *cm/sec.*)

2.7   Consider the American Creosote Works site described in Example 2.8.

[a] Calculate a clean-up level for carcinogenic PAHs in surface soil, using a CSF for B(a)P of 7.3 (mg/kg/d)$^{-1}$ and the equations in this chapter.[80] Assume residential use of the land and focus on the ingestion pathway. Use the default assumptions provided in the examples in this chapter if you think they are appropriate; if not, develop and justify your own exposure factors. How does this clean-up level compare to the concentration selected in the ROD?

[b] Perform a similar calculation for dioxin (as TCDD-TE), based on a CSF of 150,000 (mg/kg/d)$^{-1}$.

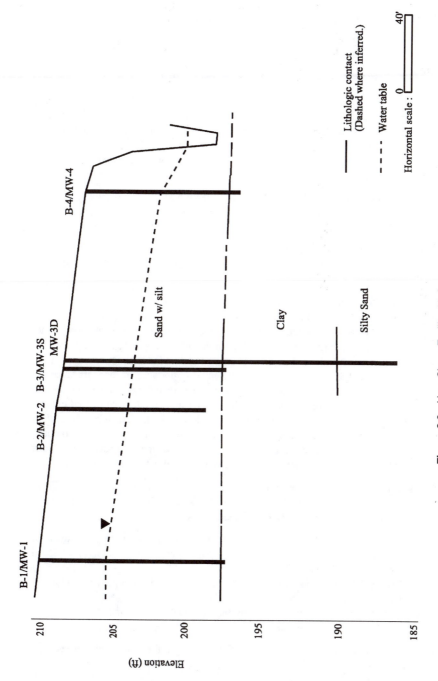

Figure 2.9 HypoChem Facility, N–S cross section.

Table 2.16   Half-Life Data
Problems 2.8 and 2.10

Compound	Half-life in groundwater (days)	Ref.
Benzene	~7	123
Phenol	0.5 to 7	124
Chlorobenzene	136 to 300	125

2.8   Consider a hypothetical site where the silty soil is contaminated with chlorobenzene, which has a k_{oc} of 219 L/kg and a k_H of 0.152 (unitless).[122] Contamination extends through the unsaturated zone within an area roughly 50 ft (width perpendicular to groundwater flow) by 100 ft (length parallel to groundwater flow). The aquifer is 35 ft thick. The downgradient edge of this area of soil is located approximately 100 ft from a well used for water supply.

The soil has a porosity of 0.40. The aquifer has a hydraulic conductivity of 2.3×10^{-5} cm/sec and a hydraulic gradient of 0.025. The mean annual infiltration rate is approximately 20.2 in./year.

A remediation goal for this site is to ensure that the concentration of benzene in that well does not exceed the MCL of 0.10 mg/L. Based on Equations 2.32 through 2.34,

[a] Calculate a clean-up level for chlorobenzene in unsaturated soil based on the potential leaching to groundwater.

[b] Derive Equation 2.33 based on a mass balance.

[c] Sophisticated models account for contaminant attenuation throughout a plume. The model comprising Equations 2.32 through 2.34, however, accounts only for dilution beneath the source area. It does not allow for attenuation between the source and the receptor. Gauge the potential importance of biodegradation by increasing C_w to account for biodegradation between the downgradient edge of the area of contaminated soil and the water supply well. Assume that chlorobenzene biodegrades in a first-order reaction with the half-life indicated in Table 2.16. (Further assume that conditions in the aquifer do not limit biodegradation.) Recalculate the soil clean-up level based upon the revised C_w. Does this level vary significantly from the clean-up level calculated in part [a]?

2.9   For each of the three exposure scenarios below, develop a clean-up level for PCE in the soil at the ASR site described in Chapter 1. Use a CSF of 5.2×10^{-2} (mg/kg/d)$^{-1}$ and a URF of 5.8×10^{-7} (ng/kg/d)$^{-1}$,[80] and assume:

[a] Industrial use of the property in the future. Further assume that an industrial worker would ingest surface soil (i.e., top 6 in.) at a rate of 50 mg/d, for 250 d per year, for 25 years.

[b] Residential use of the property in the future. Use the assumptions provided in the examples in this chapter.

[c] Protection of groundwater. Assume that the goal of remediation is to achieve the MCL in the groundwater.

2.10   [a] For each of the compounds in Table 2.16, calculate the time required to aerobically biodegrade 100 mg/L (in water) to 1 mg/L. Assume a first-order reaction.

[b] What does this calculation indicate regarding the effect of chemical structure on reactivity (biological oxidation)?

REFERENCES

1. Peters, D. G., Hayes, J. M., and Hieftje, G. M., *Chemical Separations and Measurements: Theory and Practice of Analytical Chemistry*, W. B. Saunders, Philadelphia, 1974, 7-8.

2. Massachusetts Department of Environmental Protection, *Guidance for Disposal Site Risk Characterization — In Support of the Massachusetts Contingency Plan (Interim Final)*, Policy WSC/ORS-95-141, July 1995, 2:68–69.

3. Peters, J. A., Quality Control Infusion into Stationary Source Sampling, in *Principles of Environmental Sampling*, Keith, L. H., Ed., American Chemical Society, Washington, D.C., 1988, 330–331.

4. U.S. EPA, *Test Methods for Evaluating Solid Waste*, SW-846, Volume 1B, 8240-5, 8240-31, 8250-19.

5. Morrison, R. T. and Boyd, R. N., *Organic Chemistry, 3rd ed.*, Allyn and Bacon, Boston, 1973, 22.

6. Keely, J. F., *Performance Evaluations of Pump-and-Treat Remediations*, EPA/540/4-89/005, U.S. EPA, Robert S. Kerr Environmental Research Laboratory, Ada, OK, October 1989, 5.

7. U.S. EPA, *Handbook — Ground Water, Volume II: Methodology*, EPA/625/6-90/016b, U.S. EPA, Office of Research and Development, Washington, D.C., July 1991, 50–51.

8. Piwoni, M. D. and Keely, J. F., *Ground Water Issue: Basic Concepts of Contaminant Sorption at Hazardous Waste Sites*, EPA/540/4-89/005, U.S. EPA Superfund Technology Support Center for Ground Water, Robert S. Kerr Environmental Research Laboratory, Ada, OK, October 1990, 2–4.

9. Treybal, R. E., *Mass Transfer Operations, 3rd ed.*, McGraw-Hill, New York, 1980, 278–279.

10. U.S. EPA, *Cleaning Up the Nation's Waste Sites: Markets and Technology Trends*, EPA 542-R-92-012, Office of Solid Waste and Emergency Response, Washington, D.C., April 1993, 17.

11. U.S. EPA, *Final Guidance: Presumptive Remedies Strategy and Ex-Situ Treatment Technologies for Contaminated Ground Water at CERCLA Sites*, Office of Solid Waste and Emergency Response, Washington, D.C., EPA/540/R-96/023, Prepublication Copy, October 1996, A:4-5.

12. Massachusetts Department of Environmental Protection, *Policy for the Assessment, Investigation, and Remediation of Petroleum Releases, Interim Investigation Protocol Document*, DEP Publication #WSC-401-91, Boston, MA, April 9, 1991, 34.

13. North American Manufacturing Corporation, *North American Combustion Handbook*, 2nd ed., 1983.

14. Howard, P. H. and Meylar, W. M., Eds., *Handbook of Physical Properties of Organic Chemicals*, CRC Lewis Publishers, Boca Raton, 1997.

15. U.S. EPA, *Subsurface Contamination Reference Guide*, EPA/540/2-90/011b, Office of Emergency and Remedial Response, Washington, D.C., 1990, Table 3.

16. U.S. EPA, *Development of Advisory Levels for Polychlorinated Biphenyls (PCBs) Cleanup*, OHEA-E-187, Office of Health and Environmental Assessment, Washington, D.C., May 1986, 7:3.

17. Adriano, D. C., *Trace Elements in the Terrestrial Environment*, Springer-Verlag, New York, 1986, 46–49.

18. Adriano, D. C., *Trace Elements in the Terrestrial Environment*, Springer-Verlag, New York, 1986, 57–58.

19. Clement Associates, Inc., *Chemical, Physical and Biological Properties of Compounds Present at Hazardous Waste Sites*, prepared for U.S. EPA, September 27, 1985.
20. Adriano, D. C., *Trace Elements in the Terrestrial Environment*, Springer-Verlag, New York, 1986, 148.
21. Adriano, D. C., *Trace Elements in the Terrestrial Environment*, Springer-Verlag, New York, 1986, 106.
22. Adriano, D. C., *Trace Elements in the Terrestrial Environment*, Springer-Verlag, New York, 1986, 157, 176.
23. Adriano, D. C., *Trace Elements in the Terrestrial Environment*, Springer-Verlag, New York, 1986, 219–221, 253–254.
24. Adriano, D. C., *Trace Elements in the Terrestrial Environment*, Springer-Verlag, New York, 1986, 298–300.
25. Agency for Toxic Substances and Disease Registry, U.S. Department of Health and Human Services, *Toxicological Profile for Cyanide, Public Comment Draft*, ATSDR, Atlanta, GA, August 1995, 1–3, 4–5.
26. U.S. EPA, *Soil Screening Guidance: User's Guide*, Publication 9355.4-27, Office of Solid Waste and Emergency Response, Washington, D.C., July 1996, 19.
27. U.S. EPA, *Handbook — Ground Water, Volume I: Ground Water and Contamination*, EPA/625/6-90/016a, U.S. Environmental Protection Agency, Office of Research and Development Center for Environmental Research Information, Cincinnati OH, 1990, 82–84.
28. Fetter, C. W., *Applied Hydrogeology*, 3rd ed., Macmillan College Publishing, New York, 1994, 86.
29. Fetter, C. W., *Applied Hydrogeology*, 3rd ed., Macmillan College Publishing, New York, 1994, 98.
30. Freeze, R. A. and Cherry, J. A., *Groundwater*, Prentice-Hall, Englewood Cliffs, NJ, 1979, 29.
31. Freeze, R. A. and Cherry, J. A., *Groundwater*, Prentice-Hall, Englewood Cliffs, NJ, 1979, 36–39.
32. Feenstra, S., Cherry, J. A., and Parker, B. L., Conceptual Models for the Behavior of Dense Non-Aqueous Phase Liquids (DNAPLs) in the Subsurface, in *Dense Chlorinated Solvents and other DNAPLs in Groundwater*, Pankow, J. F. and Cherry, J. A., Eds., Waterloo Press, Portland, OR, 1996, 53–88.
33. Powers, J. P., *Construction Dewatering: New Methods and Applications*, 2nd ed., John Wiley & Sons, New York, 1992, 5.
34. U.S. EPA, *Handbook — Ground Water, Volume I: Ground Water and Contamination*, EPA/625/6-90/016a, U.S. Environmental Protection Agency, Office of Research and Development Center for Environmental Research Information, Cincinnati, OH, 1990, 78.
35. Freeze, R. A. and Cherry, J. A., *Groundwater*, Prentice-Hall, Englewood Cliffs, NJ, 1979, 49.
36. Freeze, R. A. and Cherry, J. A., *Groundwater*, Prentice-Hall, Englewood Cliffs, NJ, 1979, 25.
37. Darcy, H., *Les fountaines publiques de la ville de Dijon*, V. Dalmont, Paris, 1856, 674 p.
38. Freeze, R. A. and Cherry, J. A., *Groundwater*, Prentice-Hall, Englewood Cliffs, NJ, 1979, 72–73.
39. Powers, J. P., *Construction Dewatering: New Methods and Applications*, 2nd ed., John Wiley & Sons, New York, 1992, 39.

40. U.S. EPA, *Handbook — Ground Water, Volume I: Ground Water and Contamination*, EPA/625/6-90/016a, U.S. Environmental Protection Agency, Office of Research and Development Center for Environmental Research Information, Cincinnati, OH, 1990, 89-90.

41. Fetter, C. W., *Applied Hydrogeology*, 3rd ed., Macmillan College Publishing, New York, 1994, 356.

42. Kueper, B., Pitts, M., Wyatt, K., Simkin, T., and Sale, T., *Technology Practices Manual for Surfactants and Cosolvents*, AATDF Report TR-97-2, The Advanced Applied Technology Demonstration Facility Program, Rice University, Houston, TX, Prepared for the U.S. Department of Defense, February 1997, 6–11.

43. McWhorter, D. B. and Kueper, B. H., Mechanics and Mathematics of the Movement of Dense Non-Aquesous Phase Liquids (DNAPLs) in Porous Media, in *Dense Chlorinated Solvents and Other DNAPLs in Groundwater*, Pankow, J. F. and Cherry, J. A., Eds., Waterloo Press, Portland, OR, 1996, 89–128.

44. U.S. EPA, *Dense Nonaqueous Phase Liquids — A Workshop Summary, Dallas, Texas, April 16–18, 1991*, EPA/600/R-92/030, Robert S. Kerr Research Laboratory, Ada, OK, 3.

45. U.S. EPA, *Handbook — Ground Water, Volume II: Methodology*, EPA/625/6-90/016b, U.S. EPA, Office of Research and Development, Washington, D.C., July 1991, 47–48.

46. Newell, C. J., Acree, S. D., Ross, R. R., and Huling, S. G., *Ground Water Issue: Light Nonaqueous Phase Liquids*, EPA/540/S-95/500, U.S. EPA, Office of Solid Waste and Emergency Response, Washington, D.C., 1995, 4.

47. Pankow, J. F., Feenstra, S., Cherry, J. A., and Ryan, M. C., Dense Chlorinated Solvents in Groundwater: Background and History of the Problem, in *Dense Chlorinated Solvents and other DNAPLs in Groundwater*, Pankow, J. F. and Cherry, J. A., Eds., Waterloo Press, Portland, OR, 1996, 14.

48. Feenstra, S. and Cherry, J. A., Diagnosis and Assesssment of DNAPL Sites, in *Dense Chlorinated Solvents and other DNAPLs in Groundwater*, Pankow, J. F. and Cherry, J. A., Eds., Waterloo Press, Portland, OR, 1996, 395–474.

49. MacKay, D. M. and Cherry, J. A., Groundwater contamination: limits of pump-and-treat remediation, *Environ. Sci. Technol.*, 23, 630–636, 1989.

50. Mackay, D. M., Roberts, P. V., and Cherry, J. A., Transport of organic contaminants in groundwater, *Environ. Sci. Technol.*, 19, 384–392, 1985.

51. Keely, J. F., *Ground Water Issue: Performance Evaluations of Pump-and-Treat Remediations*, EPA/540/4-89/005, U.S. Environmental Protection Agency, Robert S. Kerr Environmental Research Laboratory, Ada, OK, 1989, 2-5.

52. U.S. EPA, *Handbook — Ground Water, Volume I: Ground Water and Contamination*, EPA/625/6-90/016a, U.S. Environmental Protection Agency, Office of Research and Development Center for Environmental Research Information, Cincinnati OH, 1990, 89,109-111.

53. U.S. EPA Office of Solid Waste and Emergency Response, *Groundwater Currents: Developments in Innovative Ground Water Treatment*, Issue No. 16, September 1996, 1.

54. Freeze, R. A. and Cherry, J. A., *Groundwater*, Prentice-Hall, Englewood Cliffs, NJ, 1979, 404–405.

55. U.S. EPA, *Contaminants and Remedial Options at Wood Preserving Sites*, Office of Research and Development, Cincinnati, OH, 1994, Table 3.

56. U.S. EPA, *Risk Assessment Guidance for Superfund, Volume 1, Human Health Evaluation Manual (Part A), Interim Final*, EPA/540/1-89/002, Office of Emergency and Remedial Response, Washington, D.C., 1989, 4:8.

57. Massachusetts Department of Environmental Protection, *Guidance for Disposal Site Risk Characterization, Interim Final Policy*, WSC/ORS-95-141, Bureau of Waste Site Cleanup and Office of Research and Standards, Boston, MA, July 1995, Table 2.1.

58. Shacklette, H. T. and Boerngen, J. G., *Element Concentrations in Soils and Other Surficial Materials of the Conterminous United States*, United States Geological Survey, Professional Paper 1270, U.S. Government Printing Office, Washington, D.C., 1984, 4–6.

59. Adriano, D. C., *Trace Elements in the Terrestrial Environment*, Springer-Verlag, New York, 1986, 109.

60. U.S. EPA, *EPA Region III Risk-Based Concentration Table*, July 1996, http://www.epa.gov/reg3hwmd/riskmenu.htm?=Risk+Guidance, July 1996.

61. U.S. EPA, *ECO Update: Ecotox Thresholds*, Office of Solid Waste and Emergency Response, Washington, D.C., EPA/540/F-95/038, Vol. 3, No. 2, January 1996, Table 2.

62. U.S. EPA, *Soil Screening Guidance: User's Guide*, Publication 9355.4-23, Office of Solid Waste and Emergency Response, Washington, D.C., July 1996, 1.

63. U.S. EPA, *ECO Update: Ecotox Thresholds*, Office of Solid Waste and Emergency Response, Washington, D.C., EPA/540/F-95/038, Volume 3, Number 2, January 1996.

64. U.S. EPA, *Drinking Water Regulations and Health Advisories*, EPA 822-B-96-002, Office of Water, Washington, D.C., October 1996.

65. U.S. EPA, *ECO Update: Ecotox Thresholds*, Office of Solid Waste and Emergency Response, Washington, D.C., EPA/540/F-95/038, Vol. 3, No. 2, January 1996, Table 2.

66. U.S. EPA, *Risk Assessment Guidance for Superfund, Volume 1, Human Health Evaluation Manual (Part A), Interim Final*, EPA/540/1-89/002, Office of Emergency and Remedial Response, Washington, D.C., 1989, 7:11.

67. U.S. EPA, *Risk Assessment Guidance for Superfund, Volume 1, Human Health Evaluation Manual (Part A), Interim Final*, EPA/540/1-89/002, Office of Emergency and Remedial Response, Washington, D.C., 1989, 7:6-11.

68. National Contingency Plan, 40 CFR 300.430(e)(2)(i)(A)(2).

69. U.S. EPA, *Risk Assessment Guidance for Superfund, Volume 1, Human Health Evaluation Manual (Part A), Interim Final*, EPA/540/1-89/002, Office of Emergency and Remedial Response, Washington, D.C., 1989, 8:6-7.

70. U.S. EPA, *Risk Assessment Guidance for Superfund, Volume 1, Human Health Evaluation Manual (Part A), Interim Final*, EPA/540/1-89/002, Office of Emergency and Remedial Response, Washington, D.C., 1989, 8:11.

71. U.S. EPA, *Risk Assessment Guidance for Superfund, Volume 1, Human Health Evaluation Manual (Part A), Interim Final*, EPA/540/1-89/002, Office of Emergency and Remedial Response, Washington, D.C., 1989, 6:1.

72. U.S. EPA, *Risk Assessment Guidance for Superfund, Volume 1, Human Health Evaluation Manual (Part A), Interim Final*, EPA/540/1-89/002, Office of Emergency and Remedial Response, Washington, D.C., 1989, 6:2-50.

73. U.S. EPA, *Risk Assessment Guidance for Superfund Volume I: Human Health Evaluation Manual, Supplemental Guidance, "Standard Default Exposure Factors"* Interim Final, OSWER Directive: 9285.6-03, Office of Emergency and Remedial Response, Washington, D.C., March 25, 1991.

74. U.S. EPA, *Risk Assessment Guidance for Superfund, Volume 1, Human Health Evaluation Manual (Part A), Interim Final*, EPA/540/1-89/002, Office of Emergency and Remedial Response, Washington, D.C., 1989, 6:21.

75. U.S. EPA, *Risk Assessment Guidance for Superfund, Volume 1, Human Health Evaluation Manual (Part A), Interim Final*, EPA/540/1-89/002, Office of Emergency and Remedial Response, Washington, D.C., 1989, Appendix A.

76. Anderson, P. D., Ruffle, B., and Gillespie, W., A Monte Carlo Analysis of Dioxin Exposures and Risks from Consumption of Fish Caught in Freshwaters of the United States Affected by Bleached Chemical Pulp Mill Effluents in *Proceedings of the 1992 Environmental Conference,* TAPPI, 879–880.

77. U.S. EPA, *Risk Assessment Guidance for Superfund, Volume 1, Human Health Evaluation Manual (Part A), Interim Final,* EPA/540/1-89/002, Office of Emergency and Remedial Response, Washington, D.C., 1989, 8:19-20.

78. U.S. EPA, *Proposed Guidelines for Ecological Risk Assessment, Fed. Regist.,* Vol. 61, No. 175, September 9, 1996, 47579.

79. U.S. EPA, *Soil Screening Guidance: User's Guide,* Publication 9355.4-23, Office of Solid Waste and Emergency Response, Washington, D.C., July 1996, 21–28.

80. U.S. EPA, *Soil Screening Guidance: User's Guide,* Publication 9355.4-23, Office of Solid Waste and Emergency Response, Washington, D.C., July 1996, Attachment D.

81. Cowherd, C., Muleski, G., Englehart, P., and Gillette, D., *Rapid Assessment of Exposure to Particulate Emissions from Surface Contamination.* Prepared for U.S. EPA, Office of Health and Environmental Assessment, Washington, D.C., 1985.

82. U.S. EPA, *Risk Assessment Guidance for Superfund, Volume 1, Human Health Evaluation Manual (Part A), Interim Final,* EPA/540/1-89/002, Office of Emergency and Remedial Response, Washington, D.C., 1989, 6:45–47.

83. Smith, R. L., *EPA Region III Risk-Based Concentration Table, Background Information,* U.S. EPA Region III, Philadelphia, PA (updated semiannually via http://www.epa.gov/reg3hwmd/riskmenu.htm?=Risk+Guidance), April 30, 1996.

84. U.S. EPA, *Risk Assessment Guidance for Superfund, Volume 1, Human Health Evaluation Manual (Part A), Interim Final,* EPA/540/1-89/002, Office of Emergency and Remedial Response, Washington, D.C., 1989, 6:35–38.

85. Agency for Toxic Substances and Disease Registry, *ToxFAQs: Methylene Chloride,* U.S. Department of Health and Human Services, Public Health Service, Atlanta, April 1993.

86. Clement Associates, *Chemical, Physical, and Biological Properties of Compounds Present at Hazardous Waste Sites, Final Report,* prepared for the U.S. Environmental Protection Agency, September 27, 1985.

87. U.S. EPA, *Record of Decision, American Creosote Works, Inc.,* EPA/ROD/R04-89/055, September 28, 1989.

88. U.S. EPA, Proposed guidelines for ecological risk assessment, *Fed. Regist.,* September 9, 1996, Vol. 61, No. 175, 47604–47605.

89. ENSR, *Feasibility Study Report, Former Koppers Site, Charleston SC, Volume I, Technical Report,* Document No. 0845-042-570, prepared for Beazer East, Inc., December 1996, Section 1.

90. U.S. EPA, *Superfund Proposed Plan Fact Sheet, Koppers Co., Inc. (Charleston Plant) NPL Site, Charleston, Charleston County, South Carolina,* U.S. EPA Region IV, Atlanta, GA, March 1997, 27 pp.

91. U.S. EPA, *Soil Screening Guidance: User's Guide,* Publication 9355.4-23, Office of Solid Waste and Emergency Response, Washington, D.C., July 1996, 28–32.

92. Massachusetts Contingency Plan, 310 CMR 40.0006.

93. Grady, C. P. L. and Lim, H. C., *Biological Wastewater Treatment — Theory and Applications,* Marcel Dekker, New York, 1980, 23–40.

94. Lin, K. H., Van Ness, H. C., and Abbott, M. M. Reaction Kinetics, Reactor Design, and Thermodynamics in *Perry's Chemical Engineers' Handbook,* 6th ed., Green, D. W., Ed., McGraw-Hill, New York, 1984, 4:5–27.

95. Grady, C. P. L. and Lim, H. C., *Biological Wastewater Treatment — Theory and Applications,* Marcel Dekker, New York, 1980, 128.

96. U.S. EPA, *About Innovative Treatment Technologies*, fact sheet last modified 9/12/96, http://www.clu-in.com, 4 pp.

97. The Comprehensive Environmental Response, Compensation, and Liability Act of 1980 (Superfund)(P. L. 96-510) as amended by The Superfund Amendments and Reauthorization Act of 1986 (P. L. 99-499), §121(b).

98. U.S. EPA, *Guidance for Conducting Remedial Investigations and Feasibility Studies Under CERCLA -Interim Final*, EPA/540/G-89/004, Office of Emergency and Remedial Response, Washington, D.C., 1988, 6:15.

99. U.S. EPA, *CERCLA Compliance with Other Laws Manual: Draft Guidance*, EPA/540/G-89/006, Office of Emergency and Remedial Response, Washington, D.C., 1988, xiii–xiv.

100. U.S. EPA, *The Role of Cost in the Superfund Remedy Selection Process*, EPA 540/F-96/018, Office of Solid Waste and Emergency Response, Washington, D.C., September 1996, 2.

101. U.S. EPA, *Cleaning Up the Nation's Waste Sites: Market and Technology Trends, 1996 Edition*, EPA 542-R-96-005, Office of Solid Waste and Emergency Response, Washington, D.C., April 1997, 9:2.

102. U.S. EPA, *Cleaning Up the Nation's Waste Sites: Market and Technology Trends, 1996 Edition*, EPA 542-R-96-005, Office of Solid Waste and Emergency Response, Washington, D.C., April 1997, 1:6.

103. U.S. EPA, *CERCLA Compliance with Other Laws Manual, Draft Guidance*, EPA/540/G-89/006, Office of Emergency and Remedial Response, Washington, D.C., 1988, A-13.

104. U.S. EPA, *CERCLA Compliance with Other Laws Manual, Draft Guidance*, EPA/540/G-89/006, Office of Emergency and Remedial Response, Washington, D.C., 1988, 4:3.

105. U.S. EPA, *CERCLA Compliance with Other Laws Manual, Draft Guidance*, EPA/540/G-89/006, Office of Emergency and Remedial Response, Washington, D.C., 1988,1:15.

106. U.S. EPA, *Remedial Action Costing Procedures Manual*, EPA/600/8-87/049, U.S. EPA Office of Solid Waste and Emergency Response, Washington, D.C., 1987, 3:2.

107. U.S. Department of Energy, *1996 Baseline Environmental Management Report*, U.S. DOE, Washington, D.C., 1996, Appendix E.2.

108. U.S. EPA, *The Role of Cost in the Superfund Remedy Selection Process*, Office of Solid Waste and Emergency Response. Washington, D.C., EPA/540/F-96/018, September 1996, 2.

109. Lindenburg, M. R., *Civil Engineering Reference Manual, 4th ed.*, Professional Publications, Belmont, CA, 1986, 2:2-4 and 2:19.

110. Rast, R. R., Senior Editor, *Environmental Restoration Unit Cost Book,* published annually by R.S. Means Company, Kingston, MA and Delta Technologies Group, Littleton, CO.

111. *Means® Site Work Cost Data*, published annually by R.S. Means Company, 100 Construction Plaza, P.O. Box 800, Kingston, MA 02364-0800.

112. *The Richardson Rapid System: Process Plant Construction Estimating Standards*, published annually by Richardson Engineering Services, 1742 S. Fraser Drive, P.O. Box 9103, Mesa, Arizona 85214-9103.

113. *Vendor Information System for Treatment Technologies (VISITT)*, U.S. EPA Technology Innovation Office, Washington, D.C., updated periodically.

114. Rast, R. R., Senior Editor, *Environmental Restoration Unit Cost Book 1996*, R. S. Means Company, Kingston, MA and Delta Technologies Group, Littleton, CO, 1996, 1:5.

115. Peters, M. S. and Timmerhaus, K. D., *Plant Design and Economics for Chemical Engineers, 4th ed.*, McGraw-Hill, New York, 1991, 164–169.

116. Holland, F. A., Watson, F. A., and Wilkinson, J. K., Process Economics, in *Perry's Chemical Engineers' Handbook*, 6th ed., Perry, R. H., Green, D. W., and Maloney, J. O., Eds., McGraw-Hill, New York, 1984, 25.

117. U.S. EPA. *Innovative and Alternative Technology Assessment Manual*, EPA 430/9-78-009, Office of Water Program Operations, Washington, D.C., February 1980, A:5.

118. Peters, M. S. and Timmerhaus, K. D., *Plant Design and Economics for Chemical Engineers, 4th ed.*, McGraw-Hill, New York, 1991, 167–173.

119. U.S. EPA, *Remedial Action Costing Procedures Manual*, EPA/600/8-87/049, U.S. EPA Office of Solid Waste and Emergency Response, Washington, D.C., 1987, 3:10–11, 18.

120. Rast, R. R., *Environmental Remediation Estimating Methods*, R. S. Means Company, Kingston, MA, 1997, 548–558.

121. U.S. EPA, ECO Update, Office of Solid Waste and Emergency Response, Washington, D.C., Volume 3, No. 2, January, 1996, 12.

122. U.S. EPA, *Soil Screening Guidance: User's Guide*, Publication 9355.4-23, Office of Solid Waste and Emergency Response, Washington, D.C., July 1996, Attachment C.

123. Mackay, D., Shiu, W. Y., and Ma, K. C., *Illustrated Handbook of Physical-Chemical Properties and Environmental Fate for Organic Chemicals, Volume I*, Lewis Publishers, Chelsea, MI, 1992, 142.

124. Howard, P. H., Boethling, R. S., Jarvis, W. F., Meylan, W. M., and Michalenko, E. M., *Handbook of Environmental Degradation Rates*, Lewis Publishers, Chelsea, MI, 1991, 414.

125. Howard, P. H., Boethling, R. S., Jarvis, W. F., Meylan, W. M., and Michalenko, E. M., *Handbook of Environmental Degradation Rates*, Lewis Publishers, Chelsea, MI, 1991, 412.

CHAPTER 3

Groundwater Remediation

The general objectives for groundwater remediation, subject to site-specific considerations, are to:[1]

- Prevent exposure to groundwater contaminated at unacceptable levels;
- Contain the plume (i.e., minimize or prevent migration of contaminants);
- Control the source (i.e., minimize further groundwater contamination as a result of leaching from soils or NAPL); and
- Restore the aquifer to the expected beneficial use, to the extent practicable. The area over which an aquifer is to be restored to ARARs or health-based clean-up levels is defined by the *point of compliance*. The U.S. EPA defines the point of compliance as the boundaries of a waste management unit, if waste is left in place on a site. Otherwise, clean-up goals are to be met throughout a plume. An alternative point of compliance may be appropriate under certain circumstances.[2] Some state regulations define the point of compliance differently.

Figure 3.1 illustrates the effect of site characteristics on the ease of remediation. Source control and aquifer restoration goals may be *technically impracticable* at sites where DNAPL or certain hydrogeologic conditions exist. The U.S. EPA has developed a procedure to formally determine when remediation is technically impracticable at a site.[3]

Contaminated groundwater can be remediated in several ways, depending on the type of contaminant, hydrogeology, clean-up goals, and regulatory requirements:

- Natural attenuation is the "biodegradation, dispersion, dilution, sorption, volatilization, and/or chemical and biochemical stabilization of contaminants to effectively reduce contaminant toxicity, mobility, or volume to levels that are protective of human health and the environment.[4] When natural attenuation is selected as the remedy for a site, conditions are monitored regularly to track progress. If those data indicate that conditions could present a new threat to human health and the environment, more active remediation is required.
- *In situ* treatment. Groundwater can sometimes be treated in place, or *in situ. In situ* treatment can take several forms (see Section 3.5). For example, nutrients and an oxygen supply can be introduced into an aquifer to stimulate biodegradation of

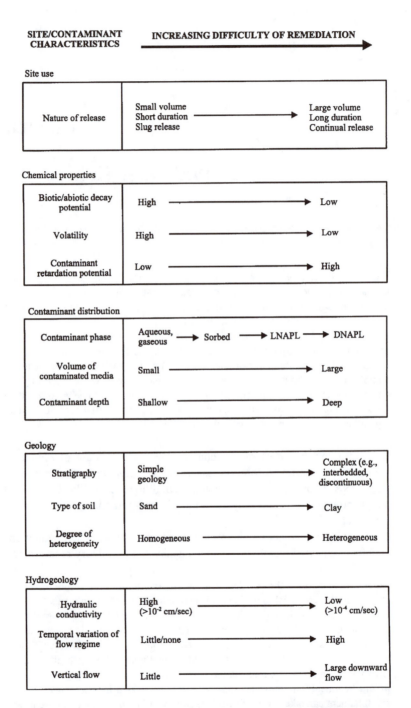

Figure 3.1 Factors affecting groundwater remediation potential. (Adapted from U.S. EPA, *Final Guidance: Presumptive Response Strategy and Ex-Situ Treatment Technologies for Contaminated Ground Water at CERCLA Sites*, EPA/540/R-96/023, Office of Emergency and Remedial Response, Washington, D.C., Prepublication copy, October 1996, 2.)

contaminants; air can be injected into an aquifer to strip volatile compounds from the groundwater; or physical barriers can be used to direct groundwater flow through a treatment cell, as in a reactive iron wall.

- Containment. Physical barriers (Section 3.1) or hydraulic gradients induced by pumping wells (Section 3.2) can be used to contain a plume and prevent contamination from spreading through an aquifer. While some mass of contamination may be removed from an aquifer, mass reduction is not the primary goal.
- Extraction and treatment. Contaminated groundwater and/or NAPL can be removed from an aquifer (Sections 3.2 and 3.6), treated (Sections 3.4 and 3.7), and discharged (Section 3.3). Also called *pump and treat*, these groundwater remedies are typically designed both to contain a plume, and to remove a mass of contamination from an aquifer and restore the original groundwater quality.

The distinctions between these various remediation strategies become blurred for some technologies. For example, dual-phase extraction of LNAPL incorporates both extraction and treatment, and *in situ* treatment (biodegradation of residual NAPL in soil).

Treatment technologies used to effect these remediation strategies can be classified according to the fate of the contaminant(s). Treatment may *separate* a contaminant from the groundwater, *immobilize* the contaminant, or *destroy* the contaminant. Contaminants which are separated from the groundwater or immobilized require further treatment or disposal, or may be recycled. The physical and chemical properties of the contaminant molecule determine how it may be separated, immobilized, or destroyed.

This chapter includes common rules of thumb for applying or designing various technologies. These rules are general guidelines, often based on the cost-effective application of commonly available equipment. When remediation options are limited or in certain other circumstances, it may be most cost effective to apply a technology despite a rule of thumb which suggests otherwise. For example, natural constituents of groundwater such as hardness can foul certain types of treatment equipment, causing increased maintenance costs. Rules of thumb have been developed for concentrations at which pretreatment to remove hardness is necessary. At a site where the level of hardness just exceeds this rule of thumb criterion, the additional capital cost of installing pretreatment equipment may far exceed the added maintenance cost which would result from equipment fouling if the groundwater was not pretreated. In such a case, adhering to the rule of thumb would not be cost effective.

Certain makes and models of equipment are mentioned in this chapter for illustration. Mention of a product in this book does not constitute endorsement.

3.1 PHYSICAL BARRIERS

Vertical barriers are used to contain contaminated groundwater or direct its flow. They are sometimes used to encircle an area in order to contain concentrated contamination. Groundwater must be pumped from inside the area to prevent it from filling up with water (the "bathtub effect"). An impermeable cap is sometimes placed over the area to minimize the quantity of water that must be pumped. A recent application of subsurface barriers is the construction of *funnel and gate* treatment systems. In such a system, vertical barriers are used to direct groundwater flow to a permeable "gate" in the barrier where *in situ* treatment can be effected. (See Section 3.5.)

Common types of vertical barriers include slurry walls and sheet piling. Vertical barriers may be *keyed* into a confining layer: installing the barrier some 2 to 3 ft into a confining layer prevents groundwater from flowing under the wall. When the confining layer is quite deep, or the objective is to contain shallow LNAPL or groundwater, the barrier may only partially penetrate into an aquifer. This configuration is called a *hanging* wall. Groundwater can flow under a hanging wall.

A *slurry wall* is constructed in a vertical trench excavated under a slurry. The trench is filled with a soil–bentonite mixture or a cement–bentonite mixture to form a barrier to groundwater flow.

The trench, typically 0.5 to 1 m in width, can be installed using a backhoe, clamshell, or other equipment. A backhoe can be used to install a slurry wall up to 80 ft deep; deeper depths (e.g., 120 ft or more) can be achieved using different equipment.[5] The excavator opens the trench under a slurry of bentonite and water. This slurry shores the trench to prevent collapse. As the water in the slurry leaches into the surrounding soils, the slurry forms a filter cake on the walls of the trench. This filter cake forms a low-permeability layer on the walls of the trench.

A soil–bentonite slurry wall is constructed by mixing the excavated soils with the bentonite slurry and gradually backfilling the trench. Backfilling begins by lowering the fill into the bottom of one end of the trench using a clamshell. When the mound of backfill reaches the surface, the remainder of the trench is filled by pushing fill down the slope of the mound into the trench using a bulldozer. Operators gradually move down the trench until the entire trench is backfilled.

The ideal soil for a soil–bentonite slurry wall is silty sand. Chunks of clay, rocks bigger than 6 in., and roots or other organic material must be removed from the soil before mixing with the bentonite slurry. Under ideal conditions, a soil–bentonite slurry wall can have a permeability as low as 10^{-7} cm/sec.[6]

Installation of a soil–bentonite slurry wall requires level ground and a large work area to mix the soil with the bentonite slurry. Soil–bentonite slurry walls are susceptible to attack by incompatible chemicals. NAPL can cause dessication and cracking. Strong acids, bases, or salt solutions will attack the slurry wall, as can high concentrations of certain organic compounds (particularly alcohols).[5]

A cement–bentonite slurry wall is similar to a soil–bentonite slurry wall, except that the trench is excavated using a slurry of Portland cement, bentonite, and water. This slurry is left in place to form the barrier. The excavated soil must be disposed of elsewhere; if the soil is contaminated, this can represent a significant expense.

Cement–bentonite construction has several other advantages and disadvantages, compared to a soil–bentonite slurry wall.[5] One advantage is that a cement–bentonite wall can be installed in an area where the ground surface is not level, because the slurry sets up into a semirigid solid. A cement–bentonite wall is also stronger, which may be important when installing a slurry wall next to a road or building. Finally, because the soil is not mixed with the bentonite slurry, installation of a cement–bentonite slurry wall requires less space for construction. Cement–bentonite slurry walls have several disadvantages compared to soil–bentonite slurry walls: slightly higher permeability, higher cost, and less resistance to chemical attack.

A variation on the slurry wall is the vibrated beam wall. An I-beam is pounded or vibrated into permeable soil. As the beam is pulled back out of the soil, a mixture of bentonite/cement or asphalt is pumped into the hole. The holes are overlapped to create a wall approximately 0.1 m thick.[7]

Steel *sheet piling* is also used to create a subsurface barrier. A series of interlocking steel sheets are driven into the ground, typically to depths of 25 ft or less. Different configurations of sheet piling are used. In cross section, a section may be linear, shaped like a flattened U, or Z-shaped. Each sheet is typically between 18 and 30 in. wide. Steel sheet piling cannot be installed in rocky soil: boulders or large pieces of rubble will damage the sheet piling as it is driven into the soil, or prevent it from being driven to the desired depth.

The effectiveness of successfully installed sheet piling at containing groundwater depends on the seal between each section. Initially, sheet piling leaks at the seams. A seal gradually forms as fine soil particles wash into the interlocking joints. Alternatively, the joints may be sealed at installation by injecting bentonite, cement, or epoxy.

As discussed further below, steel can be corroded by certain contaminants or natural constituents of groundwater. The design life of a sheet-pile wall is generally in the range of 7 to 40 years.[8]

3.2 GROUNDWATER EXTRACTION

Groundwater extraction and treatment, commonly called *pump and treat*, can be used to remove contaminant mass and restore an aquifer to clean-up goals, or to manage contaminant migration by controlling hydraulic gradients and thus the spread of contamination. Groundwater is commonly pumped from extraction wells, well points, or an extraction trench. Each of these extraction methods is discussed below. Treatment and discharge of the extracted groundwater are discussed in Sections 3.4 and 3.5.

3.2.1 Extraction Wells

The design of an extraction well system depends on the aquifer characteristics, the purpose of the system (gradient control vs. aquifer restoration), and the size of the plume. The aquifer characteristics determine how much water can and should be pumped from each well. The siting of the wells and, as a result, the total quantity of water to be pumped, depends on the purpose of the system and the size of the plume.

3.2.1.1 Well Construction and Pumps

Figure 3.2 shows a typical extraction well in a water table aquifer. The well shown in Figure 3.2 is oriented vertically; horizontal wells are occasionally used. The major components of a typical well are the casing, screen, filter pack and seal, and pump.[9-11]

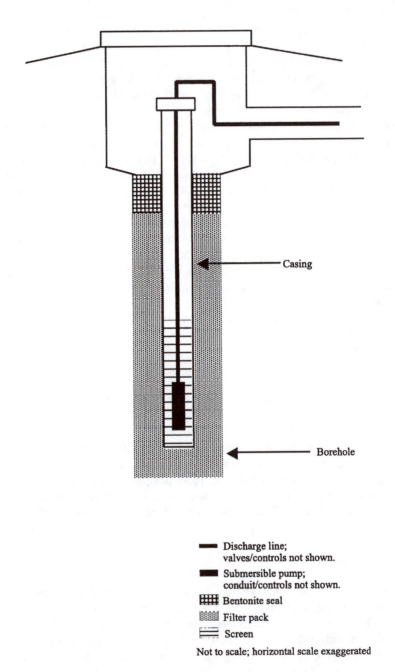

Discharge line;
valves/controls not shown.

Submersible pump;
conduit/controls not shown.

Bentonite seal

Filter pack

Screen

Not to scale; horizontal scale exaggerated

Figure 3.2 Schematic of extraction well in water table aquifer.

Groundwater flows into the well through the *screen*. The screen is commonly made of slotted polyvinyl chloride (PVC) or stainless steel pipe or continuous-slot shaped steel wire. The slots must be small enough to prevent excess soil particles or filter pack material from entering the well. However, the total slot area must be large enough so that the velocity of water through the slots is less than a critical value related to head loss through the screen. The length of the screen is based on the type, stratification, and thickness of the aquifer and the available drawdown.

The *casing* is sized to accommodate the pump. Typically, the casing is 4 in. in diameter or larger, although pumps which fit into 2-in.-diameter wells are available. However, it is more difficult to install equipment into and maintain the 2- and 4-in.-diameter wells. In general, a 4-in. well can accommodate a submersible pump with a capacity of up to 70 gallons per minute (gpm); a 6-in. well can accommodate a pump with a capacity up to 120 gpm.[11] Even larger-diameter wells are used in some dewatering applications, or where two pumps are to be installed in the well for recovery of both NAPL and water.

The *filter pack*, typically coarse to medium sand or fine gravel, surrounds the screen and casing. The filter pack has several functions. It holds the borehole open, prevents excessive soils from entering the well, and allows water to flow freely into the well. The filter pack material is selected based on the particle size of the aquifer materials.

The *pump* provides the mechanical energy needed to draw water into the well and convey it up to the ground surface and through the header (piping) to the treatment plant. Two types of pumps are commonly used to extract groundwater: *pneumatic pumps* and electric *submersible pumps*. Other types of pumps are sometimes used.

Pneumatic pumps are commonly used in low- to moderate-flow applications (up to, say, 10 gpm). A pneumatic pump uses compressed air to displace water. The pump comprises a cylinder with an air intake hose and a water discharge hose. An air compressor at the surface supplies the air. Water flows into the cylinder through a check valve until the chamber fills and triggers a level sensor. (A *check valve* is a one-way valve that allows a fluid to enter, but not to flow back out.) The pump control system then opens a valve on the air line. As air enters the pump chamber, it pushes the water out a check valve and up through the discharge line. When all of the water is displaced, the cycle begins again. In the simplest pneumatic pumps, the contact between air and water can strip volatile compounds from the groundwater into the air stream. Aeration of the water can also cause metals such as iron to oxidize and precipitate. Certain pneumatic pumps are designed to eliminate the direct contact between air and water. Pneumatic pumps can pump solids in groundwater and can pump dry without damage. They are generally less prone to fouling than electric submersible pumps.

Electric submersible pumps can be used in low- to high-flow applications. A *centrifugal pump* contains one or more impellers driven by an electrical motor. The motor is mounted on top of the cylindrical pump unit. The entire unit is suspended in the well by a metal cable. As the impeller rotates, it imparts energy to the water. Groundwater is pulled in through the intake screen and forced up through the discharge pipe or riser pipe. Because the motor on an electric submersible pump is water cooled, the motor will overheat if an electric pump draws air instead of water for a prolonged period. As a result, control systems shut down pumps when the water level in the aquifer drops and the pumps begin to pump air instead of water. The motor may be controlled to cycle on when the water level in a well reaches a certain level, and to cycle off when the water level drops to a predetermined low level. Alternatively, a load monitor may be used to control the pump operation by sensing the change in the load on the motor when the pump stops pumping water and begins to pump air. The control system will then shut off the motor for a preset period. Submersible pumps were used in water-supply wells long before they were used in remediation applications; in an uncontaminated well, a submersible pump can last for 10 to 20 years.

Pumps are sized based on the energy (head) required, which is calculated from the amount of water to be pumped and the resulting velocity through the piping; the projected friction losses through the piping; the vertical distance (elevation head) to which the water must be lifted out of the well; and the efficiency of the pump. Engineering handbooks and other references provide the calculation procedure (e.g., see Reference 12). Manufacturers summarize the capabilities of a particular pump on a *pump curve*, which plots the achievable flow rate against the total head (i.e., energy required to pump the water).

The *materials of construction* used for the pumps, screens, and other well components must be compatible with the contaminants. Certain contaminants can break down or corrode the materials used to construct wells, pumps, and piping. At elevated concentrations, low-molecular weight ketones, aldehydes, and chlorinated solvents attack PVC. Chloride ions corrode stainless steel. Teflon® is the most chemically resistant material used in well construction.[10]

The natural constituents of groundwater can foul or damage the well screen or pump. When corrosive compounds are present, the design must incorporate materials which are not susceptible to corrosion. Free carbon dioxide (CO_2) at concentrations greater than 10 to 15 mg/L can corrode steel and cast iron. Hydrogen sulfide (H_2S), encountered in tidal estuaries and oceanfront marshes, corrodes steel, cast iron, brass, and bronze at concentrations over 1 to 3 mg/L. As noted above, chloride ions corrode stainless steel, particularly at levels above 500 mg/L. Certain compounds can precipitate on well screens, sand packs, piping, and pumps; at high concentrations, these precipitates restrict flow and can damage pumps. At concentrations over 2 to 3 mg/L in groundwater, iron deposits can become a severe problem, either from precipitates of iron salts or from an iron-laden slime produced by iron-fixing bacteria. Some precipitation may occur at concentrations as low as 0.25 mg/L. Hardness over 200 mg/L can also cause problems from encrustation with calcium and magnesium carbonates. When fouling occurs, precipitates are removed by treating the well with a dilute solution of inhibited hydrochloric acid or another acid. Chlorine or hypochlorite are used to kill iron-fixing bacteria.[13,14]

3.2.1.2 Modeling the Flow to a Single Extraction Well

Pumping induces groundwater flow to a well, imposing an artificial hydraulic gradient and creating a *cone of depression* in the water level around the well, as shown in Figure 3.3. The decrease in the water table or head caused by pumping is called *drawdown*. The pumping rate and the *radius of influence* of each well can be estimated by mathematical models of groundwater behavior and/or by pumping tests. Two- and three-dimensional simulations of the effects of a network of monitoring

NO HYDRAULIC GRADIENT

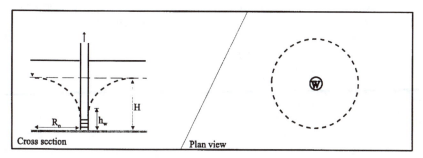

SLIGHT HYDRAULIC GRADIENT

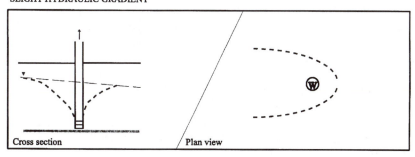

STEEP HYDRAULIC GRADIENT

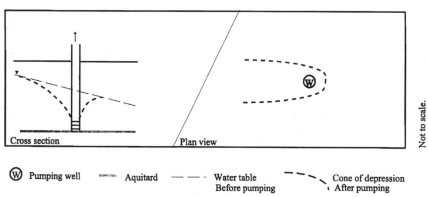

Figure 3.3 Effects of pumping. (Based on an original drawing by Kim M. Henry and used with her permission.)

wells require computer models. "Real world" systems contain many complicated variables which cannot be readily modeled. Simplifying assumptions must be built into groundwater models in order to make the models manageable. In general, models of the flow from an extraction well assume equilibrium conditions and assume ideal, homogenous aquifer conditions. As a result, they only approximate the performance of a pumping system under actual (nonequilibrium, nonideal, non-homogeneous) conditions. Developing a practical design from model projections requires considerable professional judgment.

Designing a groundwater extraction system begins with estimating the flow rate of water to be pumped from a single well and the corresponding radius of influence of that well. Hydrogeologists determine the number and spacing of extraction wells in the system based upon that radius of influence.

The flow of groundwater to a single extraction well is described by the series of equations which follows.[15] These equations necessarily contain assumptions that are rarely satisfied in "real world" applications: the aquifer materials are homogenous, of uniform thickness, extend horizontally in all directions, and do not encounter recharge or barrier boundaries; the hydraulic conductivity is the same throughout the aquifer, both vertically and horizontally (i.e., the aquifer is *isotropic*); a limitless source of groundwater is available and is instantaneously released from storage when the head is reduced; the extraction well is frictionless, small in diameter, and fully penetrates the aquifer; and the system is at equilibrium.

The radius of influence (R_o, ft) of a pumping well is ideally based on the results of a pumping test. Less accurately, it can be approximated from the aquifer properties. The order of magnitude of the radius of influence (R_o, ft) can be estimated from the transmissivity (T, gal/day/ft) and storativity (S, unitless) of an aquifer (see Section 2.3.3). Without recharge, the radius of influence is approximately[15]

$$R_o = r_w + \sqrt{\frac{Tt}{4790\,S}} \qquad (3.1)$$

where r_w is the radius of the well (ft) and t is the time since pumping began (min). The radius of influence cannot increase indefinitely with time, but increases until equilibrium conditions are reached. Thus, applying Equation 3.1 requires some judgment about when equilibrium will be reached. In a confined aquifer, equilibrium conditions can be reached in a pumping test in about a day; in an unconfined aquifer, within several days to a week.[16,17]

An empirical relationship describes R_o in terms of the drawdown (H-h) and the hydraulic conductivity (K, gal/day/ft²):[15,16]

$$R_o = 3(H - h_w)\sqrt{0.47\,K} \qquad (3.2)$$

where H is the initial head (ft) and h_w is the head at the pumping well.

The rate of flow from a pumping well depends on the radius of influence and the drawdown. The drawdown must be assumed based on the geometry of the well (i.e., screen length and depth). For an unconfined aquifer, the maximum efficiency of well operation occurs at ~67% of the maximum possible drawdown, as the yield from a well decreases with drawdown.[16]

For flow (Q, in gal/day or gpd) to a well in an extraction well that fully penetrates a confined aquifer,

$$Q = \frac{2\pi Kb\left(H - h_w\right)}{\ln\left(R_o/r_w\right)} \tag{3.3}$$

where b is the aquifer thickness [ft], H is the head at R_o [ft], and h_w is the head [ft] at the extraction well of radius r_w [ft]. The drawdown $(H - h)$ at any radius r is then

$$H - h = \frac{Q\left(\ln \dfrac{R_o}{r}\right)}{2\pi Kb} \tag{3.4}$$

The flow to a well which fully penetrates an unconfined, or water table, aquifer can be approximated by Equation 3.5:

$$Q = \frac{\pi K\left(H^2 - h_w^2\right)}{\ln\left(R_o/r_w\right)} \tag{3.5}$$

The height (h) of the water table at a distance r from the well can be predicted by Equation 3.6 when r > ~1.5H.

$$h = \sqrt{H^2 - \frac{Q}{\pi K}\ln\frac{R_o}{r}} \tag{3.6}$$

Finally, Equation 3.7 models the flow to a well in a mixed aquifer. This model would apply when pumping from a confined aquifer to the extent that the cone of depression extended below the confining layer.

$$Q = \frac{\pi K\left(2bH - b^2 - h_w^2\right)}{\ln\left(R_o/r_w\right)} \tag{3.7}$$

Equations 3.3 through 3.7 are based on the assumption that the extraction well fully penetrates the aquifer. The drawdown close to a partially penetrating well is greater than the drawdown to a fully penetrating well and, for a given drawdown, the flow rate is lower. Beyond a radius of approximately 1.5B, the drawdown is approximately the same.[15,16] The effects of partial penetration can be modeled.

3.2.1.3 Siting Extraction Wells

Groundwater remediation usually requires a network of extraction wells rather than a single well. The location of the wells and the total amount of water to be pumped depend on the goal of remediation in addition to aquifer characteristics. This goal may be to restore aquifer conditions, to control the groundwater gradient and prevent a plume from expanding, or both.

Aquifer restoration requires the removal of the contaminant mass dissolved in the groundwater and adsorbed to saturated soils. Extraction wells are placed though-out the plume to withdraw contaminated groundwater and induce flushing with clean water. These wells produce a stream of groundwater for treatment that contains relatively high concentrations of contaminants.

Gradient control requires a system of extraction wells at or near the leading edge of a plume to prevent contaminated groundwater from migrating further. Because gradient control plumes are pumping relatively clean water, the costs of treating the water can be lower than the costs of treating water from wells in the center of a plume, depending on the flow rate of water to be treated.

Thus, the goal of the remediation system determines the placement of the well network within a plume, at the leading edge, or both. Placement of the individual wells within that network depends on the radius of influence of individual wells. Wells are sited so that the radii of influence of adjacent wells overlap and no flow occurs between the wells. (This method of spacing wells is reasonably effective except in aquifers with naturally high flow velocities. In such cases, velocity distributions must be simulated using a computer model to ensure that the plume is effectively captured.) The drawdown within a network of wells in a confined aquifer can be estimated by *superposition* or *cumulative drawdown*: the drawdown at any point is the sum of the drawdowns that would have been caused by each well in the network. While the theory behind this method does not apply to unconfined aquifers, it can be applied with reasonable results when the drawdown is less than 20% of the initial saturated thickness of the aquifer.[18] Computer models are often used to simulate the flow patterns induced by different well configurations and pumping rates in order to optimize the design.

Stagnation zones can develop at points between wells or downgradient of an extraction system. A stagnation zone occurs where the hydraulic gradient is quite low. Water in that zone is not captured by an extraction well, and as a result, contaminants in a stagnation zone are not effectively removed. Extraction well networks may be designed or operated to minimize stagnation.

At the periphery of a plume, wells are sited to minimize — to the extent possible — the quantity of clean water drawn into the well. Higher flow rates require higher costs for handling and treating the water. Again, groundwater models are often used to optimize the locations of wells.

Literally hundreds of computer models are available to simulate groundwater pumping systems.[19,20] Regardless of the model, the user must be aware of the underlying assumptions, calibrate the model to site conditions using data from a field investigation, and apply the results with the knowledge that they are projections, not absolutes.

3.2.1.4 Practical Considerations

Designing and operating a groundwater extraction system involves many practical considerations in addition to the theoretical considerations discussed above. Briefly, these include:

- Groundwater treatment units are sometimes sized to a relatively small range of flow rates. Groundwater pumping rates estimated from models are approximate and may vary by an order of magnitude or more for an extraction system, depending on the model(s) and assumptions used. As a result, sizing groundwater treatment units can be a challenge. Oversizing the equipment will increase capital costs and may impair treatment effectiveness; undersizing the equipment will restrict the remediation system. The design of a treatment system must account for the effects of the underlying assumptions (is the estimate of flow likely to be high? low?), and the probability that the extraction system will expand or contract in the near future. The effects of the uncertainty in pumping rates on the cost of a treatment system can justify the costs of additional field work (such as pumping tests) or design work to narrow the range of pumping rates.
- Predictive models of pumping rates assume equilibrium conditions. However, conditions are not always at equilibrium. A greater quantity of water than predicted may be pumped initially while the water in the cone of influence is pumped out. During long-term pumping, conditions in a water table aquifer will vary as a result of precipitation.
- Drawdown of the water table can affect nearby structures and land features. Settlement of soils in the area of a groundwater pumping system can occur when compressive silts or clays or loose sands consolidate as a result of dewatering. While the effect is often minimal, under certain conditions it can be significant enough to damage the foundations of buildings or other structures. Drawdown of the water table can also dewater wetlands or cut off the water supply to trees and other vegetation in upland areas.[21]

3.2.2 Well Point Systems

Groundwater can be extracted from a system of well points through a header using a common vacuum pump mounted at the surface, rather than using a water pump in each well. The pump lifts the water from the well points and draws it through the header by suction. A valve on the hose between the header and each well point regulates the flow from each well point. Adjusting the flow from different well points is called *tuning*. A well point system differs from an extraction well network in the use of a common vacuum pump, rather than a pump in each well, and in the construction of the wells.

A well point is a screen made of wire mesh, slotted plastic, or other material, and is typically 1.5 to 3.5 in. in diameter. It can be installed by three methods. Well points can be driven into the ground. A protective steel cone is attached to the bottom of the well point and a series of connected lengths of pipe to the top. The pipe is driven into the ground using a sledge hammer, drop hammer, or air hammer. Well points can also be installed by jetting, essentially using a high-velocity stream of water to cut the borehole. Finally, a well point can be installed into drilled boreholes, similar to an extraction well. Filter sand is placed around a well screen installed by jetting or drilling.[22]

In general, well point systems are limited to a suction lift of about 15 ft, with a practical maximum of 28 ft. Well points are typically spaced 3 to 10 ft apart.[23] The normal pumping capacity is 0.1 to 25 gpm/well, although higher capacities are possible.[24]

Well point systems are commonly used in construction dewatering. They are used less commonly than extraction well networks in long-term groundwater remediation systems. All else being equal, a well point system costs less than a network of extraction wells with a pump in each well. A well point system may also be the only effective method for recovering groundwater from low-yielding formations.

However, well point systems have several limitations that do not apply to extraction wells. Well point systems are limited to shallower depths and lower pumping capacities than extraction well networks. Well points can be particularly susceptible to fouling, depending on the operation and the site geochemistry. If the well point screen is only slightly below the water table, air will be drawn into the system with the groundwater. Aeration can cause some ions, such as iron, to precipitate, and can stimulate the growth of aerobic bacteria. The induced vacuum can also enhance the precipitation of certain compounds, such as calcium bicarbonate or certain iron carbonates.[25]

3.2.3   Interceptor Trenches

Contaminated groundwater is sometimes recovered from an interceptor trench or subsurface drain. Figure 3.4 shows a typical design. In permeable soils, the trench is ideally excavated down to an impermeable layer, to prevent contaminated groundwater from flowing under the trench. A perforated pipe is wrapped in filter fabric and lowered into a trench. The trench is then backfilled with gravel or crushed stone. The backfill is more permeable than the surrounding soil and therefore transmits water more readily. The pipe conveys groundwater to a sump, where the groundwater is pumped to a treatment system.[26]

This basic design has many variations, most commonly:

- A *french drain* does not contain a pipe to convey the water; the water is simply transmitted through the gravel in the trench.
- The downgradient side of the trench may be lined with a geomembrane (e.g., high-density polyethylene, a type of plastic) to prevent inflow of water from the downgradient side.
- A plastic drainage material may be placed in the trench rather than gravel backfill. Sheets of corrugated plastic are placed vertically in the trench to provide drainage.
- Filter fabric can exclude LNAPL from the trench, and as a result may be omitted from the design of an interceptor trench designed to collect LNAPL.

An interceptor trench is oriented perpendicular to the direction of groundwater flow to prevent the migration of contaminated groundwater. A *relief drain* is constructed similarly, but oriented differently: it is placed parallel to the direction of groundwater flow in order to lower the groundwater table.

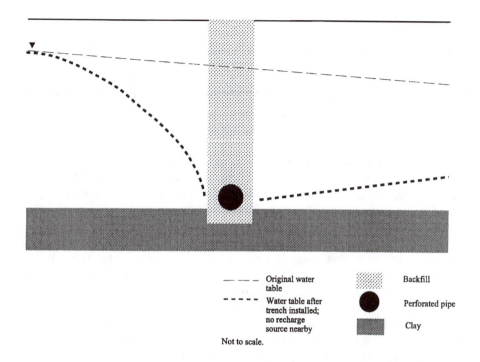

Figure 3.4 Schematic of interceptor trench.

An interceptor trench functions like an infinite line of extraction wells which lowers the water table to the depth of the drain. The influence of an interceptor trench, drawdown, and flow rate depend on the hydraulic conductivity and saturated thickness of the aquifer and the design of the trench.[26-28]

Interceptor trenches are used in relatively low-permeability soils, where wells or well points would have to be spaced quite closely. Trench construction has several limitations which do not apply to installation of extraction wells:

- Construction costs generally limit trench construction to 40 ft or less using conventional equipment, although deeper trenches have been built.[26]
- Excavation of a trench at a hazardous waste site using conventional equipment can generate a large volume of contaminated soil. Treatment and/or disposal of this soil can be quite costly.
- Underground utilities are common on industrial or urban sites. It is more difficult to construct a trench under utilities or to reroute utilities away from a trench than it is to site wells away from utility lines.

Interceptor trenches can also be more difficult to maintain than extraction wells, particularly where natural groundwater constituents such as iron or hardness are high enough to cause fouling.

3.2.4   Effectiveness of Pump-and-Treat Remedies

Water level data indicate how effectively a pumping system controls hydraulic gradients. Contaminant levels in the groundwater are monitored to determine whether or not the plume is fully contained and to measure progress in restoring the aquifer quality.[29]

Contaminant levels typically drop rapidly when a pump-and-treat system begins operating. The rate at which the concentration drops decreases with time, however, eventually reaching an asymptote at many sites. At the asymptote, the slope of a plot of concentration (C) vs. time (t) (i.e., dC/dt) is not significantly different from zero.[29,30]

When the asymptotic concentration exceeds the clean-up level, it may not be possible to restore the aquifer to the original remediation goals. In EPA jargon, remediation may be *technically impracticable*. The pump-and-treat system may be modified to improve its performance if possible (Section 3.2.5), or the remediation goals may have to be rethought.

When the asymptotic concentration is below the clean-up level, the pump-and-treat system can be shut down. However, monitoring must continue to verify that the remediation goal has truly been reached, as the concentration can *rebound* when the pumps are shut off, or increase above the asymptotic level.

This phenomenon apparently results from mass transfer from a continuing source such as NAPL residuals or contaminants sorbed to soils. If groundwater velocities during pumping are too high to allow dissolution or desorption to reach equilibrium, groundwater will flow through pore spaces without removing an appreciable amount of the contaminant. When pumping stops, groundwater in the pore spaces can reach equilibrium with the NAPL residual and/or sorbed contaminants and the dissolved-phase concentration will increase.[31]

Studies of data from pump-and-treat systems suggest several factors that can prevent a remediation system from achieving its goals:[32,33]

- Extraction systems cannot effectively circulate groundwater throughout a contaminated aquifer when the aquifer materials are highly heterogeneous soils containing lenses or strata of low-permeability materials, or comprising fractured bedrock.
- The ability to remove contaminant mass from the system and reduce contaminant concentrations is limited by continuing sources of contamination. These sources may include contamination which leaches from unsaturated soils, contamination adsorbed to aquifer solids, or NAPL. DNAPL is the most problematic continuing source.
- The design of the extraction system may not suffice. Systems may be underdesigned based on a lack of initial characterization data. In some cases, natural conditions make it difficult to design a system that can adequately control groundwater gradients. These natural conditions include very high groundwater flow rates, or heterogenous aquifer materials/fractured bedrock.

These factors can vary across a hazardous waste site. As a result, it may be possible to achieve remediation goals in one portion of an aquifer, but not in another.

3.2.5 Enhancements

Basic pump-and-treat systems may be augmented with *in situ* treatment or the components of the system varied to improve performance. For example, additional extraction wells can be installed to extend the capture zone of the system, or injection wells could be added to improve gradient control. While these augmentations may be part of the initial design, they may also be added to the system based on the initial performance of a system.

Gradient control can be augmented using a vertical barrier, such as a slurry wall or sheet piling, or by injecting groundwater. Vertical barriers may be placed down-gradient of a source area or used to encircle a source area. The barrier helps to contain the source while groundwater is pumped and treated. Alternatively, a vertical barrier can be placed upgradient of a source area to divert clean groundwater around the source area and minimize further groundwater contamination. Reinjection of treated groundwater creates a *mound of impression* at each well, essentially the inverse of a cone of depression; a line of injection wells can create a so-called *pressure ridge*. The gradients induced by groundwater reinjection can help to contain a plume or to increase flushing through a contaminated zone. (See the discussion of water flooding in Section 3.6 for application to DNAPL remediation.)

Pulsed pumping can sometimes be used to enhance the performance of an extraction well system when its performance levels off.[34] By varying pumping rates at different wells, sometimes called *adaptive pumping*, operators can increase the flow through stagnant zones between wells where little flow previously occurred. This increased flow will remove contamination from the stagnant zones. Pulsed pumping can also entail shutting off and restarting the pumps in one or more extraction wells. This process addresses residual soil contamination in the immediate vicinity of the well. When a well starts pumping and a cone of depression forms around the well, the soil in that unsaturated cone contains contaminants that will not be removed by the groundwater extraction system. When the well stops pumping, the cone of depression flattens and soils near the well that had been dewatered become saturated again. When pumping begins again, groundwater removed from the cone of depression will carry some of the contamination that was sorbed to the soils. Repeatedly cycling the pumps on and off, allowing the groundwater to rebound, will gradually flush contamination from the soils in the cone of depression. Finally, pulsed pumping may also enhance the desorption of contaminants from saturated soils. As noted previously, when pumping stops groundwater can equilibrate with contaminants adsorbed to aquifer solids. Equilibrium concentrations in groundwater exceed the concentrations which can desorb into rapidly flowing groundwater under pumping conditions. Pulsed pumping must be applied carefully to maintain control of hydraulic gradients and prevent the uncontrolled migration of a plume when the pumps cycle off.

Pump-and-treat systems can be augmented by changing the permeability of aquifer materials. Hydraulic or pneumatic fracturing can increase the permeability of aquifer materials. These techniques have been considered and tested on unsaturated and saturated soils. They are discusssed further in Section 4.3.1.

Surfactants may also be used to enhance the recovery of organic contaminants by increasing their solubility. While surfactants have been used for some time in the petroleum industry to enhance oil recovery, their use in groundwater remediation has been limited. For additional information on surfactants, see the discussion of surfactant use for DNAPL remediation in Section 3.6.2.

3.3 GROUNDWATER DISCHARGE

The extent and type of treatment required for extracted groundwater depends on the limits placed on the discharge. Discharge options include:

- Surface water. Groundwater may be discharged to a stream, lake, or river. The federal Clean Water Act and analogous state laws regulate such discharges. Discharge limits are set under a National Pollution Discharge Elimination System (NPDES) permit under the Clean Water Act or the state analog. Regulatory agencies may set these limits in several ways. Discharge limits are frequently based on the levels of pollutants that can be allowed in the water without significantly harming aquatic life. These levels may be calculated from Ambient Water Quality Criteria based on dilution of the discharge by the surface water, or may be based on the results of bioassays. In certain circumstances (e.g., badly polluted surface water), no detectable concentrations of contaminants may be allowed in the discharge. Limits for industrial wastewater discharges may also be based on the source of wastewater and the technologies available to treat the wastewater; these categorical limits are sometimes applied to groundwater discharges.
- Groundwater. Treated groundwater may be reinjected into the aquifer through injection wells, a trench, a leach field, or a recharge impoundment. The hydraulic gradients resulting from injecting the water may be used to control or contain groundwater migration. Reinjection may also be used to help flush contaminated soil and/or groundwater. Reinjection may require a permit. Discharge limits typically correspond to the ultimate cleanup goal for the aquifer (e.g., maximum contaminant levels [MCLs] in an aquifer that is or may be used for drinking). Powers[35] notes that returning water to the ground is more difficult, and usually more costly, than extracting it. Recharge systems are prone to fouling from precipitates of natural groundwater constituents such as iron, suspended solids, and bacterial growth (particularly since treated groundwater is usually oxygenated). Reinjection systems usually require a greater area than that encompassed by the extraction system (assuming the hydraulic conductivity is similar in both areas). Example 3.5 describes the use of groundwater injection at one site.
- Sewer/publicly owned treatment works (POTWs). Treated or untreated groundwater may be discharged to the local sewage treatment plant. The discharge must meet the requirements of the pretreatment program of the plant. POTWs, typically designed to treat domestic sewage biologically, are reluctant to accept large flows of "dilute" groundwater.
- Industrial wastewater treatment plant. Groundwater remediation sometimes occurs at an industrial facility which already has a treatment plant for industrial wastewater from the facility. Extracted groundwater may be pretreated and discharged to the treatment plant, or discharged directly to the treatment plant, depending on the permit limitations of the plant. Discharge of groundwater requires a modification

of the existing NPDES permit of the plant; that discharge may also trigger Resource Conservation and Recovery Act (RCRA) requirements if the groundwater is considered to be a RCRA waste.

3.4 GROUNDWATER TREATMENT

Groundwater treatment technologies can separate a contaminant from the water, immobilize the contaminant, or destroy it. The first two types of treatment produce residuals which must be treated further, to destroy the contaminant, or landfilled.

Natural groundwater constituents can strongly affect treatment technologies designed to remove hazardous substances. Pretreatment to remove those natural constituents can drive the selection and operation of treatment technologies.

Specialized analyses have been developed to characterize water for treatment purposes. These tests are often omitted from remedial investigations, but may be critical to selecting and designing groundwater treatment systems. Critical parameters include:

- *Biological oxygen demand* (BOD) indicates the concentration of compounds in the groundwater that can be readily biodegraded (5-d test).
- *Chemical oxygen demand* (COD) indicates the concentration of organic compounds that can be degraded using a strong chemical oxidant. The test does not oxidize volatile straight-chain hydrocarbons, but will oxidize reduced metals (e.g., iron).
- *Total organic carbon* includes carbon compounds that cannot be readily oxidized biologically or chemically. This test does not measure inorganic compounds that can contribute to BOD or COD.
- *Alkalinity* indicates the natural buffering capacity of the water due to dissolved carbonates. (Buffering means that the water resists a change in pH upon addition of an acid or a base.)
- pH indicates the hydrogen (hydronium) ion concentration in water.
- Calcium (Ca) and Magnesium (Mg), collectively characterized as *hardness*, can precipitate and clog pipes or treatment equipment. Hardness may be measured as total hardness, or Ca and Mg can be measured separately and hardness calculated:

$$Hardness\ \left(mg\ equivalent\ CaCO_3/L\right) = 2.5\left[Ca, mg/L\right] + 4.1\left[Mg, mg/L\right] \quad (3.8)$$

- Iron (Fe) can precipitate and clog treatment equipment.
- Manganese (Mn) can precipitate and clog treatment equipment.
- *Total dissolved solids* (TDS) is a gross measure of ions in solution.
- *Total suspended solids* (TSS) is a measure of particulates in solution; it may be artificially high in groundwater samples from monitoring wells.
- Phosphorus (P) is a nutrient necessary for biodegradation.
- Nitrogen is also required for biodegradation. Nitrogen levels are measured in three ways: ammonia–nitrogen (NH_3-N), Nitrate–Nitrogen (NO_3-N), and total-Kjeldahl nitrogen (TKN), which measures organic nitrogen plus ammonia nitrogen.

Methodologies for these tests are found in *Standard Methods for the Examination of Water and Wastewater*.[36]

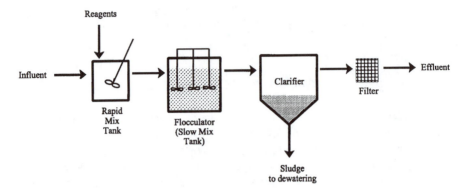

Figure 3.5   Metals precipitation system.

3.4.1   Immobilization Technologies

Metals dissolved in groundwater can be removed by adding chemicals that cause the metals to form insoluble compounds. These solid *precipitates* can be removed by *flocculation, sedimentation,* and/or filtration, as described below. Contaminants may also be removed by *coprecipitation.* Coprecipitation is an adsorption phenomenon wherein a trace constituent is adsorbed onto a bulk-precipitated solid, or coagulated and enmeshed in the bulk precipitate. Coprecipitation can remove contaminants to levels below the solubility level.

Precipitation is used to treat metals such as cadmium, trivalent chromium, lead, and mercury. Some forms of metals — such as hexavalent chromium and some forms of arsenic — may require pretreatment before they can be removed by precipitation. Hexavalent chromium, which commonly exists in highly soluble anions (e.g., $Cr_2O_4^{2-}$) is reduced to trivalent chromium before precipitation as the oxide. Precipitation and sedimentation are also used to remove solids (TDS, TSS) and nonhazardous metals that can otherwise foul treatment equipment.

Chemicals are added to the water to adjust the pH and to react with the metals to form insoluble precipitates. As shown in Figure 3.5, these chemicals are typically added in a *rapid-mix tank.* As the name implies, this unit is sized to allow for nearly instantaneous dispersion of the reagents. The mixing detention period is approximately 30 sec. Water then flows to a *slow-mix tank,* where *coagulation* and *flocculation* occur. The detention time in a slow-mix tank is at least 30 min.[37] Coagulation of suspended solids refers to the neutralization of the charge of the particles to overcome the repulsion between particles and enable the formation of microflocs ranging in size from micrometers to millimeters. *Flocculation* is the formation of interparticle bridges to form macroflocs, which range in size from millimeters to centimeters. The flocs are removed from the water by sedimentation to form a sludge which must be dewatered and landfilled. Sedimentation and sludge handling are described further below.

The degree of treatment depends on the chemicals added, the type of precipitate formed, and the pH. A variety of chemicals can be added to treat the water. The type and concentration of additives are typically determined through *jar tests* during the feasibility study or at the beginning of the design phase. *Hydroxide* and *sulfide* precipitation are commonly used to remove metals from water.

In hydroxide precipitation, lime (CaO), caustic (NaOH), or, less commonly, magnesium hydroxide ($Mg(OH)_2$) are added to the water to adjust the pH and precipitate the metal as the hydroxide. When caustic is added, for example,

$$M^{n+} + nNaOH \rightarrow M(OH)_n(\downarrow) + nNa^+$$

where M^{n+} is the metal ion. (Recall that the symbol ($\downarrow$) indicates that the compound forms a solid which can precipitate.)

The solubility of a metal depends on the pH of the water, as shown in Figure 3.6. Note that the minimum solubility occurs at a relatively precise pH level. As a result, chemical dosages must be carefully controlled. Note too, that the minimum solubility occurs at a different pH for different metals. As a result, treatment to a series of pH levels may be required to adequately remove several metals from groundwater.

The solubility of a metal in groundwater also depends on the concentrations of organic material and of other ions in the water; high concentrations of organics or of anions such as Cl^- or SO_4^{2-} will *complex* with the metal and cause it to be more soluble than it would be in pure water at the same pH. As a result, the solubilities shown in Figure 3.6 represent the minimum solubilities observed. Treatment systems relying on precipitation can rarely reach these low levels (except, in certain cases, when coprecipitation occurs).

As noted above, either lime or caustic soda may be used for hydroxide precipitation. Lime is less expensive than caustic soda (by 40% to 60%). However, more sludge is typically formed from treatment by lime than by caustic soda, so sludge disposal costs are higher.

Other chemicals may be added, or pretreatment steps may be required. For example:

- Soda ash (Na_2CO_3) is sometimes used with lime in water softening (removal of hardness).
- Alum ($Al_2(SO_4)_3 \cdot 18H_2O$) may be added to improve coagulation, particularly for lime/soda ash softening; alum generates voluminous sludge.
- Polymers may be added to improve coagulation/flocculation.
- Cr(VI) must be reduced to Cr(III) before it can be precipitated as the hydroxide. Reduction can be achieved by lowering the water pH to 2.0 to 3.0 with sulfuric acid, then adding a chemical reducing agent such as sulfur dioxide, sodium bisulfite, metabisufite, or hydrosulfite, or ferrous sulfate. The reduction reaction requires a retention time of approximately 20 to 30 min.[38]
- Iron can be removed from water by aeration (to convert Fe^{2+} to Fe^{3+}), formation of the hydroxide at pH 7 to 10, and sedimentation/filtration.
- Arsenic does not precipitate as readily as the hydroxide, but can be removed by coprecipitation with iron. Iron is added at a pH between 5 and 6 (to achieve a ratio of iron to arsenic of approximately 8:1), then the pH is increased to 8 to 9 using lime.[38]

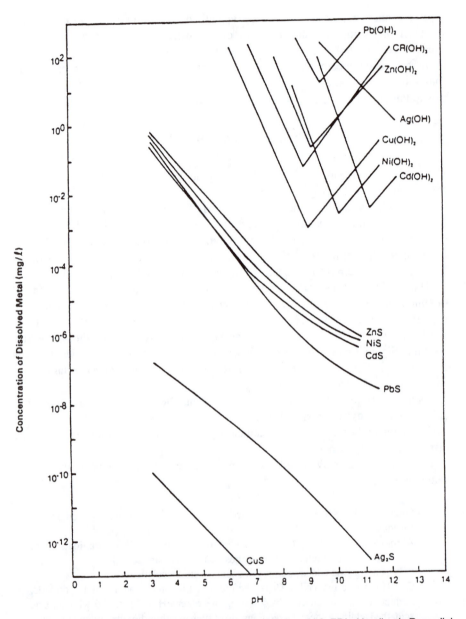

Figure 3.6   Solubilities of metal hydroxides and sulfides. (From U.S. EPA, *Handbook: Remedial Action at Waste Disposal Sites (Revised)*, EPA/625/6-85/006, Office of Emergency and Remedial Response, Washington, D.C., October 1985, 10:23–25.)

Sulfide precipitation can be achieved by two processes: the insoluble sulfide process and the soluble sulfide process.

In the *insoluble sulfide precipitation process* a freshly prepared slurry of ferrous sulfate (FeS) is added to wastewater (or groundwater) to precipitate heavy metals as the sulfide salts. Ferrous ions dissociated from FeS will precipitate as ferrous hydroxide at the operating pH of 8 to 9.

$$FeS \rightarrow Fe^{2+} + S^{2-}$$

$$M^{2+} + S^{2-} \rightarrow MS(\downarrow)$$

$$Fe^{2+} + 2OH^- \rightarrow Fe(OH)_2$$

Only those metals whose sulfides are less soluble than ferrous sulfide will precipitate; most priority pollutant metals will precipitate, but (for example) manganese will not.

The *soluble sulfide precipitation process* is similar, except that sodium bisulfide is used rather than ferrous sulfide.

$$M^{2+} + HS^- + OH^- \rightarrow MS(\downarrow) + H_2O$$

Metal sulfides are less soluble than hydroxides, as shown in Figure 3.6. As a result, sulfide precipitation may be the preferred form of treatment when discharge limits are very low or there are concerns over the classification of the waste sludge (i.e., nonhazardous or hazardous by characteristic). Sulfide sludge can be more difficult to flocculate than hydroxide sludge; however, the sludge can generally be thickened more easily. A final operational concern is the level of soluble sulfide remaining in the effluent.

Dewatering and disposal of the sludge generated by hydroxide or sulfide precipitation can represent a significant cost. The amount of sludge generated from treating contaminated groundwater may be predicted from jar tests or, if hydroxide precipitation is used and a nearby municipality or industry draws a water supply from the aquifer, from the results at their drinking-water treatment plant. Note that much of the sludge may result from removing natural groundwater constituents (e.g., hardness, iron) rather than the contaminants themselves. Lacking site-specific test data, rules of thumb and general ranges of chemical dosages and sludge generation are available. The rules of thumb which follow pertain to the removal of hardness or iron and manganese, not to the removal of heavy metals per se. However, these natural groundwater constituents may comprise most of the sludge generated by a treatment unit primarily intended to remove heavy metals. In addition, natural groundwater constituents must often be removed in a pretreatment step to prevent fouling of treatment equipment intended to remove organic contaminants.

- A survey of 23 municipal water-softening plants using lime softening found that treatment required 336 to 2200 lb lime per million gallons of water treated. The process generated 1.5 to 3.8 lb solids generated per pound of lime dose, or 0.0004 to 0.02 lb sludge per gallon water. The report did not indicate the sources of the water or provide information on influent water quality.[39]

- EPA has also suggested that the amount of lime required can be estimated as about three times the stoichiometric amount.[40]
- For each milligram per liter of iron or manganese in solution, 1.5 to 2.0 mg/L sludge is produced from oxidation and sedimentation.[38]

3.4.2 Separation Technologies

Treatment technologies may separate contaminants from groundwater based on the density, volatility, or ability of the contaminants to sorb to solids.

3.4.2.1 Sedimentation and Sludge Handling

Sedimentation refers to the separation of solids from water by gravitational settling. This process commonly follows precipitation and flocculation in order to remove metals or solids from water. Sedimentation typically occurs in a *conventional clarifier*, or a *lamella (or tube) clarifier*.

A conventional clarifier is simply a circular or rectangular basin, or a tank. A wide range of tank configurations and sizes have been used. The clarifier is designed to allow water to flow through slowly enough, and with minimal agitation, to allow solids to settle. A sloped bottom directs settled sludge to a collection point. Sludge collected from the bottom of the clarifier typically has a relatively low solids content, e.g., 2% to 15% solids from lime–soda ash softening of groundwater.[38,41]

Weber[42] provides some general guidelines for sizing a clarifier based on conventional drinking water treatment: for aluminum and iron floc, a surface loading rate of 600 to 1800 gpd/ft^2 and a detention period of 2 to 8 h; for calcium carbonate precipitates, a surface loading rate of 600 to 2880 gpd/ft^2 and a detention period of 1 to 4 h.[42] (The dimensions of the clarifier can be calculated from the flow rate, the surface loading rate, and the detention time.) Portable or prefabricated units are used for relatively small flows (up to approximately 40 gpm).[43]

Example 3.1: Determine the clarifier size for a flow rate of 100 gpm, a surface loading rate of 800 gpd/ft^2, and a detention time of 6 h.

Calculate the surface area and radius for a circular clarifier:

$$\frac{(100 \ gal/min)\cdot(1440 \ min/day)}{800 \ gal/day/ft^2} = 180 \ ft^2$$

$$\frac{\pi d^2}{4} = 180$$

$$d = 15.1 \ ft, \ \text{say 15 ft}$$

Then the cross-sectional area is 177 ft^2.

Determine the depth of the clarifier:

$$Detention\ Time = \frac{tank\ volume}{flow\ rate}$$

$$Volume = (240\ min) \cdot \left(100\ \frac{gal}{min}\right) \cdot \left(0.1337\ \frac{ft^3}{gal}\right) = 3209\ ft^3 \qquad (3.9)$$

$$\frac{3209\ ft^3\ volume}{177\ ft^2\ area} = 18\ ft\ height$$

The clarifier would be 15 ft in diameter and 18 ft high (water depth) plus freeboard.

A lamella clarifier consists of a series of inclined plates or tubes inside a conical-bottom tank. The inclined plates are used to increase surface area available for settling and thus sedimentation efficiency. As a result, a lamella clarifier can handle up to ten times the flow rate of a similarly sized conventional clarifier at a similar removal efficiency.[43] The plates slope at 45–60° to allow gravity settling.[42]

Sludge exiting a clarifier must be dewatered before disposal. Options include a *plate and frame filter press, belt filter press, centrifuge,* or *sludge drying beds.* As noted above, sludge handling and disposal requirements determine the cost of water treatment by precipitation: the present value of sludge disposal costs can far outweigh the capital equipment costs. The costs of increased dewatering are weighed against the cost of disposal of the sludge weight resulting from the water content. Sludges which must be landfilled in a hazardous waste landfill cannot contain free water.

A plate and frame filter press consists of a series of vertical plates which are held in a frame and are pressed together between a fixed and moving end. The face of each plate is covered with a filter cloth or precoated with a filter aid to enhance solids retention.

A plate and frame filter press operates on a batch basis. The sludge is fed into the press under pressure and passes through feed holes in each tray along the length of the press. The water drains through, but the solids are retained and form a *filter cake* on the plates. The feed is stopped when the spaces between the plates are full. When filtrate essentially stops draining from the plates, the plates are separated and the solids allowed to drop (or are scraped) into a hopper. When the sludge has been removed, the plates are slid back together again and the cycle starts over. From a feed sludge containing 1% to 10% solids, a plate and frame filter press can capture up to 99% of the solids and achieve up to 50% solids in the filter cake. A plate and frame filter press is particularly well suited to dewatering hard-to-dewater sludges.[44]

When a belt filter press is used to dewater sludge, the sludge is conditioned with polymers, then fed between two continuous belts. Water is removed from the sludge by three mechanisms: drainage, pressure filtration between the rollers, and shear-pressure filtration. Belt filters can produce sludge containing up to 25%–40% solids and capture up to 90%–99% of the solids.

A centrifuge separates solids from water by centrifugal force. Centrifugation can produce sludge containing 15% to 40% solids, and capture 50% to 95% of the influent solids (the higher solids capture is achieved with polymer addition). Centrifuges are used more commonly in large wastewater treatment plants than in groundwater treatment systems.

3.4.2.2 Filtration

The effluent from a sedimentation basin may require filtration to remove remaining suspended solids. Filters are also commonly placed before activated carbon or other treatment units to remove solids (absent precipitation and sedimentation).

Filters are distinguished by the type of flow (gravity or pressure) and the filter medium. Filter media commonly include sand, mixed-media, and cartridge or bag filters. Sand and mixed-media filters are classically used in drinking water treatment, but can be used as part of a system to treat contaminated groundwater.

A *slow sand filter* relies on gravity to filter water through a sand bed. Such filters are used in drinking water treatment in Europe, less commonly in the U.S. A *rapid sand filter* comprises an 18- to 30-in.-thick layer of sand on an underdrain system. If the filter is in a closed tank, pressure can be applied to the water to force the water through the filter. Typical filtration rates are on the order of 4 to 5 gpm/ft^2 filter area.[45]

In a mixed-media filter, water is filtered through two layers of material in a steel or concrete vessel: a top layer of coarse anthracite (12 to 24 in.), and a bottom layer of sand (6 to 16 in.). Head loss is generated at a lower rate than for typical sand filters; mixed-media filters are used instead of sand filters after biological treatment or for water containing high TSS. A typical filtration rate is 2 to 8 gpm/ft^2 filter area. For example, one vendor's single-compartment gravity filter with a filter area of 20 ft^2 can treat a flow range from 100 to 160 gpm. The tank dimensions are 5.5 × 5.5 × 11.5 ft (height).[46]

As solids accumulate, it becomes more difficult to pass water through the filter (the "head loss" referred to above). Sand filters and mixed media filters are *backwashed* periodically to remove accumulated solids. Clean water is run backward through the filter to expand the medium and flush out the solids. This stream of water, which contains a high concentration of solids, is returned to the head of the plant for treatment. (Filters are typically backwashed at 15 to 25 gpm/ft^2 filter area for 5 min.)

Bag filters and cartridge filters are commonly used in groundwater treatment, particularly in relatively low-flow systems. Particulates are removed as water flows through the filter, which is mounted in a steel or plastic housing. A bag filter is essentially a nylon, polypropylene, or polyester sock suspended in the housing. The filter medium used depends on the contaminants in the groundwater and the size of particle to be removed (e.g., nominal particle size 1.5, 3.5, or 23 μm). Bag filters are available in various sizes to accommodate flows up to on the order of 150 gpm. A cartridge filter comprises a corrugated polyester medium (which looks like an elongated version of the air filter on a car) or a hollow cylinder comprised of wrapped polyester string; for a given kind of cartridge filter, the medium varies to filter different sized particles (e.g., nominal 0.35, 1, 10 μm, etc.). Cartridge filters are available in various sizes, with capacities ranging from a few gallons per minute to a few hundred gallons per minute. Bag filters and cartridge filters are usually simply replaced and thrown away when spent.

Microfilters or *ultrafilters* are designed to remove extremely small particles from water by pressure filtration through a membrane or other material containing very small pores. Microfilration systems can remove particles an order of magnitude smaller than the particles removed by bag or cartidge filters. Several vendors offer microfiltration systems. For example,[47]

- Oberlin Filter Company developed a membrane microfiltration system which uses a Tyvek® fabric for filtration. The fabric has openings about one ten-millionth of a meter (or 0.1 µm) in diameter.
- EPOC Water, Inc. markets the EXXFLOW microfilter. Filter modules are made from a proprietary tubular woven polyester. Water is pumped into the tube for filtration of particles larger than 0.2 µm. Treatment produces a concentrate stream which is then dewatered.

3.4.2.3   Oil–Water Separators and Dissolved Air Flotation

Recall that LNAPL may be removed from an aquifer as a separate phase, using skimmer pumps, or may be recovered with groundwater using total-fluid pumps. If LNAPL and groundwater are recovered together, they must be separated above ground. LNAPL and water can be separated by gravity due to the difference between the density of the two fluids. Two types of processes are generally used: simple *gravity separation* and *dissolved air flotation.*

Gravity separation is achieved by retaining contaminated water in a holding tank where oil and other materials with a specific gravity less than or equal to water float to the surface. The organic phase is skimmed off for incineration or recovery. Solids in the influent may partition into the organic phase, remain in the water phase, or settle to the bottom of the tank. Chemicals — polymers and salts — may be added to break oil–water emulsions. Emulsions can also be broken by lowering the pH.

Retention times are on the order of 20 min or more. The removal efficiency increases with the retention time; removal efficiencies generally range from 60% to 99%.

Gravity separators can be constructed in various sizes and shapes (e.g., horizontal cylindrical decanters, vertical cylindrical decanters, and cone-bottomed settlers). Various vendors have added *coalescing media* (e.g., a series of corrugated plates) to their oil–water separators. Small oil droplets coalesce into larger droplets on the surface of the media, thereby improving phase separation. Figure 3.7 shows an example of an oil–water separator containing coalescing media. The American Petroleum Institute has developed specifications for gravity separators used for treating refinery wastewater; as a result, gravity separators are frequently referred to as *API separators.*

Dissolved air flotation (DAF) may also be used to remove LNAPL from extracted groundwater. The water is saturated with air under pressure. When the pressure is reduced, the air comes out of solution as micron-sized bubbles. Oily or suspended material attaches to the bubbles and rises to the surface of the tank. The floating layer is skimmed off for recycling or incineration. DAF may be used instead of gravity separation if the LNAPL has a specific gravity very close to that of water. Air emissions of volatile organic compounds may be of concern

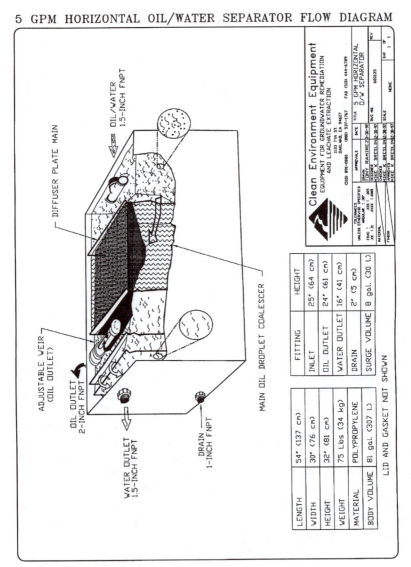

Figure 3.7 Oil/water separator. (Reprinted with permission of Clean Environment Equipment. Copyright 1998 by Clean Environment Equipment, Oakland, CA.)

3.4.2.4 Air Stripping

Air stripping refers to a phase change of contaminants, from liquid to gas, that occurs when water containing dissolved volatile organic compounds (VOC) is intimately contacted with air. In other words, contaminants are "stripped" from the water into the air.

Air stripping effectively removes VOCs such as light hydrocarbons, benzene and analogous compounds, and chlorinated ethylenes. Highly soluble compounds or compounds of low volatility are removed poorly or not at all. As discussed in Section 2.2.1.5, the Henry's law constant (k_H) characterizes the volatility of a compound, and thus the tendency of a compound to be removed from groundwater by air stripping. In general, compounds with k_h greater than 0.01 (unitless) can be effectively removed by air stripping. Compounds with a lower k_h can be stripped at a relatively low efficiency.[48]

Air stripping can generally remove 95% to 99% of volatile contaminants from groundwater. The actual degree of treatment depends on the volatility and solubility of the contaminant(s), stripper design, and operational efficiency (or loss of efficiency due to fouling, for example).

Two types of air strippers are commonly used: packed towers and tray strippers. Aeration tanks and other designs are also used. Air strippers are usually operated continuously, but may be used to treat groundwater on a batch basis.

In a *packed tower*, air and water flow countercurrent through a cylindrical tower constructed of fiberglass, aluminum, or stainless steel. The tower is packed with plastic or ceramic media. A blower supplies air to the bottom of the tower. Contaminated groundwater is distributed across the top of the packing through a spray nozzle. As the water flows down through the packing, water contacts the air flowing up through the tower. The packing provides a large surface area for the contact to occur. Contact between the air and water causes contaminants to move from the liquid phase into the gas phase. "Contaminated" air exits the top of the column and treated water flows out the bottom.

Design variables include the air to water ratio, the type of packing and depth in the column, the tower diameter, and the temperature:

- The necessary air to water ratio depends on the volatility of the contaminant.[49] The air to water ratio is generally between 25:1 to 250:1. Higher ratios may provide higher stripping efficiency, but require more energy (thus increasing operating costs) for the blower.
- The shape and materials of construction of the packing varies. The packing is designed to maximize the surface area for gas–liquid contact, resist fouling, and resist chemical attack by the groundwater contaminants. *Random packing* consists of a large number of packing pieces dumped into a tower. The individual pieces are typically up to 4 in. in diameter, and may comprise perforated spheres, cylinders, saddles, or other shapes. Packing may be made of metal, plastic, or ceramic. The height to which the packing is piled in the tower determines, in part, the stripping efficiency.[49]
- The tower diameter depends on the flow rate of water to be treated.[49] For a given air to water ratio, the tower diameter determines the velocity of the gas stream and the pressure drop through the tower. *Flooding* occurs when the air velocity is so high that the water cannot disperse through the tower.[50,51]

• Temperature may also be a consideration, as volatility increases with temperature. Preheating the groundwater above the natural temperature of approximately 55°F or preheating the air can improve the stripping efficiency, particularly for compounds with a low Henry's law constant. Steam is also used, in lieu of air, to strip recalcitrant compounds. The energy costs of heating the influent or generating steam can make other forms of groundwater treatment more cost effective.

The primary design variables are related by the following mass-balance equation:[49,52]

$$Z = \frac{L}{K_L a} \cdot \frac{R}{R-1} \cdot \ln \frac{\frac{C_i}{C_e} \cdot (R-1) + 1}{R}$$ (3.10)

where: Z = packing height,
      R = stripping factor; $R = k_H \cdot G/L$ (dimensionless),
      G = gas (air) flow rate,
      L = liquid (air) flow rate,
      C_e = target effluent concentration in water,
      C_i = influent concentration in water,
      k_H = Henry's law constant, and
      $K_L a$ = mass transfer coefficient, which depends on the tower design and type of packing. For a particular application, this coefficient is determined by field testing, values provided by a vendor for their product, values provided in standard references such as Perry and Chilton's *Chemical Engineer's Handbook*,[53] or other literature values.

Computer models (and/or field experience) are generally used to design a stripper for a particular application.

Packed towers are advantageous when high air to water flow ratios are required, or when foaming may be a problem. However, they are more susceptible to fouling than a tray stripper or aeration tank.

Example 3.2: The Verona Well Field supplies potable water for the city of Battle Creek, MI. In 1981 the Calhoun County Health Department found that the groundwater in a portion of the well field contained trichloroethylene (TCE) and other VOCs. As a result, the site was proposed for the NPL in 1982 and listed in 1983. In 1983, bottled water and portable showers were provided to affected residences and businesses until water main connections from the city system were completed.

Subsequent investigations identified several sources of contamination: a leaking underground storage tank at a nearby solvent recycling company, spills at a railroad loading spur at that company, and spills at a section of the railroad line southeast of the well field. New water-supply wells were installed in a clean portion of the well field and 12 gradient control wells were installed in 1984. Groundwater was extracted from the gradient control wells at approximately 2000 gpm and, after an initial period of treatment using activated carbon, treated in an air stripper to remove VOCs. The project team initially estimated that it would take 5 years to restore groundwater quality.

The 10-ft-diameter air-stripping tower was 60 ft high and constructed of PVC-wrapped fiberglass. It was packed to a height of 40 ft with 3.5-in.-diameter polypropylene pall

Figure 3.8   Photograph of counterflow plate stripper. (Reprinted with permission of Carbtrol®
Corporation, Westport, CT.)

rings. With an air to water ratio of 20:1, the air stripper reduced VOC concentrations by
approximately 82.9% from 131 µg/L. The operators controlled iron fouling by circulating
a solution of sodium hypochlorite through the stripper about four times per year.

The off gases from the air stripper were heated by approximately 30°F to reduce the
relative humidity to 40%, then treated using activated carbon. Two activated carbon
units were used in parallel. Each carbon bed was 4 ft deep and 10 ft in diameter, and
contained 19,000 pounds of carbon.

As of 1996, groundwater was still contaminated with tetrachloroethylene, TCE, toluene,
xylene, and vinyl chloride. Groundwater extraction (from six of the original gradient
control wells) and treatment continued. In the 1990s, the initial system was supplemented with additional groundwater extraction and treatment systems in three source
areas. Soil in those source areas was remediated using soil vapor extraction as described
in Example 4.4.

This summary is based on information from several EPA sources.[48,54,55]

In a *tray stripper*, which may be cylindrical or rectangular, water flows downward
through a series of perforated plates or trays. Air flows upward through the perforations
in the trays. As the air contacts the contaminated water, contaminants move from the
liquid phase into the vapor phase. Figures 3.8 and 3.9 show one type of tray stripper.

The number of trays depends on the necessary removal efficiency and the volatility and solubility of the contaminant. The diameter or cross-sectional area of the
stripper depends on the air and water flow rates. These design parameters can be
calculated for a particular application;[56] most groundwater treatment problems fall
within a range that can be handled by off-the-shelf units. Tray strippers tolerate
higher levels of solids than packed towers, are less prone to fouling, and can be
easier to clean when they do foul.

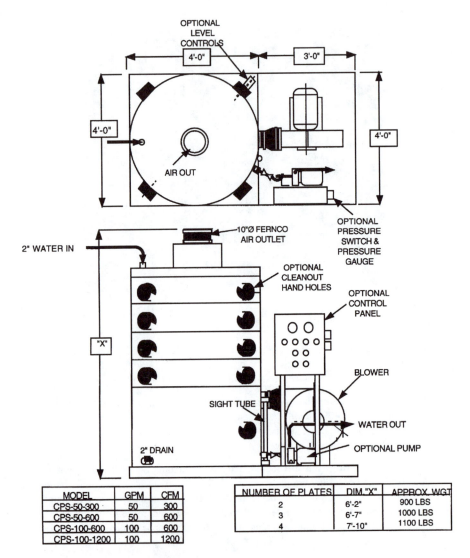

MODEL	GPM	CFM
CPS-50-300	50	300
CPS-50-600	50	600
CPS-100-600	100	600
CPS-100-1200	100	1200

NUMBER OF PLATES	DIM."X"	APPROX. WGT
2	6'-2"	900 LBS
3	6'-7"	1000 LBS
4	7'-10"	1100 LBS

Figure 3.9 Drawing of counterflow plate stripper. (Reprinted with permission of Carbtrol® Corporation, Westport, CT.)

An *aeration tank* or basin is the simplest and least efficient type of air stripper. A blower supplies air to the tank through a perforated pipe or porous plates (aerators) in the bottom of the tank. Volatile contaminants are stripped from the groundwater as the air bubbles rise through the tank.[57] Baffles may be used to direct the water flow through the tank and increase the effective contact time. This design does not include the packing or perforated trays, as described above, which enhance the contact between liquid and vapor phases. Vendors provide skid-mounted package units.

For any type of stripper, air emissions control may be required by health and environmental concerns and/or by regulations. Three types of emission control are commonly used with air strippers, following a demister to remove water droplets: vapor-phase carbon adsorption, catalytic incineration, and thermal incineration. See Section 3.7 for a discussion of these technologies.

Air strippers can foul as a result of bacterial growth or precipitation of naturally occurring metals in groundwater:

- Bacteria may grow on the packing/trays if water high in biodegradable organics is treated. Eventually, biological growth will clog the packing/trays. This concern may be significant enough to require a different treatment technology instead of air stripping. Relatively low levels of bacterial growth can be cleaned off the packing or trays as part of routine maintenance.
- High levels of calcium and magnesium (hardness) can precipitate out of groundwater, fouling piping and clogging air strippers. Fouling can be a significant problem when the hardness is greater than 800 mg/L.[58] Hardness can be removed by precipitation or ion exchange.
- Iron and manganese may precipitate in the stripper and clog the unit. A packed column can typically tolerate up to 5 mg/L (total) iron and manganese before fouling becomes an operation and maintenance problem.[58,59] A tray stripper can tolerate higher levels: according to one vendor, concentrations on the order of 25 mg/L iron and manganese in the influent.[60]

Pretreatment may be required to remove iron and manganese. Options for pretreatment include precipitation, manganese greensand filters, and the addition of sequestering agents. Fe(II) can be removed by aerating the water, to oxidize the iron to Fe(III). Ferric (Fe(III)) hydroxides precipitate readily at a pH between 7 and 10.[61] Aeration will not effectively oxidize manganese below pH 9.5.[61] Chemical oxidation (by chlorination or potassium permanganate) may be required to precipitate manganous oxides. Iron and manganese are also removed by the lime softening process used to remove hardness.[61] A *manganese greensand filter* contains a combination of New Jersey glauconite (greensand), manganese sulfate, and potassium sulfate. This mixture contains high levels of manganese oxides. When manganous ion in groundwater contacts the bed, an oxidation-reduction reaction forms insoluble manganese dioxide. The manganese dioxide remains on the filter bed. The bed is periodically reoxidized with permanganate, and backwashed to remove solids. For high levels of manganese in the groundwater, the necessity for frequent maintenance can make the use of manganese greensand filtration infeasible. For additional discussion of iron and manganese removal, see *Water Treatment Plant Design*.[62] *Sequestering agents* are chemicals that can be added to the groundwater to chelate the metal ions and keep them in solution. Sequestering agents are only effective at relatively low concentrations of iron.

3.4.2.5 Carbon Adsorption and Related Technologies

Charcoal filtration was reportedly used nearly 4000 years ago to purify water.[63] Today, engineers commonly use carbon adsorption to separate organic compounds and some inorganic compounds from contaminated groundwater. Contaminants are attracted to and then held on the surface of activated carbon particles. Compounds that can be treated include organic compounds such as:

- VOCs (e.g., benzene, toluene, ethyl benzene, xylenes [BTEX], chlorinated ethanes, and ethenes),
- petroleum hydrocarbons (TPH),
- polynuclear aromatic hydrocarbons (PAHs), and
- PCBs (subject to certain handling restrictions on the spent carbon).

Treatment is particularly effective for organic contaminants that are nonpolar, of low solubility, or have a high molecular weight.

While activated carbon is most commonly used to remove organic contaminants, certain heavy metals and other inorganics also adsorb onto carbon. These include chromium, copper, lead, nickel, and zinc.[64,65] Activated carbon can also remove mercury from water.[66] The removal efficiency for metals depends strongly on pH. Activated carbon has also been used to remove cyanide.[67]

Activated carbon may be derived from a variety of carbon sources, commonly from coal, sometimes from other sources such as wood, coconut shells, or nut shells. The performance of activated carbon depends on the original carbon source and the way in which it is manufactured. Activated carbon is manufactured by crushing the source material, roasting it in an oxygen-free atmosphere to make charcoal, then roasting it a second time in the presence of steam. The steam etches away the surface of the carbon particles, creating highly porous granules. These pores provide the extremely high surface area that makes activated carbon an effective adsorbent. The surface area of a grain of activated carbon can vary between 500 and 1000 square meters per gram, or more than a square mile of surface area in six pounds of granular activated carbon.[68]

Activated carbon is classified by the size of the carbon particles. Granular activated carbon (GAC) particles have a diameter greater than 1 mm.[69] Powdered activated carbon (PAC) is also used in groundwater treatment, although less commonly than GAC. PAC particles have a diameter of less than 200 mesh.[69]

GAC is packed in a column constructed of stainless steel, lined steel, or polyethylene. Typically, contaminated water flows down through the column under pressure. As the water flows through the unit, contaminants adsorb onto the carbon. Two carbon columns in series are frequently used to treat groundwater. Multiple units may be used in parallel to provide hydraulic capacity.

Carbon units are sized to provide sufficient time for the contaminated groundwater to contact the carbon and contaminants to adsorb to the carbon. The typical contact time is 10 to 30 min per unit. The size is also based on the necessary bed depth; room needed for expansion during backwashing (discussed below); and the bed life, or frequency of carbon replacement. Liquid-phase carbon treatment units are available from many vendors, in a variety of sizes. For example, flows up to 10 gpm and 7.5 psi can be treated in a unit the size of a 55-gallon drum (24 in. diameter × 34 in.

high). This size typically contains up to 200 lb of carbon and provides 15 min of contact time at 10 gpm. These small units are not designed for backwashing or replacing carbon on site.[70,71] Flows up to approximately 50 gpm can be treated in a larger unit, approximately 45 in. diameter × 88 in. high, containing 1800 lb of carbon.[72] Units are available in larger sizes (e.g., can treat up to 700 gpm in a single unit).

Solids in the influent gradually accumulate on the carbon bed, causing a pressure drop across the carbon. Large carbon adsorption units intended for long-term use are designed to permit periodic *backwashing* to remove particulates when the pressure drop becomes too high. Backwashing is generally necessary at a solids loading of roughly 1 to 3 lb/sf.[59] When backwashing is to occur, treatment of contaminated groundwater must stop. Backwashing is accomplished by reversing the flow through the carbon unit, using clean water to expand the carbon bed and remove the solids. The wastewater generated from backwashing must be treated.

When the adsorption capacity of the carbon is used up, or *spent*, contaminant *breakthrough* occurs and contaminants are detected in the effluent from the carbon bed. The use of two columns in series to treat groundwater is intended to prevent discharge of unacceptable levels of contaminants when breakthrough occurs: when contaminants break through the first column (Column A) in series, the second column (Column B) adsorbs the contaminants. At that time, the carbon in the first bed (Column A) is replaced or regenerated to provide fresh carbon. The flow pattern is then reversed, so that the water flows first through Column B, then through Column A. (The cleanest carbon bed is always last.)

Spent carbon may be landfilled or *regenerated*. Regeneration means that the carbon is thermally treated to remove and destroy adsorbed organic contaminants. (Metal contaminants are not removed.) The initial adsorptive capacity is partially, although not completely restored. About 5 to 10% of the carbon is destroyed during regeneration.[69] Carbon can be regenerated on or off site. On-site regeneration is generally not cost effective unless more than 2000 lb of carbon are used per day.[73]

The potential performance of activated carbon in a specific application is best predicted by performing bench-scale tests. Lacking site-specific test data, literature data can be used to estimate carbon usage. The adsorption of a particular compound onto activated carbon is indicated by an adsorption isotherm. The capacity of carbon to adsorb a contaminant varies with the concentration of that compound in solution; adsorption *isotherms* describe this variation at a constant temperature. *Freundlich Isotherms* are frequently used to characterize the use of carbon adsorption in a particular application. As indicated in Section 2.1.1.4,

$$\frac{x}{m} = K_p C^{\frac{1}{n}} \tag{3.11}$$

or

$$\log\left(\frac{x}{m}\right) = \log\left(K_p\right) + \frac{1}{n}\log C \tag{3.12}$$

Table 3.1　Adsorption Isotherms for Liquid-Phase Carbon

Compound	Range of equilibrium concentrations (μg/L)	Isotherm parameters[a] K_p $(\mu g/g)(L/\mu g)^{1/n}$	1/n
Benzene[b]	462–3.2	1,260	0.533
Toluene[b]	104–2.3	5,010	0.429
Ethylbenzene[b]	565–39.4	9,270	0.415
p-Xylene	32.7–1.6	12,600	0.418
Chlorobenzene	732–15.4	9,170	0.348
Pentachlorophenol	3410–0.4	42,600	0.339
Methylene chloride	715–18.1	6.25	0.801
Chloroform	226–13.2	92.5	0.669
Carbon tetrachloride	429–9.1	387	0.594
1,1-Dichloroethane	559–13.4	64.6	0.706
1,2-Dichloroethane	392–6.9	470	0.515
1,1,1-Trichloroethane	860–14.6	335	0.531
1,1-Dichloroethylene	392–6.9	470	0.515
trans-1,2-Dichloroethylene	415–13.5	618	0.452
Trichloroethylene	442–7.7	2,000	0.482
Tetrachloroethylene	421–3.6	4,050	0.516

[a] Data were obtained from tests with distilled deionized water, unless otherwise noted. In this series of laboratory experiments, natural water isotherms generally had a reduced capacity compared with the isotherms generated using distilled deionized water. Tests performed using Filtrasorb 400®, Calgon Corp.
[b] Ethanol cosolvent was used (<1:750 v/v); not expected to affect adsorption of other compounds.

From Speth, T. F. and Miltner, R. J., *Technical Note: Adsorption Capacity of GAC for Synthetic Organics*, J. Am. Water Works Assoc., February 1990, 72-75. With permission.

where:　x/m = mg contaminant adsorbed per kg carbon (or μg/g),

K_p = experimentally determined constant for a particular contaminant and type of carbon,

1/n = experimentally determined constant for a particular contaminant and type of carbon, and

C　= inlet concentration of contaminant (mg/L or μg/L).

Isotherms are frequently reported as tabulated values of K_p and 1/n for different contaminants. Table 3.1 presents EPA-derived data for some of the most common groundwater contaminants. Isotherms are also presented as linear plots of log (C) vs. log (x/m).

Because the isotherms depend on the type of carbon and groundwater characteristics, as well as the contaminant, results can vary. Table 3.2 illustrates the range of isotherms which can be generated for a single compound.

Estimates of carbon usage calculated from isotherm data provided by a vendor without site-specific testing or in the literature do not account for site-specific groundwater chemistry. Isotherms are developed for a single contaminant in solution; they do not account for the competition between multiple contaminants for adsorption sites. In addition, isotherms are developed for batch solutions at equilibrium,

Table 3.2 Isotherm Data for Trichloroethylene

Carbon	k	1/n	C Units	x/m units
Westvaco WV-G	3260	0.407	µg/L	µg/g
Westvaco WV-W	1060	0.5	µg/L	µg/g
Hydro darco-3000	713	0.47	µg/L	µg/g
Filtrasorb-300	28	0.62	mg/L	mg/g
Filtrasorb-400	36.3	0.592	mg/L	mg/g
Filtrasorb-400	45	0.625	mg/L	mg/g
Filtrasorb-400	3390	0.416	µg/L	µg/g

From U.S. EPA, *Treatability Database, Version 5.0*, Risk Reduction Engineering Laboratory, Cincinnati, OH, April 1994.

not for flowing systems. As a rule of thumb: when estimating carbon usage without site-specific testing, *double* the carbon usage estimated from isotherms to account for the differences between laboratory isotherm tests and field conditions.

Example 3.3: Estimate the amount of carbon used annually to treat a 10-gpm groundwater stream containing 500 µg/L benzene.

From Table 3.1, $K_p = 1260$ $(µg/g)(L/µg)^{1/n}$ and $1/n = 0.533$. Substituting these values into Equation 3.11,

$$\frac{x}{m} = (1260)(500 \ µg/L)^{0.533} = \frac{34600 \ µg \ benzene}{g \ carbon}$$

The mass of benzene in extracted groundwater, per year (assuming that the concentration is constant) is

$$(500 \ µg/L)(10 \ gal/min)(3.785 \ L/gal)(1440 \ min/day)(365 \ day/yr)$$
$$= 9.9 \times 10^9 \ µg \ benzene$$

The estimated amount of carbon used each year is

$$2 \cdot \frac{9.9 \times 10^9 \ µg \ benzene}{34600 \ µg \ benzene/g \ carbon} \cdot 10^{-3} \frac{kg}{g} \cdot 2.205 \frac{lb}{kg} = 1300 \ lb \ carbon$$

Activated carbon units can foul as a result of bacterial growth or precipitation of naturally occurring metals in groundwater. They can also foul or lose capacity due to naturally occurring organics or contaminants other than the target compounds. Operating concerns include the following conditions:

- Suspended solids ≥50 ppm require backwashing so frequently as to be impractical.[73,75] Pretreatment (e.g., by sedimentation or filtration) would be required.
- High levels of calcium and magnesium (hardness) can precipitate out of groundwater, fouling piping and clogging carbon beds. Hardness levels greater than 100 mg/L as $CaCO_3$ may cause problems.[75] Hardness can be removed by precipitation or ion exchange.

- Iron and manganese may precipitate and clog the carbon unit. Pretreatment may be required to remove iron and manganese if their total concentration exceeds ~10 mg/L.[59] Options for pretreatment were discussed in Section 3.4.2.4.
- Bacteria may grow on the carbon bed if the groundwater contains high levels of biodegradable organics. This biological growth can clog the carbon bed.
- Oil and grease at levels ≥10 mg/L may foul carbon.[75]
- High levels of contamination (e.g., 1000 mg/L) may exhaust carbon capacity quickly, making treatment by carbon adsorption impractical or not cost effective. Naturally occurring organic compounds may exhaust carbon capacity and/or interfere with the adsorption of the target compounds, e.g., humic and fulvic acids (found in some surface waters) at concentrations of 10 to 100 mg/L.[73]

Other adsorbants are used besides GAC, including:

- peat moss;
- regenerable synthetic carbonaceous absorbents. Rohm and Haas developed Ambersorb® 563 adsorbent to remove organic contaminants from groundwater. An EPA pilot test indicated that Ambersorb® 563 could remove trichloroethylene and other chlorinated organics to levels below the MCL; the Ambersorb® 563 adsorbent could treat approximately two to five times the bed volumes as did GAC, while operating at five times the flow rate loading.[76]
- Mixtures of clay and anthracite, used to absorb hydrocarbons and break emulsions, are available from several vendors.[77,78]

3.4.2.6 Ion Exchange

This form of treatment is used to remove metal ions from water. When in contact with contaminated water, certain minerals and synthetic resins will exchange "their own" ions with ions in the water. Ion exchange is used to remove hardness (Ca^{2+} and Mg^{2+}) or toxic metals from water. It can also be used to remove inorganic pollutants such as sulfate (SO_4^{2-}). It is sometimes used as a polishing step rather than a primary treatment step.

The ion-exchange matrix can be a *zeolite* (a naturally occurring aluminosilicate) or a synthetic polymeric resin. Recently developed products include ion-exchange sponges and biological ion-exchange media.[79-81] For simplicity, the remainder of this discussion will refer to ion-exchange *resins*. As discussed further below, ion-exchange resins can be classified as *cation-* or *anion-exchange* resins. *Cation-exchange resins* are used to remove positively charged ions, such as Pb^{2+} (lead), Cr^{3+} (trivalent chromium), Ca^{2+}, or Mg^{2+}. *Anion-exchange resins* are used to remove negatively charged ions, such as SO_4^{2-}. In aqueous solution (such as groundwater), hexavalent chromium (Cr^{6+}) forms anions such as chromate (CrO_4^{2-}), dichromate ($Cr_2O_7^{2-}$), and bichromate ($HCrO_4^-$); these anions can be removed using anion-exchange resins.

Typically, ion-exchange resins are placed in columns. As the water flows downward through the column, ions in the water exchange with ions on the resin (see Figure 3.10). The column is sized based on the flow rate, the capacity of the resin

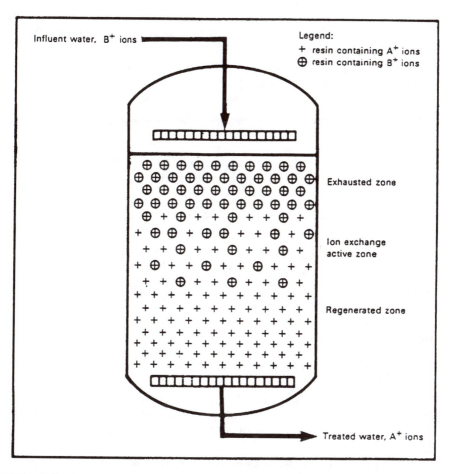

Figure 3.10   Ion exchange column in service. (From U.S. EPA, *Summary Report: Control and Treatment Technology for the Metal Finishing Industry — Ion Exchange*, EPA 625/8-81-007, U.S. EPA, June 1981.)

under site-specific conditions, and the intended frequency of regeneration. It typically consists of a vertical cylindrical pressure vessel made of steel. The ion-exchange resin is supported on a screen. Influent water is distributed across the top of the resin bed. The unit may be plumbed to allow for regeneration through either downward flow or upward flow of a regenerant solution.[82]

When the ion-exchange capacity of the resin is exhausted, the column is backwashed if necessary to remove trapped solids and then *regenerated*. Cation-exchange columns are regenerated with a strong acid solution. This solution removes the metals adsorbed onto the resin and "replaces" them with hydrogen ions. The resin is then ready for reuse. Anion-exchange columns are regenerated with a strong base solution. The regeneration process produces a highly acidic or basic waste stream that contains concentrated contamination. This stream typically requires further treatment. In some applications, the metals can be recovered from the regenerant stream as a resource. The alternative to regeneration is to simply replace the resin with fresh material and dispose of the spent resin.

An ion-exchange resin may be characterized by its capacity, i.e., the number of functional groups per gram of resin (generally, capacity ranges from 1 to 10 milliequivalents per gram, or meq/g). It may also be classified by its acid or base strength.

Cation-exchange resins can be classified as *strong-acid resins*, *weak-acid resins*, or *chelating resins*.

Strong-acid resins have sulfonic acid functional groups ($-SO_3H$). Cationic contaminants exchange with a hydrogen ion on the resin.

$$M^{2+} + 2\,R-SO_3-H \;\rightarrow\; \left(R-SO_3\right)_2 - M + 2\,H^+$$

where M^{2+} is a divalent metal ion (contaminant), and R is the resin matrix.

If several cations are in the groundwater, the resin will preferentially adsorb certain cations. This tendency can limit the application of ion exchange. For example, consider groundwater contaminated with cadmium (Cd^{2+}) that also contains very high levels of hardness (Ca^{2+} and Mg^{2+}). An ion-exchange resin would preferentially adsorb Ca^{2+} over Cd^{2+}.[83] Very large volumes of resin might be required to remove both the hardness and the cadmium, with correspondingly high waste generation.

These resins hold the metal contaminants far more strongly than hydrogen ions. As a result, regeneration requires several times the stoichiometric amount of acid.

Weak-acid resins are similar to the strong-acid resins, but have carboxylic acid groups ($-COOH$) rather than sulfonic acid groups. Because of the functional group, they do not operate below pH ~4. Weak-acid resins are more easily regenerated than strong-acid resins.

Chelating resins are similar to weak-acid resins; the functional group is iminodiacetate. These resins are particularly effective for mercury, copper, and lead. Due to their relatively slow kinetics, use of chelating resins requires a large resin volume and low flow rate (thus a long retention time). Chelating resins are relatively expensive compared to other resins.

Anion-exchange resins can be classified as *strong-base resins* or *weak-base resins*. Strong-base resins have a quaternary amine functional group ($-CH_2N(CH_3)_3^+Cl^-$). Weak-base resins have tertiary ($-CH_2NH(CH_3)_2^+$) or secondary ($-CH_2NH_2(CH_3)^+$) amine functional groups.

Operating concerns include:

- treatment/disposal of concentrated regenerant;
- fouling due to high levels of suspended solids in the water (>10 mg/L), which will clog or blind the resin;[84]
- high levels of organic contaminants, particularly oil and grease, which can coat and foul the resin;
- interference from natural groundwater constituents (e.g., hardness); and
- a potential concern may be the concentration of exchanged ions in the treated water (e.g., sodium levels in drinking water).

3.4.2.7 Reverse Osmosis

Reverse osmosis (RO) separates inorganic contaminants from water by forcing pure water through a semipermeable membrane under pressure.[85] The contaminants cannot cross the membrane. The treatment process generates clean water and a wastewater containing concentrated contaminants.

Membranes may be made of cellulose acetate, aromatic polyamides, or thin-film composites (comprised of a thin layer of membrane on a porous support polymer). The equipment may take several forms. For example:

- A tubular module consists of a porous, membrane-lined tube. Contaminated water is pumped through the tube, and the pure water flows radially outward. The concentrated solution of contaminants exits through the other end of the tube.
- A plate and frame module is analogous to a plate and frame filter press. It consists of a stack of circular plates sandwiched or wrapped with the RO membrane.

RO can achieve 50% to 99% removal of inorganics in one stage.

Operating concerns include:

- treatment/disposal of concentrated wastewater;
- clogging membrane with particulates;
- acceptable pH range (varies with the membrane); and
- fouling the membrane with organics.

3.4.3 Destruction Technologies

Depending upon their chemical structure, chemical compounds can be destroyed by biological oxidation, chemical oxidation, or reduction.

3.4.3.1 Biological Treatment

Engineered biological treatment of sewage dates back over 100 years.[86] Traditional sewage-treatment processes have been adapted to treat contaminated groundwater. Extracted groundwater can be treated in a bioreactor to destroy organic contaminants and certain inorganic contaminants. Biological treatment can also be applied *in situ* (Section 3.6). Compounds which can be treated biologically include: ammonia, nitrates; ketones; alcohols, phenols; hydrocarbons (TPH); monoaromatics (BTEX) and polynuclear aromatic hydrocarbons (PAH); cyanides; and, under certain conditions, chorinated ethenes (PCE, TCE). Metals — which as elements cannot be biodegraded — may adsorb to the biological mass and thereby be removed from the groundwater.

Biological wastewater treatment may be *aerobic* or *anaerobic*. Aerobic treatment requires oxygen, which serves as the *electron acceptor* in a redox reaction. Contaminants are oxidized to form carbon dioxide, ammonia, microbial cell tissue, and other byproducts. Aerobic processes are generally used to treat wastewater containing between 50 and 4,000 mg/L BOD,[87] although they can be used to treat lower concentrations (e.g., 10 mg/L BOD).[88]

Anaerobic treatment does not require oxygen. The microorganisms use electron acceptors other than oxygen, such as nitrate, sulfate, and carbon dioxide, to oxidize organic compounds. Treatment may depend upon a consortium of microorganisms, e.g., nonmethanogenic microorganisms to hydrolyze large organic molecules or break down smaller molecules into organic acids, and methanogenic microorganisms to convert certain compounds such as hydrogen, carbon dioxide, methanol, or organic acids to methane.[89] Anaerobic processes have traditionally been used to treat highly concentrated industrial or food-processing wastewater, e.g., 4,000 to 50,000 mg/L BOD.[87] They are also used to *digest* the sludge generated by aerobic biological wastewater treatment, in order to decrease the volume of sludge that must eventually be disposed of. More recently, anaerobic treatment processes have been used to degrade relatively low concentrations of certain organic contaminants in groundwater.

The general theory and design of bioreactors is described below, followed by a brief discussion of some of the applications to groundwater treatment.

3.4.3.1.1 Model for Bioreactors — Bacterial growth in a single population in a batch culture follows a general pattern.[90]

(1) *Lag phase* — represents the time required for the microorganisms to acclimate to their environment. The microorganisms begin to reproduce.
(2) *Log-growth phase* — rapid increase in the number and mass of microorganisms; food is so plentiful that the microorganisms reproduce as fast as physiologically possible.
(3) *Declining growth phase* — the food supply has decreased. The rate of microbial reproduction decreases until the number of viable microorganisms is stable. (At the peak of the growth curve, the microbial population is stable: the growth of new cells is offset by the death of old cells.)
(4) *Endogenous phase* — the bacterial death rate exceeds the production of new cells. The original food supply is exhausted or nearly so.

Operators of biological treatment systems generally seek to maintain a bacterial population at the end of the declining growth phase or the beginning of the endogenous phase for optimum treatment. Operation of a treatment system during the log-growth phase has two problems: less-than-optimum removal of BOD, and poor settling of microbial mass in suspended growth systems. The *food to microorganism ratio* (F/M) is used to characterize the operation of a biological treatment system. A high F/M ratio implies that the system is in the log-growth phase; a low F/M ratio implies that a system is in the endogenous phase.

The growth of microorganisms in a batch culture is described by the empirical *Monod equation*. The degradation of a substrate can be calculated by:[89,91]

$$\frac{dS}{dt} = \frac{-kSX}{K_s + S} \qquad (3.13)$$

where S is the concentration of the growth-limiting substrate, t is time, k is the maximum use of substrate per unit weight of microorganisms (time^{-1}), X is the concentration of microorganisms, and K_s is the Monod half velocity coefficient, equal to the substrate concentration when $dS/dt = 0.5k$ (units of mass/volume). When K_s is large with respect to the substrate concentration, Equation 3.13 reduces to an equation describing a first-order reaction:

$$\frac{dS}{dt} = \frac{-kSX}{K_s} \tag{3.14}$$

Various other first-order and pseudo first-order kinetic expressions have been used to describe biological treatment processes.[89] The kinetics of bioremediation must be evaluated on a site-specific basis.

3.4.3.1.2 General Design Considerations — Bioreactors are traditionally designed to optimize natural biodegradation using equations based on F/M, reaction kinetics, and/or hydraulic loading and retention time. The design of a treatment system ultimately requires site-specific treatability testing at bench and/or pilot scale.

Design considerations include:

- the type of contamination, concentrations, and potential variations in concentrations over time;
- kinetics of degradation reactions and temperature dependance (reaction slows with temperature);
- potential toxicity of groundwater constituents;
- requirements for macro- and micronutrients (biological treatment requires roughly 1 g of nitrogen per 30 g COD removed and 1 g of phosphorus per 140 g COD removed, although actual requirements are site specific);[92]
- pH, which is optimally between 6 and 9;
- the need for supplemental food, or a cometabolite;
- type of reactor, as described below;
- air supply, for aerobic treatment;
- potential for air emissions of VOCs; and
- sludge generation, dewatering, and disposal.

Suspended-growth, fixed film, or natural treatment systems may be used to treat extracted groundwater.

In *suspended-growth systems*, the microorganisms grow in solution, or are suspended in the water being treated. Suspended growth systems include activated sludge and sequencing batch reactors (Figure 3.11).

Activated sludge is the classic form of aerobic biological wastewater treatment: most municipal sewage treatment plants are activated sludge units. The term "activated sludge" refers to the biological mass (sludge) maintained in a continuous flow stirred tank reactor. Air is sparged into the bottom of the tank to provide oxygen and mixing. The effluent from the treatment tank flows to a clarifier, where biological sludge settles by gravity. Some of the biological sludge (roughly 30%) is recycled to the aeration tank. As a result of this recycle, the *mean cell residence time* of the microorganisms differs from the hydraulic residence time. Some of the sludge is

Activated Sludge

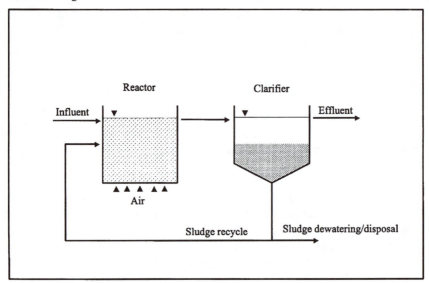

Sequencing Batch Reactor

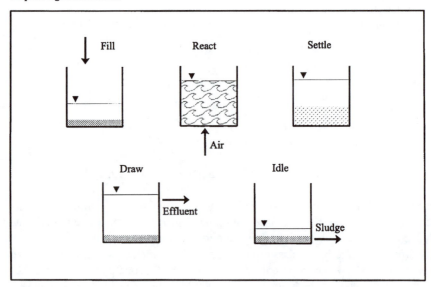

Figure 3.11 Suspended growth biological reactors.

wasted from the clarifier to remove organic material from the system and maintain the F/M ratio in the desired range (or in other words, to keep the population of microorganisms in the desired growth phase). The wasted sludge must be dewatered and disposed of.

There are several variations of the activated sludge process. Most variations pertain to the type of air supply. Design variations include: step aeration, contact stabilization, extended aeration, and pure oxygen systems.

Conventional activated sludge systems have a hydraulic retention time in the aeration basin of 4 to 8 h. The *mean cell residence time* of biological solids is typically 5 to 15 d, although it may be longer for certain organics.

Sequencing batch reactors (SBRs) are batch suspended-growth treatment units. Two units in parallel may be used to treat a continuous flow of groundwater. Each treatment "batch" includes four steps, followed by an idle period:

(i) *Fill* — the tank is filled with groundwater.
(ii) *React* — the contents of the tank are aerated and mixed to treat the groundwater.
(iii) *Settle* — the air supply and mixers are shut off and the biological mass allowed to settle.
(iv) *Draw* — treated groundwater is decanted off the top of the tank. A slurry of biological solids remains in the tank for the next batch treatment. (Excess solids are periodically wasted.)
(v) *Idle* — in a multiple-tank system, one tank may be temporarily idle while the other tank is in use.

As batch treatment units, SBRs offer flexibility in treating influent at varying concentrations and flow rates. Retention time and oxygen supply can easily be varied to suit varying influent concentrations and flows. SBRs can also achieve better solids separation than continuous-flow systems, since settling occurs under quiescent conditions. There are fewer design standards for SBRs than activated sludge systems.

In fixed-film systems, as the name implies, a film of microbial growth is attached (or *fixed*) to an inert support medium. As the water flows over the support medium, the contaminants are biodegraded. Designs include trickling filters, rotating biological contactors, fluidized bed reactors, and proprietary systems.

In a trickling filter, classically used to treat sewage, the water is sprayed over and trickles through a bed of crushed stone approximately 4 in. in diameter. Biological growth is attached to the stone. An underdrain system collects treated water and conveys it out of the trickling filter. Excess biological growth sloughs off the trickling filter. As a result, a trickling filter is followed by a clarifier to settle out the excess biological mass. Biological solids may be recycled (similar to an activated sludge system).

Biological towers or *packed bed reactors* represent a variation on the tricking filter design. Somewhat similar in construction to an air stripper, a biological tower is filled with a synthetic packing that provides the surface area for biological growth. Biological towers enable control of VOC emissions. They also tolerate higher flow rates than trickling filters.

A *rotating biological contactor* consists of a shaft of rotating circular plastic plates partially immersed in a tank containing the water to be treated. Biological growth attaches to the disks. As the disks rotate into the tank, they are coated with water. As the disks rotate out of the tank, they are exposed to oxygen in the air.

Fixed-film systems offer several advantages over suspended-growth systems. In general, they are simpler to operate, can tolerate lower influent concentrations, and, depending on the design, may generate less sludge. However, they also have several disadvantages compared to suspended-growth systems. In general, they are not as efficient, there is less control over the treatment process, and they are subject to icing in cold weather unless enclosed.

So-called *natural treatment systems* or *phytoremediation* systems use the capabilities of natural plants or ecosystems to treat waste. Such systems include constructed wetlands, floating aquatic plants, and agricultural plants or trees, as described below.

3.4.3.1.3 Applications to Groundwater Treatment — The following discussion — by no means comprehensive — describes some of the equipment and techniques used to treat extracted groundwater. Example 3.5 in Section 3.6 describes another form of treatment. These examples show how the traditional wastewater techniques described above have been adapted for use on hazardous waste sites.

- The ZenoGem™ process was developed by Zenon Environmental Systems, Inc. and tested by the U.S. EPA under the Superfund Innovative Technology Evaluation program. The trailer-mounted unit utilizes a 1000-gallon continuous-stirred tank reactor for aerobic suspended-growth biological treatment unit, followed by ultrafiltration. Ultrafiltration occurs in a unit containing a series of connected tubes which contains an ultrafiltration membrane. That membrane filters some dissolved compounds and particulates over 0.003 to 0.1 μm in size, depending on the membrane used. The solids are recycled to the biological treatment unit. After filtration, the groundwater is treated with activated carbon. The EPA has tested a trailer-mounted system on groundwater containing methyl methacrylate, with a COD of 1500 mg/L and higher, at a rate of 350 to 500 gpd. Groundwater was effectively treated. The ZenoGem™ process has also reportedly been used to treat oily wastewater, cleaning solutions containing detergents and alcohols, landfill leachate, and other waste streams.[93]
- Biotrol, Inc. is developing a two-stage methanotrophic bioreactor to treat chlorinated compounds such as TCE. In general, aerobic bacteria cannot use chlorinated ethanes or ethenes as a primary carbon source. One microorganism, *Methylosinus trichosporium* OB3b (abbreviated OB3b), can oxidize those compounds while using a single-carbon compound such as methane as its primary substrate. In BioTrol's two-stage bioreactor, a highly concentrated microbial culture of OB3b grown in a culture medium contacts contaminated groundwater in a plug flow reactor. Air and methane are supplied to the culture vessel.[94]
- BioTrol has also developed a fixed-film bioreactor for aerobic or anaerobic treatment. The support medium comprises a series of corrugated polyvinyl sheets. The unit successfully treated groundwater containing pentachlorophenol under the U.S. EPA Superfund Innovative Technology Evaluation program. The system has also reportedly been used to treat water contaminated with gasoline, phenol, and creosote.[95]
- Allied Signal has developed the Immobilized Cell Bioreactor (ICB). The ICB is a fixed-film reactor that can be used for aerobic or anaerobic treatment. The support medium is a proprietary mixture of polyurethane foam coated with activated carbon and plastic supports.[96,97]

• Wetlands can remove contaminants by several means, including: filtration of suspended and colloidal material; uptake into plants; adsorption onto organic materials; neutralization and precipitation through the generation of HCO_3^- and NH_3 by bacterial decay of organic matter; and destruction or precipitation of contaminants catalyzed by either aerobic or anaerobic bacteria. A constructed wetland is designed to maximize the specific removal process necessary to treat the target contaminant. The design is highly site specific. A constructed wetland was used to raise the pH and remove heavy metals from acid mine drainage in a 2-year test. The microbial reduction of sulfate to sulfide followed by precipitation of heavy metal sulfides removed over 90% of the copper and zinc, and raised the pH from 3 to 6.[98]

3.4.3.2 Chemical Oxidation

Strong chemical oxidants, such as peroxide (H_2O_2) or ozone (O_3), are used to degrade organic contaminants in groundwater. The use of chemical oxidants alone frequently has kinetic limitations. As a result, ultraviolet (UV) light is frequently used with chemical oxidants to overcome kinetic limitations and enhance the degradation of the target organics. Chemical oxidation is frequently referred to as UV/oxidation.

Chemical oxidation can be used to destroy organic compounds with double bonds, such as chlorinated ethylenes (PCE, TCE), mononuclear aromatics (BTEX, phenols), PAHs, and cyanides. VOCs may also be inadvertently removed from groundwater by air stripping in a UV/ozone unit.[99] UV/oxidation is often used in lieu of carbon adsorption or air stripping when destruction is preferred to separation; the target compounds are not readily treated by other means; or the operating costs of other treatment methods (e.g., due to high carbon loading) balance the relatively high capital costs of a UV/oxidation unit.

Water to be treated is dosed with the chemical reagent and exposed to UV light (if UV light is used) in a reactor. Although typically continuous flow, treatment can be performed on a batch basis. If ozone is used, it must be generated on site.

Several reactions contribute to the destruction of organic contaminants using UV/oxidation. The type of reactions depends on the reagents and equipment used. Three types of reactions can occur:[99]

(1) Oxidants such as ozone or peroxide can react directly with the organic molecules. Organic molecules are oxidized to carbon dioxide, water, and chlorides (if the original molecule was chlorinated). These reactions are often very slow.

(2) *Hydroxyl radicals* (HO·) generated from peroxide (H_2O_2) and/or ozone (O_3) in the presence of UV light (*hv*) attack and destroy organic compounds. The hydroxyl radical is a stronger oxidant than hydrogen peroxide or ozone. One or more of the following reactions may occur, depending on the type of equipment, reagents used, and presence of naturally occurring iron:[99-101]

$$H_2O_2 + hv \rightarrow 2\ HO^\cdot$$

$$2\ O_3 + H_2O_2 \rightleftarrows 2\ HO^\cdot + 3\ O_2$$

$$Fe^{2+} + H_2O_2 \rightarrow HO^\cdot + Fe^{3+} + OH^-$$

(3) Absorbtion of UV light, if supplied at the proper wavelengths, can break the chemical bonds in the contaminant molecules by *photolysis*.

Several design variables must be considered, including:

- type of reagent (peroxide, ozone, or peroxide plus ozone), and dose,
- wavelength and intensity of UV light, and number of lamps (note that UV light, high-intensity lamps in particular, add heat to the water),
- use of catalysts, and
- retention time.

These variables are manipulated to ensure adequate treatment of the target compounds. Insufficient treatment resulting from incomplete oxidation of target compounds can leave the target compounds or degradation intermediates in the effluent. Treatability testing is required to develop the design for a specific application.

UV/oxidation units have several operating concerns:

- Storage and handling of strong oxidants require safety precautions.[99]
- Iron and manganese can precipitate and foul the UV lamps. (In recognition that certain compounds can coat UV lamps and reduce their effectiveness, certain models incorporated scrapers to keep the lamps clean.) UV/oxidation units can generally tolerate an influent of up to approximately 5 to 10 mg/L iron, depending on the unit. Iron also absorbs ultraviolet light and consumes oxidant; as a result, high levels of iron can also interfere with oxidation of the target compounds.[102]
- Trivalent chromium, if present in the water, will be oxidized to the more toxic hexavalent form.[99]
- Bicarbonate and carbonate ions (alkalinity) will scavenge or consume oxidants, increasing reagent usage and cost. If bicarbonate and carbonate ions are present at a concentration greater than approximately 400 mg/L as $CaCO_3$, adjustment of the pH to 4 to 6 (to change the chemical species) may be required.[99,101,102]
- A high level of COD, potentially resulting from groundwater constituents other than the target compounds, can consume reagents and interfere with the oxidation of the target compounds. One vendor suggests that a level of COD on the order of 100 to 200 mg/L will interfere with the treatment of target organic compounds.[102]
- "Scavengers" which can interfere with treatment also include sulfides, nitrites, and bromides.[99]
- Visible, free, or emulsified oil and grease can foul the UV lamps.[99,101,102]
- Suspended solids can reduce the transmission of UV light and adsorb organics; water containing greater than 30 mg/L TSS will generally require pretreatment to remove suspended solids.[99,101,102]
- Emissions of excess reagent — ozone in air, peroxide in water — may be of concern due to their toxicity. UV/ozone treatment units include an ozone decomposition unit to limit ozone emissions.
- Air stripping and resultant emissions of VOC.

While the preceeding discussion focused on UV oxidation using peroxide or ozone, other chemical oxidation systems are used to treat groundwater. A brief description of several options follows.

Fenton's reagent consists of Fe^{2+} and hydrogen peroxide. Rather than being a nuisance which can foul UV lamps, as described above, iron is used to catalyze the oxidation reaction. (A *catalyst* accelerates the rate of reaction without being appreciably changed during the reaction.)

$$Fe^{2+} + H_2O_2 \rightarrow HO^{\cdot} + Fe^{3+} + OH^-$$

The free radical ($HO^{\cdot}$) produced by this reaction oxidizes the target compounds, initially to carboxylic acids and ultimately to carbon dioxide and water. Treatment is carried out at acidic pH (2 to 5) to prevent precipitation of $Fe(OH)_3$. Several reactions have been postulated for the observed regeneration of Fe^{2+} from Fe^{3+} (e.g., References 103 and 104). Fenton's reagent has been used to treat wastewater containing phenols,[105] chlorophenols,[103] chlorobenzene, PCBs, PAHs, cyanides,[106] benzene, alcohols,[107] and other compounds. Fenton's reagent has also been used to treat groundwater and NAPL residual *in situ* (see Section 3.6.2) and soils.

Chlorine in various forms (e.g., hypochlorite) is also used as a chemical oxidant. However, treatment of organics using hypochlorite can produce toxic chlorinated byproducts, such as chloroform. Finally, potassium permanganate is used in certain applications.

3.4.3.3 Reduction by Zero-Valent Iron (Reductive Dehalogenation)

Zero-valent iron can destroy a variety of compounds in a redox reaction where iron is oxidized and the contaminant is reduced. When applied to chlorinated organic compounds, this form of treatment is known as *reductive dehalogenation*. Corrosion of iron by chlorinated solvents was first observed as an operating problem in the chemical process industry.[108] The earliest environmental application was reported in 1972, but reductive dehalogenation was not applied to groundwater treatment until the 1990s.[109] Because of the potential to destroy chlorinated organic compounds, which are otherwise difficult to degrade, research and development of the technology have expanded rapidly in the 1990s. Field experience is limited, but growing. As of this writing in 1997, a bibliography of research on contaminant remediation with zero-valent iron metal can be accessed through a web site maintained by the Tratnyek Group at the Oregon Graduate Institute of Science and Technology, at http://www.ese.ogi.edu/ese_docs/tratnyek/ironrefs.html. EnviroMetal Technologies, Inc. (ETI), of Guelph, Ontario, Canada, holds the license on the patent on the use of zero-valent iron to treat groundwater contaminants.

Reductive dehalogenation can destroy chlorinated compounds such as halogenated methanes, chlorinated ethanes and ethenes, and vinyl chloride.[110] The process has also been used to degrade PCBs at high temperature in bench-scale experiments.[111] Reductive dehalogenation can also degrade other organic compounds, including Freon-113, ethylene dibromide, certain nitro-aromatics, and *N*-nitrosodimethylamine.[112] Zero-valent iron can reduce hexavalent chromium to the less toxic, less soluble trivalent ion.[109]

Reductive dehalogenation can be used in a treatment plant or *in situ*. Groundwater pumped out of an aquifer can be passed through a tank containing a bed of iron filings mixed with sand, sometimes called "magic sand", or a bed of iron filings. The groundwater first passes through an air eliminator and a filter to remove particulates which could clog the iron bed.[113] The treatment unit must be sized to allow sufficient hydraulic retention time to allow the contaminants to reach the desired effluent concentration. (See the discussion of reaction kinetics below.) Alternatively, reductive halogenation is used *in situ* by creating a permeable gate of iron filings in a funnel-and-gate system (see Section 3.5.4).

As an alternative to iron filings, CERCONA Inc. has developed iron-based foams comprising iron and aluminosilicates. The foam can incorporate catalysts such as palladium or copper. It has a relatively high hydraulic conductivity and surface area; as a result of the high surface area, unpublished studies have shown relatively high reaction rates.[114] Other researchers have examined the use of Fe° colloids.[115]

Reductive dehalogenation is an abiotic pseudo first-order reaction. The reaction is relatively independent of contaminant concentration.[116] It is strongly dependent on the surface area of the iron catalyst. The rate constant has been observed to decrease over time in batch studies, potentially due to the accumulation of reaction products (hence, "pseudo" first order).[110] A study of available data determined that first-order rate constants from a variety of reported batch and column studies varied widely; however, once the data were normalized to account for the iron surface area concentration, the reaction constants for a given compound were relatively consistent.[117] In practical terms, this means that the reaction rate constants or half lives developed for one type of iron medium (e.g., iron filings, iron filings mixed with sand, iron foam) should not be applied to a design based on a different medium without accounting for the difference in surface area.

The design must account for the degradation of chlorinated daughter products, in addition to the degradation of the target compound. In general, dechlorination is more rapid at saturated carbon centers than unsaturated centers, and high degrees of halogenation favor rapid reduction. In other words, PCE will degrade more quickly than TCE, which will degrade more quickly than daughter products such as dichloroethylene or vinyl chloride. As a result of these kinetics, the degradation of daughter products to meet discharge or clean-up levels may drive the design.

In a report on a field test of an above-ground reactor, EPA indicated that flow conditions and reaction rates in the iron bed may decrease over time due to precipitation.[113] Example 3.4, which describes a funnel-and-gate system, provides an example of the use of reductive dehalogenation in a reaction wall and the potential decrease in porosity over time.

3.5 *IN SITU* TREATMENT

Treating groundwater contaminants in place can offer many advantages over pump-and-treat systems. In a pump-and-treat system, potentially costly equipment and energy are needed to bring groundwater to the surface and move the water through a series of treatment units. The options for discharging treated water are sometimes quite limited. However, *in situ* treatment also has limitations. The mass transfer of reagents and contaminants can be limited in heterogeneous or low-permeability soil. *In situ* processes can be difficult to control and to monitor. Because of the potential advantages of *in situ* treatment, researchers have focused on several forms of *in situ* treatment in recent years. These include natural attenuation, biodegradation, air sparging, reductive dehalogenation, oxidation, and surfactants. (The latter two technologies are described with respect to NAPL remediation in Section 3.6.2. Similar principles apply to groundwater remediation.) Many of these technologies are adaptations of the above-ground treatment technologies discussed previously in this chapter.

Subsurface conditions must be changed to actively treat groundwater *in situ*. Three techniques are commonly used: extraction/injection wells (or leaching systems), vertical circulation wells, and funnel-and-gate systems. The latter two technologies are relatively new. Air and aqueous solutions of reagents can be added to an aquifer through injection wells and removed through extraction wells. Several vendors have developed single-well systems which can inject air or reagents and circulate groundwater vertically around each well. The groundwater circulation creates a miniature reactor around each well, providing control of mixing and hydraulic retention time. Section 3.5.2 describes such systems in more detail. Funnel-and-gate systems channel groundwater flow toward a treatment cell. A vertical barrier such as sheet piling is placed perpendicular to groundwater flow or at a slight angle to perpendicular. The vertical barrier directs groundwater flow toward a permeable treatment cell, or gate, where treatment occurs. Sections 3.5.2 and 3.5.4 describe the use of funnel-and-gate systems for *in situ* air sparging and reductive dehalogenation, respectively. Other applications have been considered or tested. These include the use of absorbant material such as activated carbon; chelating agents or precipitating agents to immobilize metals; or bioreactors in the gate.[118]

3.5.1 Natural Attenuation

Natural attenuation refers to the physical, chemical, and biological processes which act to reduce the concentration of a contaminant in an aquifer. Organic contaminants can hydrolyze or biodegrade (sometimes called *intrinsic bioremediation* or *passive bioremediation*). Advection, dispersion, and dilution from infiltration physically decrease contaminant concentrations. Volatilization can transfer groundwater contaminants to the air in soil pores. Finally, contaminants can sorb onto aquifer solids, reducing their mobility. Of these attenuation mechanisms, only volatilization and chemical and biological degradation actually reduce the *mass* of contamination in the aquifer. The other attenuation processes simply reduce the contaminant *concentration*. Section 2.2.1 briefly describes chemical and physical attenuation mechanisms. Intrinsic bioremediation and certain abiotic reactions are described further below.

Selection of natural attenuation as a remedy depends on regulatory requirements as well as site conditions.[119] Natural attenuation can be an appropriate remediation choice when contaminants readily degrade or disperse and will not present a significant risk to human health or the environment while they attenuate. Natural attenuation costs less than active remediation (although demonstrating that it will occur can be quite expensive) and has less effect on the use of the site. Regulators may approve natural attenuation more readily if source of further contamination are removed or contained. Natural attenuation is generally not appropriate when a site contains a significant amount of NAPL, contaminant concentrations are so high that they are toxic to microorganisms, contamination presents an unacceptable risk to human health or an ecosystem, and/or the rate of attenuation is unacceptable to regulators or the public.

3.5.1.1 *Conditions for Microbial Growth; Electron Acceptors*

An aquifer is a complex ecosystem. It contains a variety of microorganisms which compete for food and strive to reproduce. Their environment reflects a complex mixture of variables: contaminant (food) distribution; pH; the type of soil and resulting nutrient levels, hydraulic conductivity, and geochemistry; and the availability of electron acceptors. These variables affect the occurrence and rate of contaminant biodegradation.

In order to grow and reproduce, microorganisms need food to supply energy and carbon for cell growth. The microorganisms also need certain nutrients for cell growth. In a contaminated aquifer, microorganisms may use the organic contaminants as food. Microbial degradation of contaminants occurs in a series of redox reactions. As a *substrate* molecule is oxidized, it loses an electron to an *electron acceptor*. The electron acceptor is reduced. Thus, microorganisms also need a supply of electron acceptors. In an aerobic system, oxygen serves as the electron acceptor.

Several zones may develop in an aquifer at a contaminated site. These zones reflect the availability of various electron acceptors, which can change over time or over space (through a plume) as a result of changes in contaminant concentration. In order of preference, these electron acceptors are oxygen, nitrate, manganese (IV), iron (III), sulfate, and carbon dioxide or organic matter.[120-122] Different microbially mediated reactions occur in each zone.

Groundwater naturally contains some dissolved oxygen (DO) as a result of the contact between air in the atmosphere and precipitation before it infiltrates, and because of the contact between soil gases and groundwater.[120] The level of DO depends in part on the type of soil through which precipitation must infiltrate. The concentration of DO in shallow groundwater is typically relatively low in recharge areas in silty or clayey soils (i.e., less than 0.1 mg/L) and somewhat higher (i.e., greater than 0.1 mg/L) in sandy or gravelly soils.[123] The concentration of DO also depends on the extent of geochemical reactions in the aquifer which consume oxygen.[120] Finally, the concentration of DO depends strongly on the concentration of organic compounds and the resulting extent of aerobic biodegradation.

Significant aerobic biodegradation can occur when the groundwater contains 1 to 2 mg/L of oxygen or more. At the site of a large spill, biodegradation of readily oxidized compounds can quickly consume the available oxygen. Aerobic biodegradation produces organic acids and carbon dioxide, which affect the pH and alkalinity of the aquifer.[124]

When the supply of DO becomes depleted, microorganisms begin to use nitrate (NO_3^-) as the electron acceptor. When, in turn, the nitrate becomes depleted, microorganisms use manganese (IV) and then iron (III) as the electron acceptors. Redox reactions produce soluble Mn (II) and Fe (II). When those electron acceptors are used up, microorganisms turn to sulfate (SO_4^{2-}) as the electron acceptor, producing hydrogen sulfide (H_2S). Finally, when no other electron acceptors are available, methanogenic organisms use carbon dioxide or certain organic compounds as the electron acceptor and produce methane. In the degradation of certain chlorinated organic compounds, hydrogen or another organic molecule serves as the electron donor and the chlorinated organic compound serves as the electron acceptor.[125] *Methanogenic fermentation* occurs when a portion of the substrate molecule serves as the electron donor, and a portion of the substrate molecule serves as the electron acceptor.[122]

3.5.1.2 Degradation of Hydrocarbons

Hydrocarbons from fuels (including BTEX) dissolved in groundwater degrade readily. Microorganisms use the contaminant molecules as a source of both energy and carbon atoms for growth. Groundwater often contains a sufficient supply of electron acceptors for intrinsic bioremediation of fuel contamination provided free product has been removed.[124]

3.5.1.3 Degradation of Chlorinated Solvents

Chlorinated compounds tend to persist in the environment, and the daughter products of degradation are potentially toxic. As a result, many researchers and regulatory agencies have focused on the natural attenuation of chlorinated solvents such as carbon tetrachloride, trichloroethane, tetrachloroethylene, and TCE. These compounds can biodegrade, as discussed further below. Abiotic degradation is not significant, except for trichloroethane, which undergoes two abiotic reactions. Some 20% degrades to dichloroethylene (DCE), and 80% hydrolyzes to acetic acid (which can biodegrade):[125]

$$Cl_3C - CH_3 \rightarrow H_2C > CCl_2 + H^+ + Cl^-$$

$$Cl_3C - CH_3 + H_2O \rightarrow CH_3COOH + 3H^+ + 3Cl^-$$

Figure 3.12 summarizes the reactions involved in biodegrading common chlorinated organic solvents. Different reactions occur at different rates, mediated by different bacteria.

Depending on the microorganisms and site conditions, the chlorinated organic compound can serve as either an electron acceptor (and degrade by *reductive dechlorination*) or as an electron donor. Different microorganisms degrade chlorinated organic compounds as either a primary substrate or via a cometabolite.

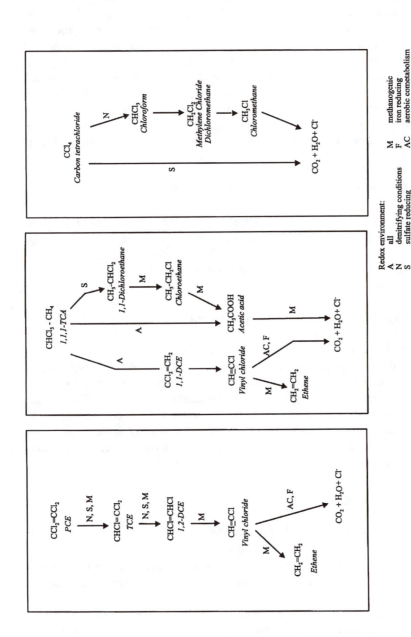

Figure 3.12 Degradation pathways for chlorinated hydrocarbons. (From McCarty, P. L., in Proceedings of the Symposium on Natural Attenuation of Chlorinated Organics in Ground Water, Dallas, TX, September 11–13, 1996, EPA/540/R-96/509, September 1996, 5–9.)

Several organisms can use chloroethylenes directly as electron acceptors. These organisms variously use hydrogen (H_2) or organic compounds as the electron donors and may use other compounds (e.g., nitrate, sulfite, or other compounds) as other electron acceptors. They also require organic matter as a carbon source. Biodegradation of the target contaminants thus requires an adequate supply of those other electron donors/receptors and a carbon source.[126] The rate of reductive dechlorination decreases with the degree of chlorination (e.g., PCE degrades more quickly than vinyl chloride). Reductive dechlorination occurs most rapidly under methanogenic conditions, although certain reactions can occur under nitrate- and sulfate-reducing conditions.[124]

Microorganisms can also use certain chlorinated organic compounds as electron donors and a source of carbon. These reactions appear to be limited to less oxidized compounds such as vinyl chloride, methylene chloride, and 1,2-dichloroethane. They occur under aerobic and some anaerobic conditions.[124]

Other aerobic and anaerobic organisms can degrade chlorinated compounds as a cometabolite. Cometabolism has been described as a process whereby the microorganisms degrade the chlorinated compounds "more or less accidentally or incidentally as the organisms carry out their normal metabolic functions; the organisms apparently derive no growth-linked or energy-conserving benefit from the reductive dechlorination."[126] For example, microorganisms can degrade TCE aerobically provided sufficient DO and an electron donor such as methane, ammonia, or phenol are available.[125]

As this summary indicates, researchers are just beginning to understand the complex biological processes which can attenuate contaminants in an aquifer. Natural attenuation depends on the chemical nature of the contaminant, its distribution throughout an aquifer as a result of the site geology and hydrogeology, and its utility as a desirable food source to microorganisms.

3.5.2 Air Sparging

Air sparging is simply *in situ* air stripping. Injection wells convey air from a compressor to the aquifer. Like air stripping, air sparging is used to transfer VOCs from groundwater to the air stream.

Air sparging is usually coupled with soil vapor extraction to capture the stream of contaminated air. Soil vapor extraction, which is discussed in Section 4.3.1, uses a series of extraction wells screened in the unsaturated zone to capture soil vapors. A blower mounted at the surface induces a vacuum on the wells and stimulates air flow through the soil to the wells. The soil vapor extraction system should be designed to collect more air than is injected through air sparging in order to prevent uncontrolled emissions. The EPA suggests that the soil vapor extraction rate should be 1.25 to 5 times the sparge rate.[127] Extracted air is treated to remove the water vapor and the contaminants.

3.5.2.1 Physical Principles

Air sparging will remove compounds with a high Henry's law constant (i.e., greater than 100 atm), a boiling point less than 250 to 300°C, or a vapor pressure greater than 0.5 mmHg most readily.[128] The increased oxygen supply can also stimulate the biodegradation of less volatile compounds. The primary challenges of air sparging are to distribute the air throughout the area of contamination and to control and capture the emissions.

The distribution of air through an aquifer depends on the homogeneity and intrinsic permeability of the soils. The *intrinsic permeability* (k [cm^2]), which indicates the ability of a soil to transmit fluids, is related to the hydraulic conductivity (K [cm/sec]):

$$k = K \cdot \frac{\mu}{\rho g} \qquad (3.15)$$

where μ is the dynamic viscosity of water [g/cm · sec], ρ is the density of water [g/cm^3], and g is the acceleration due to gravity. (At 20°C, $\mu/\rho g = 1.02 \times 10^{-5}$ cm · sec.) Air sparging is generally effective when $k \geq 10^{-9}$; it may be effective when $10^{-9} \geq k \geq 10^{-10}$, and will probably be marginally effective or ineffective when $k < 10^{-10}$.[127] Air sparging appears to work best in uniform coarse-grained soils such as sands and gravels.[129]

As with hydraulic conductivity, the intrinsic permeability can change significantly with depth and areal location on a site. Air will not be distributed evenly through a highly heterogenous aquifer, such as a silty/sand aquifer containing lenses of sand. Instead, the air will preferentially flow through higher-permeability materials.

3.5.2.2 Air Sparging Through Wells

The design of an air-sparging system is usually based on the results of a field test. Important variables include the number and spacing of injection wells, the air pressure and flow rate, the means for collecting vapors, and the monitoring system.

Air is injected through vertical or horizontal wells which are 1 to 4 in. in diameter. Each well is screened or fitted with perforated pipe some 5 to 15 ft below the zone to be treated. The well screen of a vertical well is shorter than in an extraction well, generally 1 to 3 ft long.[127,130] Injection wells are fitted with a check valve to keep groundwater from backing up the piping, a control valve to regulate the air flow, and a pressure gauge to monitor air pressures. A sampling port may also be fitted to the well head.

An oil-free air compressor supplies the air to the injection wells through a manifold. The air compressor must supply air at sufficient pressure to overcome three forces: head losses from friction in the air distribution system, the static pressure (head) of the water above the discharge point (1 psi per 2.3 ft of head), and the capillary pressure holding water in the soil pores. Air pressures are typically between 10 and 15 psig. Excessive pressure can fracture the soil formation, creating channels for preferential air flow.[127,129-131] The air flow rate must suffice to induce mass transfer from the water phase to the vapor phase. The air flow rate is typically between 3 and 25 standard cubic feet per minute (scfm) per well.[127,130]

The spacing of injection wells depends on the radius of influence (ROI), which is defined by the "greatest distance from a sparging well at which sufficient sparge pressure and air flow can be induced to enhance the mass transfer of contaminants from the dissolved phase to the vapor phase."[127] In general, the design ROI ranges from 5 ft, for fine-grained soils, to 100 ft, for coarse-grained soils.[127] A survey of the operating results of pilot- and full-scale treatment systems at petroleum-contaminated sites indicated that the ROI usually ranges between 10 and 25 ft.[131]

Both field experience and theoretical considerations suggest that air sparged into an aquifer is not distributed uniformly through the aquifer or to a great distance from the injection point. Because the buoyancy of the air exerts a significant vertical force, and the horizontal pressure gradients induced by sparging are comparatively small, air sparged into an aquifer migrates only a small lateral distance from the injection point in homogenous media.[132] The air injected into an aquifer migrates as a separate phase through channels in the aquifer material. Bubbles do not apparently form, except potentially in gravel aquifers or a gravel-filled trench. As a result of the channeling, air is not distributed thoroughly throughout aquifer materials. Oxygen does diffuse out of the channels into surrounding aquifer materials, and contaminants in the surrounding soils diffuse into the channels. However, diffusion occurs relatively slowly.[132,133]

Air-sparging wells can be distributed throughout the area to be treated based on the estimated ROI. Alternatively, a vertical barrier such as sheet piling or a slurry wall can be used to direct contaminated groundwater to a central point for treatment. The wall opens to a *gate* at the central point. This configuration is known as a *funnel-and-gate* system or an *in situ treatment curtain*. The use of air-sparging wells in a funnel-and-gate system is a developing application.

The gate could be filled with material more permeable than the aquifer materials, such as gravel, to enhance the flow of groundwater through the gate. Alternatively, the gate could comprise parallel walls of slotted or perforated sheet piling, braced to support the walls without backfilling. After the walls are installed, the soil between the walls would be excavated to create a subsurface flow-through tank for treating groundwater. Depending on the hydraulics, the gate comprises 5% to 20% of the wall. Air-sparging wells placed in the gate strip contaminants from the groundwater as it flows through the gate. Groundwater flow can be controlled most effectively when the vertical barrier is keyed into a low-permeability layer.[134,135]

3.5.2.3 Effectiveness

Parameters that may be monitored to determine the effectiveness of air sparging include air pressures, at the wellhead and in the subsurface; air flow rates, in the injection system and in the soil vapor extraction system; vapor concentrations in the air stream extracted from the ground and in the emissions from off-gas treatment unit(s); levels of DO in the groundwater; and/or concentrations of contaminants dissolved in groundwater. Air sparging can also cause mounding of the groundwater, which can be measured in piezometers or monitoring wells.

Data obtained from monitoring wells may not accurately represent aquifer conditions. If a monitoring well borehole intercepts one or more of the channels through which the sparged air moves, air bubbles can form within the well and rise to the surface. As a result, the levels of DO or dissolved contaminants in the water in the well do not accurately represent the levels in the aquifer.[133]

Air-sparging systems can be effective in a relatively short time for compounds which can be easily stripped and easily biodegraded. For example, several sites where groundwater in a sandy/silty aquifer contained BTEX at concentrations up to 30 mg/L were remediated in less than a year.[127] The actual remediation time depends on the characteristics of the contaminant, its distribution, the type of soil, and the design and operation of the remediation system.

Contaminant concentrations typically decrease to an asymptotic concentration, and may rebound after the system is shut down, as discussed for pump-and-treat systems in Section 3.2.4. If that occurs before the system achieves remediation goals, the system may be modified to enhance performance by pulsing the injection of air and/or installing additional injection or extraction wells.[127,129,136]

As with air stripping, reduced iron (Fe(II)) will oxidize and form hydroxides upon aeration in an air-sparging system. The iron hydroxide precipitate can foul injection wells. The U.S. EPA suggests that iron fouling should not be a problem when the concentration of iron is below 10 mg/L, but can be a significant problem at concentrations above 20 mg/L.[127] However, some researchers disagree that iron fouling is likely to be a problem.[129] That conclusion is based in part on the use of aeration to *prevent* iron fouling of water supply wells. In the Vyredox™ process, aerated water is injected into a series of wells constructed in a ring around a production well. Aeration causes iron and manganese to precipitate and to be fixed by bacteria at some distance from the production well. Driscoll[137] notes that the aquifer does plug or foul around the aeration wells, but "at a rate that takes much longer to become objectionable than the typical economic life of the well."

Under certain circumstances, air sparging can worsen a groundwater contamination problem. The head induced by an air-sparging system can cause LNAPL to spread.[127] The turbulance induced by air sparging may enhance the desorption of contaminants from soils. If those contaminants are not volatile enough to be stripped from the groundwater, the dissolved-phase contamination may increase.[127] Air sparging can also cause unacceptable vapor contamination in subsurface structures such as nearby basements or utility lines, if the emissions cannot be carefully captured and controlled.

3.5.2.4 In-Well Systems

The preceding discussion focused on air-sparging systems which use a series of air injection and extraction wells to effect treatment. Alternatively, air stripping can be accomplished in a single well, or series of wells, each of which provides for both injection and extraction of air within the well and induces the groundwater to circulate vertically around the well. Several patented systems are on the market, including the NoVOCs™ in-well stripping technology developed by EG&G Environmental, Inc.[138]

and the *Unterdruck-Verdampfer-Brunnen (UVB)* or *Vacuum-Vaporizer Well* technology patented by IEG mbh of Germany and licensed to IEG™ Technologies Corporation in the U.S. The UVB technology has reportedly been used at over 180 sites in the U.S. and Europe.[139]

Each UVB well is a large-diameter well (up to 2 ft diameter) with two well screens. The borehole is sealed between the two screens to prevent short circuiting through the filter pack. Several different configurations have been used to effect treatment; in general, all induce the vertical circulation of groundwater around each well. Groundwater flows in through the bottom screen and out through the top screen.

In one configuration, the top screen is immediately below the water table. A blower at the surface induces a vacuum on the water in the well. A vent pipe provides for air intake from the surface through a sieve plate located at the bottom of the first screened interval. The vacuum induced by the blower draws air in from the surface through this pipe.

The air-lift effect, supplemented by a submersible pump in the well, induces the flow of water in through the bottom screen and up through the well. As the water moves upward through the well and contacts air bubbles, volatile compounds transfer from the water to the air. A "stripping reactor" positioned above the sieve plate provides additional contact time between air bubbles and water to enhance air stripping. The air emissions are drawn to the surface by the blower and treated.

The water exiting the stripping reactor falls back down the well casing and out the upper screen. Much of this water is drawn in through the lower screen. As a result, the groundwater flowing through the well includes both "fresh" water and recirculating water that has already been treated. Recirculating water can account for up to 90% of the flow, depending on the radius of circulation and the ratio of the vertical hydraulic conductivity to the horizontal hydraulic conductivity. This recirculated water effectively dilutes the contaminants entering the well.[140,141]

The patent or license holder designs the UVB well systems. A general description of some design variables follows. The circulation of water around the well depends on the hydraulic conductivity, gradient, and homogeneity of the aquifer materials. In general, the radius of circulation around the well is approximately 2.5 times the distance between the well screens. It can be difficult to establish an adequate stripping zone in aquifers where the depth to groundwater is less than 5 ft. It can also be difficult to install an effective system in an aquifer that is less than 10 ft thick.[140] Finally, in heterogenous aquifer materials, it is difficult to throroughly treat contaminated water.

3.5.3 *In Situ* Bioremediation

Engineered *in situ* bioremediation systems seek to optimize biodegradation by providing an adequate supply of electron acceptors, and minimizing mass-transfer limitations. Various methods have been developed to enhance both aerobic and anaerobic biodegradation. Although *in situ* bioremediation has been used successfully to remediate groundwater since the early 1970s,[142] it is still a developing technology.

Lower-molecular-weight aliphatic and monoaromatic hyrdrocarbons readily bio-degrade aerobically. Higher-molecular-weight ($>C_9$) aliphatic, monoaromatic, or polynuclear aromatic hydrocarbons,[143] and highly chlorinated molecules such as PCBs, degrade more slowly or not at all. Extremely low clean-up levels of such compounds can be difficult to achieve.

Chlorinated organic compounds can be degraded aerobically through the use of a cometabolite. While not commonly applied at full scale, studies have shown that bacteria can degrade TCE aerobically if they are provided with another compound, such as toluene, phenol, or methane, as a primary substrate.[144] (See Section 3.5.1 for a related discussion.)

3.5.3.1 Aerobic Bioremediation

Aerobic biodegradation requires an adequate supply of oxygen, which can be calculated from the stoichiometry of the oxidation reaction(s). The oxygen supply can be introduced to the aquifer by injecting or infiltrating oxygen-rich water, biosparging air or oxygen, or injection of peroxide. Each of these methods has a different capability for conveying oxygen. Aerated water can hold 8 to 10 mg/L oxygen. If oxygen is sparged into the aquifer, 40 to 50 mg/L oxygen levels can be achieved. Finally, up to 500 mg/L oxygen in water can theoretically be achieved with the addition of 1000 mg/L hydrogen peroxide, from the reaction:[145]

$$H_2O_2 \rightarrow H_2O + \tfrac{1}{2}O_2$$

Each of these oxygen delivery techniques is described further below.

Water can be aerated and injected into an aquifer to supply oxygen for bioreme-diation. (Injection of aerated water can also serve other purposes, such as hydraulic control or disposal of treated water, as discussed previously.) However, because the amount of oxygen which can be delivered to the aquifer is limited, injection of aerated water is suitable only when relatively dilute concentrations of contaminants are to be treated *in situ* (<10 mg/L COD).[142]

Biosparging systems deliver air or oxygen to the aquifer through air injection wells. Emissions may be captured by a soil vapor extraction system. As biosparging with air is similar to air sparging, it is subject to the same physical constraints and design considerations described in Section 3.5.2. Injection of oxygen to enhance biodegradation is a developing technology.

Oxygen can also be supplied by injecting an aqueous solution of hydrogen peroxide into the aquifer. Three issues affect the application of hydrogen peroxide:

• The rate of hydrogen peroxide decomposition to oxygen and water, and the avail-ability of the oxygen produced from the hydrogen peroxide to the microbes in the subsurface — if hydrogen peroxide decomposes too rapidly, oxygen bubbles form, limiting the availability of DO.[146] Hydrogen peroxide can decompose so rapidly that oxygen is not effectively distributed through the groundwater.[145]

- The toxicity of hydrogen peroxide — peroxide concentrations greater than 1000 mg/L are toxic to bacteria at room temperature, while concentrations between 500 and 1000 mg/L can inhibit bacterial growth.[146]
- The extent of chemical oxidation — as discussed in Sections 3.4.3.2 and 3.6.2.4, hydrogen peroxide can be an effective chemical oxidant, particularly in combination with iron (Fenton's reagent) or other metals. Other reactions can also potentially occur upon injection of hydrogen peroxide into an aquifer.[146]

Despite these concerns, and the complicated chemical reactions which occur when hydrogen peroxide is injected into an aquifer, it has reportedly been used successfully for *in situ* bioremediation.[142,146]

Bioremediation also requires a pH favorable for growth and an adequate supply of nitrogen, phosphorus, and micronutrients. A pH between 5 and 9 is generally acceptable; a pH of 7 or slightly higher is optimum.[147,148] An adequate nutrient supply, expressed as a mass ratio between the carbon content of the degradable compounds, nitrogen and phosphorus, is generally between 100:10:1 and 100:1:0.5.[147] Many soils naturally contain sufficient nitrogen and phosphorus. Field tests of the related technique of bioventing — *in situ* bioremediation of unsaturated soils — have shown no improvement in treatment with nutrient addition.[148]

Finally, biodegradation requires a viable population of bacteria. Bacteria have been added to aquifers to enhance remediation. The level of *heterotrophic* bacteria — those which rely on organic compounds for a food source — is measured in the saturated soils at a site before designing a bioremediation system. *Plate counts* of heterotrophic bacteria indicate the number of *colony-forming units (CFUs)* per gram of soil. Plate counts of 10^4 to 10^7 CFU/g are typical. A plate count below 10^3 CFU/g could indicate the presence of toxic levels of contaminants. As a general rule of thumb, concentrations of petroleum hydrocarbons over 50,000 mg/kg, or of metals over 2,500 mg/kg, can be toxic to microorganisms.[147]

Treatability testing is often used to determine the optimum treatment conditions and to gauge the kinetics and ultimate effectiveness of treatment. Laboratory studies for *in situ* bioremediation of groundwater or soil often rely on a series of batch tests of soil/water slurries. Each test examines a different combination of treatment conditions. The results of a slurry phase test represent the optimal mass transfer of nutrients and oxygen to the contaminants. Soil column studies are also used to simulate *in situ* bioremediation of groundwater and soil. A field test is often performed before designing a full-scale system in order to evaluate and design systems for delivering nutrients and oxygen to the subsurface, and capturing emissions, if necessary. Field trials and full-scale systems are monitored to determine effectiveness by measuring DO levels, CO_2 levels, pH, and contaminant concentrations.

3.5.3.2 Anaerobic Bioremediation

Systems designed to enhance anaerobic biodegradation *in situ* are not common. ABB Environmental Services has developed a "Two-Zone, Plume Interception, *In Situ* Treatment Strategy" for treating chlorinated organic compounds.[149] Two treatment zones are created in an aquifer in sequence, either in space or in time. Nutrients and a readily degraded carbon source such as glucose are injected into the first zone. Biodegradation of the glucose depletes the oxygen supply and creates anaerobic conditions in the aquifer so that chlorinated compounds such as PCE can be biodegraded by methanogenic bacteria. In the second treatment zone, downgradient of the first zone or in the same place at a later time, oxygen is introduced into the aquifer to methanotrophic bacteria to aerobically degrade the partially dechlorinated byproducts of the first zone (vinyl chloride, dichloroethylene). Methane may be introduced if needed as a carbon source. Field testing of this system began in late 1996 at a site where PCE and its daughter products contaminated the groundwater. The system was designed to enhance first anaerobic degradation, then aerobic degradation. Preliminary results indicated that anaerobic dechlorination was occurring.[150]

3.5.4   Reductive Dehalogenation

Reductive dehalogenation, as discussed in Section 3.4.3.3, is used in a funnel-and-gate system to treat groundwater *in situ*. The gate is sized to allow sufficient contact time between the contaminants in groundwater and the iron medium to achieve the necessary degree of treatment. The gate must also be designed to allow for the hydraulics of groundwater flow which is funnelled through the gate.

Example 3.4:   The U.S. Air Force demonstrated the use of a zero-valent iron reactive wall at Lowrey Air Force Base in Colorado.[151,152] The surficial aquifer at the site contained chlorinated organic compounds at concentrations up to 107.925 mg/L TCE, as well as PCE, 1,2-DCE isomers, 1,2-DCE, and vinyl chloride.

Soil in the test area comprised interbedded sands, silts, and clays. The depth to groundwater was approximately 8 ft below ground surface (BGS), and the surficial aquifer was underlain by a low-permeability bedrock at approximately 17.5 ft BGS. The groundwater flow across the chosen region was approximately 1 ft/d. The test area was approximately 300 ft downgradient from a DNAPL source, and was just upgradient of a creek. Groundwater in the test area contained DO at 1 to 6.6 mg/L; alkalinity at 320 to 560 mg/L; and TCE at 0.16 to 1.26 mg/L; and had a pH between 6.7 and 7.44.

The project team performed bench-scale tests to evaluate the kinetics of the dehalogenation reaction. The results of these tests were used to calculate the residence time within the iron filings required to degrade the maximum concentration of each contaminant in the test area. Table 3.3 summarizes the test results. The design was based on a target residence time of 2 d.

The design team used two groundwater models, MODFLOW and Path3D, to evaluate two different conceptual designs in order to determine which would provide the optimum combination of hydraulic control and reaction kinetics. The first configuration comprised a 10-ft-wide by 5-ft-thick gate with 10-ft-long funnels on both sides of the gate oriented parallel to the width of the wall. The second configuration comprised a 10-ft-wide by 6-ft-thick gate with 14.3-ft-long funnels on both sides of the gate oriented at an angle of 45 degrees (in the upgradient direction) from a line parallel to the width of the wall. Groundwater modeling of the second configuration suggested that the

Table 3.3   Kinetics Data for Reductive Dehalogenation Lowry Air Force Base[151,152]

Compound	Half-life calculated from bench-scale tests (hours)	Half-life calculated from field results (hours)
PCE	—	1.6
TCE	0.45	0.9
cis-1,2-DCE	2.2	2.2
Vinyl Chloride	3	2.1

Note: Half-life of PCE in laboratory tests not reported.

groundwater would flow through the gate at approximately 2.60 ft/d. The team selected the second configuration as it reduced the potential for the wall to divert groundwater to discharge to the downgradient creek.

The funnel walls were constructed by installing interlocking sheet piles to a depth of approximately 18 ft BGS. Each funnel wall was approximately 15 ft long. The gate installed between these walls was approximately 11 ft long. The contractor installed sheet piling to form a box-shaped gate with sidewalls of approximately 10.5 ft, then used a clamshell excavator to remove native soils from inside the "box". Engineered supports held the box open during construction. Once the soil was removed, three layers of material were placed in the box. Oriented perpendicular to the direction of flow, these layers comprised pea gravel, iron filings, and pea gravel. (See Figure 3.13). When these layers were in place, the contractor removed the sheet piling which formed the upgradient and downgradient walls of the box in order to permit groundwater flow through the gate. Finally, the contractor installed 37 monitoring wells and 5 stream gauges to monitor the performance of the system.

Monitoring data obtained during the first 7 months indicated that:

- Oxidation–reduction potential dropped significantly with distance into the gate, potentially indicating the oxidation of Fe° to Fe^{2+};
- pH rose with distance into the gate to approximately 10, likely indicating the production of OH^-;
- total alkalinity, sulfate, and calcium concentrations decreased in the gate;
- most chlorinated compounds degraded after travelling less than 1 ft into the iron filings (approximately 9 h residence time);
- cis-1,2-DCE, 1,1-DCA, and 1,2-DCA were detected further into the gate, either because less chlorinated compounds degrade more slowly or because those compounds were produced by the dechlorination reaction;
- vinyl chloride — the ultimate degradation product — was degraded to the quantitation limit within the first 2 ft of the iron filings (residence time of 18 h);
- precipitation of carbonates and hydroxides could theoretically reduce the porosity of the wall by 13% to 14% per year (i.e., , from a porosity of 0.40 to a porosity of 0.35). This rate of porosity loss could vary based on decreasing corrosion rates and/or changes in groundwater geochemistry and velocity.

Table 3.3 indicates the half-life values calculated from the first 4 months of data.

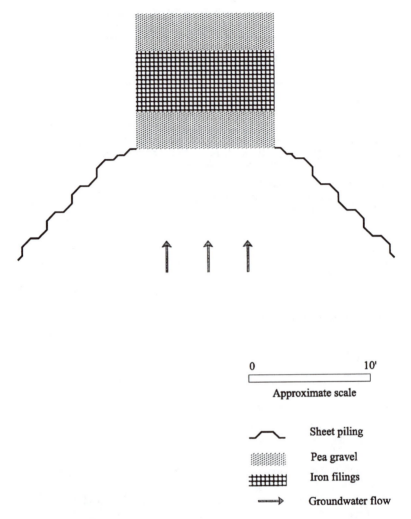

Figure 3.13   Plan view schematic of funnel-and-gate system, Lowry Air Force Base. (Adapted from Muza, R., *Reactive Wall Demonstration Project, Lowry Air Force Base, Colorado*, presentation materials provided by R. Muza, U.S. EPA, Region VIII, August 1997.)

3.6 NONAQUEOUS PHASE LIQUIDS

LNAPL and DNAPL present quite different challenges to remediation. As described in Section 2.3.4.1, LNAPL and DNAPL serve as continuing sources of dissolved contamination. Because LNAPL floats above the water table, it is usually easier to locate and remove than DNAPL. LNAPL compounds are also typically more biodegradable. LNAPL can be removed by pumping or aspiration. When DNAPL pools can be located, they can be removed. Researchers are working to develop other means to actively remediate DNAPL. Meanwhile, the remedial solution for many DNAPL sites is to contain the DNAPL and associated plume.

3.6.1 Light Nonaqueous Phase Liquids

Mobile LNAPL is often pumped to the surface for treatment using recovery systems analogous to groundwater extraction wells or well points. Alternatively, it is aspirated in a multiphase extraction system. The residual LNAPL remaining in the soil is less easy to remove.[153] Soil containing residual LNAPL is sometimes excavated or treated *in situ* using soil vapor extraction, air sparging, or bioventing. Some of the techniques described for DNAPL recovery in Section 3.6.2, such as water flooding, surfactant use, steam injection, and oxidation, have also been contemplated for or applied to remediation of separate-phase or residual LNAPL.

3.6.1.1 Extraction

The earliest methods of LNAPL recovery usually relied on pumping liquids to the surface from a series of extraction wells or an extraction trench. Two variations are used: removing LNAPL alone, or removing both LNAPL and groundwater. While these methods remove liquid organics, they do not remove LNAPL residuals held in soil pores by capillary forces. Even under optimum conditions, various estimates suggest that pumping systems remove less than half the total LNAPL volume.[153] More recently, vacuum extraction, also known as dual-phase extraction or bioslurping, became common. Vacuum extraction can remediate NAPL residuals in soil as well as removing free product. Each of these methods is described below.

3.6.1.1.1 Removal of LNAPL Only — LNAPL can be recovered passively by suspending a collection unit in the LNAPL layer in a well. Different types of collection units can be used, such as a thin cylinder comprising absorbant material, or a small bucket or bailer. These units must be manually removed from the well and emptied or replaced when full. Passive units are best used on relatively small LNAPL spills where labor is readily available to check and replace the units.

LNAPL can also be recovered with a *skimmer pump* or product-only pump installed in an extraction well or the sump of an extraction trench (see Figure 3.14 for an example). Extraction wells are typically 6 in. in diameter or larger and are screened across the LNAPL layer. The product pump is suspended in the well at a fixed position in the NAPL layer, or the pump inlet is attached to a float which allows the inlet to follow fluctuations in the NAPL layer. LNAPL is pumped to the surface for storage and ultimate disposal. The limit of LNAPL recovery depends in part on the pump used; for example, one manufacturer states that their product recovery pump can remove pure product from a layer as thin as ⅛-in. under equilibrium conditions, or down to a sheen if a small amount of water is also recovered.[154] In general, skimmer pumps have very low LNAPL recovery rates, because they cause little or no drawdown of the water table. Recovery rates tend to be higher when the LNAPL mass is very large and very mobile, and the subsurface permeability is high.

H/250/SPG-2
DEEP-WELL AUTOMATIC PRODUCT ONLY RECOVERY
SYSTEM WITH TFSO, HWSO, & SELECTIVE OIL SKIMMER

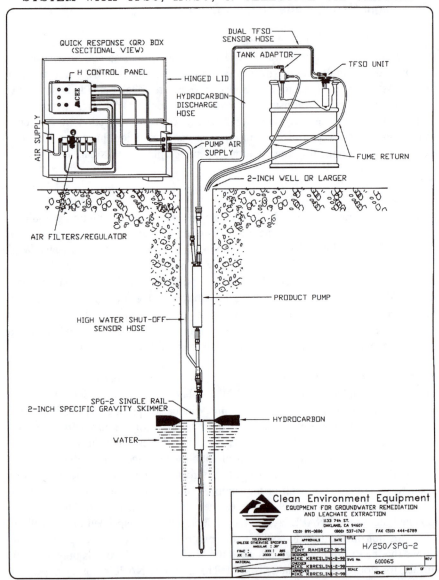

Figure 3.14 LNAPL recovery using skimmer pump. (Reprinted with permission of Clean Environment Equipment. Copyright 1998 by Clean Environment Equipment, Oakland, CA.)

Removing LNAPL with a skimmer pump can be an effective solution when a rapid response is necessary to remove LNAPL and thereby limit the ultimate extent of groundwater remediation necessary. (A remediation system that begins with recovery of LNAPL can be modified later to enable collection of groundwater.) Removing NAPL alone may be all that is necessary or when the clean-up goal is to remove free product and remediation of dissolved-phase groundwater contamination is not required.

3.6.1.1.2 Removal of LNAPL and Groundwater — LNAPL and groundwater may be recovered using a series of two pumps installed in an extraction well or sump in a collection trench (see Figure 3.15 for an example). A lower pump draws down the groundwater, forming a *cone of depression* in the water table. NAPL flows into this depression, where it is extracted by a skimmer pump. The extracted water and NAPL are then treated as discussed below. Alternatively, a single *total fluids* pump is sometimes used to extract both LNAPL and groundwater, particularly when the LNAPL layer is relatively thin (see Figure 3.16 for an example).

Extraction of both LNAPL and groundwater recovers both the product and dissolved-phase contamination, and can provide gradient control to prevent dissolved-phase contamination from spreading. However, this method can have several disadvantages depending on the site conditions. First, pumping groundwater at a high rate, which may be necessary to maintain drawdown in some soils, can entail high treatment costs. Second, as the LNAPL flows into the cone of depression, LNAPL smears over a band of soil, leaving a residual that will not be addressed by the remediation system. Third, *emulsions*, or mixtures of NAPL and water that cannot be readily separated, can form, particularly when using a single pump. The formation of emulsions complicates the treatment and disposal of the recovered liquid.

3.6.1.1.3 Vacuum Extraction/Multi-Phase Extraction — In this remediation method, a vacuum is applied to a recovery well to extract vapor-phase LNAPL and water. Vacuum extraction can remove both liquid organics and residual LNAPL in soil. In addition, the air flow induced through the unsaturated soil by vacuum extraction volatilizes contaminants and stimulates aerobic biodegradation. As a result, this technology is sometimes called *bioslurping*. Vacuum extraction is closely related to soil vapor extraction; see the more detailed discussion of soil vapor extraction in Section 4.3.1.

Vapors are recovered from a well or series of wells screened across the water table. The wells can be as small as 2 in. internal diameter, although larger diameter wells provide more working room. The spacing of the recovery wells is determined by field testing or by modeling the hypothetical air flow through the soil under a specified vacuum.

Vapors are extracted from a *suction pipe,* also called a *drop tube* or *slurp tube* (e.g., 1-in. diameter tubing) suspended in each well. The tubes from multiple wells are connected through a manifold which leads to a treatment system. Figures 3.17 and 3.18 show one off-the-shelf treatment system.

The first unit in the treatment system is typically a knock-out tank or air–water separator. As the vapors enter the tank, the velocity slows due to the increase in diameter. This velocity change causes water to condense and collect in the bottom of the tank. The water is pumped out for treatment, e.g., by carbon adsorption.

HW/250/SPG-2
DEEP-WELL AUTOMATIC HYDROCARBON AND WATER RECOVERY SYSTEM WITH SPECIFIC GRAVITY SKIMMER

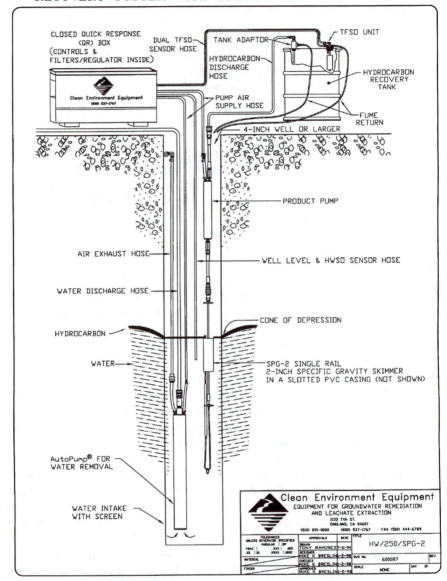

Figure 3.15   Recovery of LNAPL and groundwater using two-pump system. (Reprinted with permission of Clean Environment Equipment. Copyright 1998 by Clean Environment Equipment, Oakland, CA.)

A vacuum pump, which induces the vapor flow from the wells and into the treatment system, is typically located after the knock-out tank. The vacuum pump is sized to induce the necessary vacuum and air flow rate, taking into account the

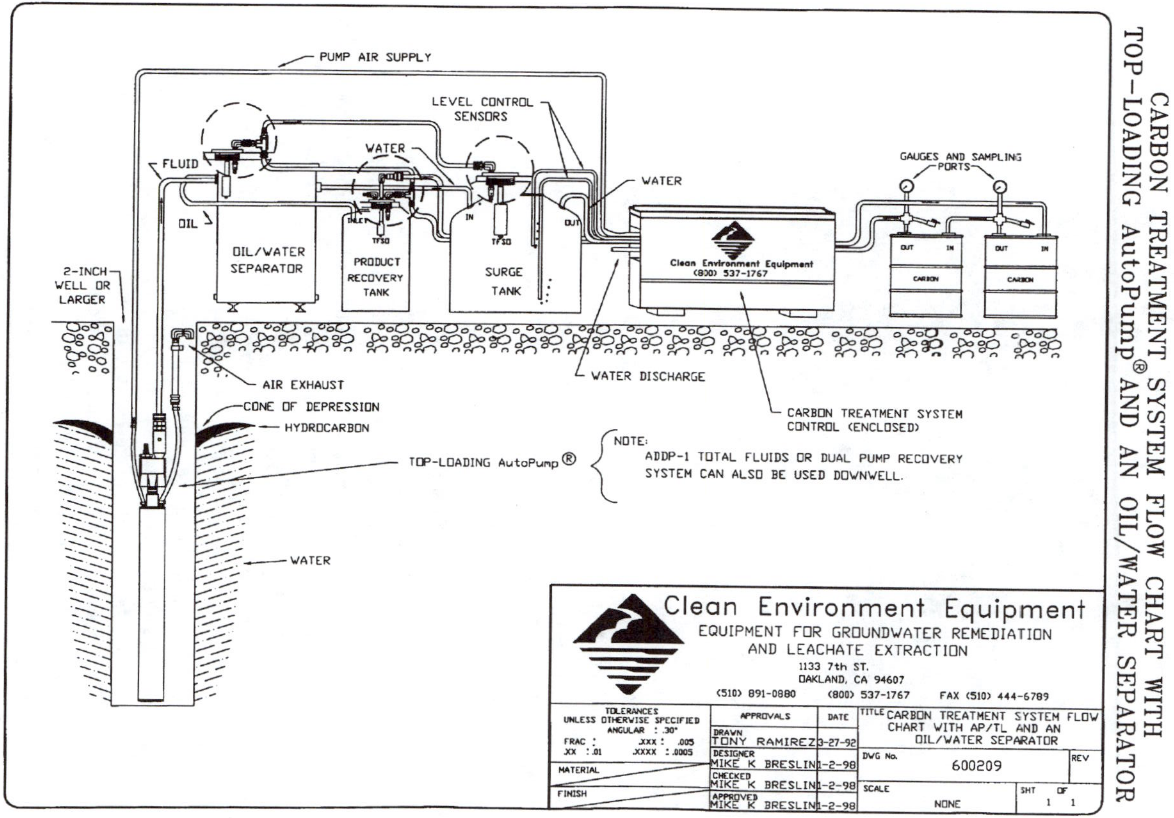

Figure 3.16 Total fluids recovery and treatment. (Reprinted with permission of Clean Environment Equipment. Copyright 1998 by Clean Environment Equipment, Oakland, CA.)

Figure 3.17   Photograph of multiphase extraction system. (Reprinted with permission of Carb-trol® Corporation, Westport, CT.)

number of wells, permeability of the soils, the air velocity in the slurp tubes necessary to suspend droplets, and the head loss in the piping. Depending on emission limits, vapors may discharge from the vacuum pump to the atmosphere or to a treatment unit such as activated carbon. (See Section 3.7 for a discussion of common emission-control devices.) The pump motor and controls must be explosion-proof in order to safely handle many LNAPLs.

The effectiveness of vacuum extraction depends on the properties of the LNAPL, particularly the tendency to sorb to soils (see Section 2.2.1.4), which can limit the effectiveness of removal, and volatility. Compounds with a vapor pressure >0.5 mmHg, a boiling point of less than about 250 to 300°C, or a Henry's law constant of less than 100 atm cannot be effectively volatilized in a vacuum extraction system. Less volatile compounds may be treated, however, by the biodegradation which is enhanced by the increased oxygen supply.[155]

The effectiveness of vacuum extraction also depends on the properties of the soil:

- In general, single-pump vacuum extraction is most effective in soils with intrinsic permeabilities (Equation 3.15) between 10^{-9} to 10^{-11} cm^2. At higher permeabilities, a single pump may not be economical because of the high air flow required to maintain sufficient vacuum. At low permeabilities, the soils do not readily transmit vapors or oxygen to enhance biodegradation.[155]
- Stratification affects the effectiveness of treatment. Air flows preferentially through fractures and zones of higher-permeability material such as sands, limiting the remediation of NAPL trapped in finer-grained material.[155]
- The moisture content of the soil also affects the flow of air and vapors through the vadose zone. In general, when the soil moisture reaches a level ≥80% of field capacity, the air permeability of soil is effectively zero. Groundwater pumping may

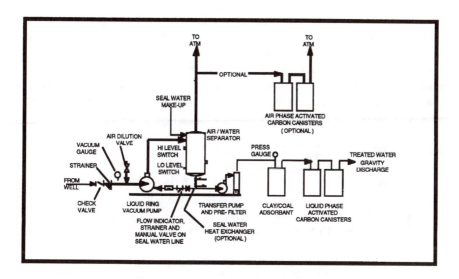

Basic Extraction System

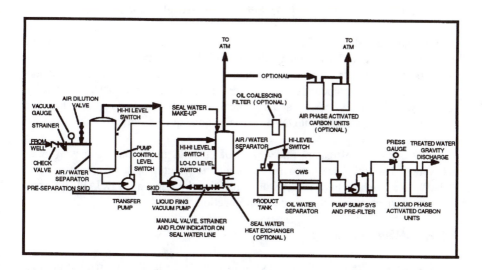

Extraction System with pre-separation capability

Figure 3.18 Diagram of multiphase extraction system. (Reprinted with permission of Carbtrol® Corporation, Westport, CT.)

be used to lower the water table and dewater the capillary fringe in order to increase the air flow.[155,156]

- A shallow system, installed at a site where the water table is relatively close to the ground surface, may preferentially draw air from the surface rather than drawing air through the zone of contamination. Sealing the surface, with pavement or a temporary layer of polyethylene, can limit this short-circuiting.

Vacuum extraction can have several advantages over liquid pumping systems. First, a relatively smaller flow of groundwater is extracted than in a LNAPL/groundwater pumping system, lowering groundwater treatment costs. This is an advantage at sites where groundwater does not need to be extracted to recover or control a plume of dissolved-phase contamination. Second, vacuum extraction can treat the soil overlying an LNAPL plume: the induced air flow removes volatile compounds by soil vapor extraction and by enhancing aerobic biodegradation.

3.6.1.2 Disposal

Recovered LNAPL is commonly incinerated. Relatively large volumes of petroleum products may be returned to a refinery for recycling.

3.6.2 Dense Nonaqueous Phase Liquids

As described in Section 2.3.4.1, investigators typically infer the presence and extent of DNAPL from indirect evidence. Site investigations rarely identify discrete pools of DNAPL. As a result, the strategy commonly used at DNAPL sites is to hydraulically contain the estimated DNAPL area, making conservative assumptions regarding the possible extent of DNAPL, and to actively remediate the dissolved-phase plume.

Full-scale remediation of DNAPL is not yet common. Technologies which are under development or have been applied at a limited number of sites include extraction, enhanced by water flooding, surfactants, or steam, and treatment by oxidation.

Many different compounds or mixtures of compounds are DNAPL. Density, solubility, volatility, viscosity, and chemical reactivity can vary greatly between DNAPLs. As a result, different DNAPLs can respond quite differently to remediation technologies.

Occasionally, investigators find a pool of DNAPL in a monitoring well or an excavation. The author has heard anecdotal reports of pumping such pools to remove DNAPL. Researchers have focused on enhancing the removal of DNAPL by water flooding, surfactant injection, and steam injection.

3.6.2.1 Water Flooding

Water flooding refers to the process of increasing the hydraulic gradient in order to flush DNAPL residual from soil pores and/or remove DNAPL from pools on an aquitard. Water is injected into the aquifer to induce the hydraulic gradient. An extraction system is used to collect and contain the mobilized NAPL. Vertical and horizontal hydraulic gradients must be carefully controlled to prevent the uncontrolled spread of DNAPL. Water flooding has not been commonly used at full scale for DNAPL remediation; however, it has been used in the petroleum industry.

The interface between the water-wet surface of soil particles and DNAPL is characterized by DNAPL–water interfacial tension. DNAPL is held in soil pores primarily by capillary forces (adhesive-cohesive forces), which are proportional to the DNAPL–water interfacial tension.[157] Recall from Section 2.3.4.1 that DNAPL penetrates the saturated zone when the pressure head of accumulated DNAPL exceeds the capillary forces which hold water in soil pores. This process must be reversed to remove DNAPL by water flooding. The movement of DNAPL depends on its viscosity, the interfacial tension between the DNAPL and the water at the site, and the permeability of the soil.

The potential to mobilize residual horizontally is proportional to the *capillary number* (N_c), which represents the ratio of viscous to capillary forces:[157,158]

$$N_c = \frac{k\rho_w g I}{\sigma} \qquad (3.16)$$

where k is the intrinsic permeability of the aquifer, ρ_w is the density of water, g is the force of gravity, I is the hydraulic gradient, and σ is the NAPL–water interfacial tension. (σ is typically 0.005 to 0.035 N/m for the NAPL-water at a waste site).[159] Residual DNAPL mobilizes in the horizontal direction at the *critical capillary number*, N_c^*. The critical capillary number can be reached by increasing the hydraulic gradient or by decreasing σ with surfactant addition, as discussed in Section 3.6.2.2. In soils finer than sand and gravel, the hydraulic gradient cannot be increased enough to effectively mobilize NAPL.[160]

The limited experience to date suggests that in practice, water flooding cannot completely remove residual DNAPL. An estimated 30% to 60% of a DNAPL can be removed, particularly if the DNAPL/water has a relatively low interfacial tension, the DNAPL has a relatively low viscosity, and the aquifer is relatively homogenous.[158]

The necessary hydraulic gradient can be induced using a horizontal drain or a groundwater extraction/injection system. Where DNAPL is pooled on a low-permeability unit, dual drains can theoretically be used to maximize DNAPL recovery. A lower drain recovers DNAPL and water, and an upper drain, which collects groundwater only, maximizes the gradient in the water phase.[158] The desired drawdown reflects a balance. Increasing the gradient on DNAPL maximizes the driving force; however, thinning the layer of DNAPL (e.g., in a cone of depression around a well or horizontal drain) minimizes the permeability of the formation to DNAPL. An upper drain (in a dual drain system) which draws down the groundwater can also cause upwelling in the DNAPL pool, theoretically minimizing the latter effect.[161]

Vertical hydraulic gradients must be carefully controlled to prevent mobilized DNAPL from moving downward, outside the range of the remediation system. Ideally, water flooding is applied at a site where a competent layer of low-permeability material underlies the DNAPL residual. Alternatively, the vertical gradient required to prevent downward migration of the DNAPL can be calculated theoretically.[162,163] The required gradient is proportional to the height of pooled DNAPL and the density difference between DNAPL and water. Containment of chlorinated solvents such as TCE and PCE may require a vertical hydraulic gradient as high as 0.6. Containment of creosote, which is less dense than such chlorinated solvents, would theoretically require an upward gradient on the order of 0.05 to 0.1.[164]

3.6.2.2 Surfactants and Cosolvents

Surfactants and cosolvents have the potential to enhance the solubility and mobility of low-solubility contaminants or DNAPL and thus their removal from an aquifer in a pump-and-treat system. The success of surfactant use depends on the chemical nature of the surfactant and the contaminant, the homogeneity and type of aquifer materials, and on maintaining hydraulic control of the surfactant-treated plume. Surfactants and alcohol cosolvents have analogous chemical structures and similar effects. Sections 3.6.2.2.1 and 3.6.2.2.2 discuss surfactants and cosolvents, respectively.

3.6.2.2.1 Surfactants — A surfactant is a bipolar molecule. One end of the molecule is polar or ionic, and as a result is hydrophilic. The polar functional group can be anionic, cationic, zwitterionic (both positively and negatively charged), or nonionic. The other end of the molecule is nonpolar, e.g., a long hydrocarbon chain, and is therefore hydrophobic. This bipolar structure lends certain characteristics to surfactants. (The word *surfactant* is a contraction of the phrase "surface active agent".) Surfactants concentrate at interfacial regions. At an oil–water interface, the polar ends of the surfactant molecules are attracted to the surface of the water layer, and the nonpolar ends of the surfactant molecules are attracted to the surface of the oil layer. A surfactant also forms *micelles* in water when the concentration of the surfacant exceeds the *critical micelle concentration* (CMC), which is typically between 10 mg/L and 2000 mg/L.[160] A micelle is a small cluster of surfactant molecules which can be spherically shaped. The molecules are oriented so that the nonpolar ends of the molecules cluster together in the center, and the nonpolar ends of the molecules point outward toward the aqueous phase.[157]

Surfactants affect DNAPL in two ways: first, surfactants increase the solubility of hydrophobic compounds, and second, surfactants can mobilize residual NAPL. Surfactants have been used to mobilize PCBs;[165] carbon tetrachloride;[166] PCE, 1,1,1-TCA, and TCE;[167] and creosote.[168] Full-scale applications to waste site remediation are rare; however, surfactants have been used in the petroleum industry since the 1960s.[160]

In a contaminated aquifer, the hydrophobic interior of surfactant micelles effectively serves as an organic pseudophase into which hydrophobic organic contaminants can partition.[157] The solubility of the contaminant increases linearly with the concentration of the surfactant (at concentrations above the CMC and below the solubility limit of the surfactant).[169] Addition of a surfactant can increase the solubility of a contaminant by up to three orders of magnitude.[167] Two parameters are used to characterize this phenomenon: the *molar solubilization ratio (MSR)* and the *micellular-water partition coefficient (K_m)*. The MSR is the ratio of the moles of contaminant to the moles of surfactant, and can be determined as the slope of a plot of aqueous contaminant concentration vs. the surfactant concentration (above the CMC). K_m is the molar ratio of the contaminant in the micellar phase divided by the molar ratio of the contaminant in the aqueous phase, and is calculated from the MSR, contaminant water solubility, and the molar concentration of water.[160,169]

A surfactant also affects the interface between DNAPL and water. Recall that DNAPL residuals are held in soil pore spaces by capillary forces (Section 2.3.4.1). The capillary forces are proportional to the interfacial tension at the NAPL–water interface. An effective surfactant lowers the interfacial tension, making it easier to flush NAPL residual from the pore space.

Surfactant systems are characterized by the *hydrophilic/lipophilic balance (HLB)*, an empirical method, and by the *Winsor system.*[157,169] The HLB number of a surfactant indicates the types of oils which it can emulsify. A high HLB number indicates that the surfactant is relatively water soluble. A contaminant which is quite hydrophobic would generally require a solvent with a low HLB. The Winsor system is based on the relationship between the interfacial tension and the formation of a microemulsion at the interface between water and an organic layer:

- In a *Winsor type I* system, the surfactant forms micelles in the water phase. Hydrophobic organic contaminants partition into the micelles, forming swollen micelles.
- In a *Winsor type II* system, the surfactant forms reverse micelles in the NAPL phase. These micelles are swollen with water molecules.
- A *Winsor type III* system is a balanced system, where a mixture of water, NAPL, and surfactant forms a separate phase microemulsion of intermediate density. DNAPL is mobilized in a Winsor type III system.

The type of surfactant and its concentration, and the salinity/hardness and temperature of the water determine whether a type I, II, or III system exists.

Selection of a single surfactant or surfactant mixture for aquifer remediation is not a simple task. Candidate surfactants for use in remediation can be identified based on the characteristics of the surfactant, contaminant, and aquifer. The final selection of a surfactant depends on the results of site-specific treatability testing. Critical factors include the following:

- The first criterion for surfactant selection is its potential toxicity. Regulatory agencies are unlikely to approve the injection of potentially harmful levels of a surfactant into an aquifer that suppies drinking water or discharges to surface water. Researchers have focused on the use of food-grade surfactants approved for human consumption by the U.S. Food and Drug Administration.[169] Certain surfactants can also inhibit biodegradation, which may be of concern in some remedial programs.
- Selection of a surfactant also depends on the target contaminant(s). As described below, two parameters are conventionally used to describe the relationship between a surfactant and an organic liquid. Ideally, a surfactant effectively lowers the interfacial tension and enhances the solubility of the contaminant(s) without forming a stable emulsion that cannot be broken in the system used to treat extracted water.
- Surfactants can react with aquifer materials. Clay particles have large negatively charged surface area which can adsorb surfactant molecules, and can form aggregates with surfactant molecules. Soils high in organic carbon content also adsorb surfactants. Ideally, soils should have less than 10% clay content and less than 5% organic carbon content. The cations in soil can also react with surfactants.[170] Food-grade surfactants are particularly prone to precipitation and sorption. Surfactant

losses on the soil can add substantially to the cost of remediation. Researchers are investigating various methods to limit surfactant losses on soils.[169]

- Surfactants can also be affected by the natural constituents of groundwater (hardness, alkalinity, pH). Treatability tests performed in a laboratory should account for natural groundwater characteristics.[169,170]
- Certain cationic surfactants can enhance the adsorption of dissolved organic contaminants to aquifer materials in a rapid and reversible reaction.[171]
- The behavior of a surfactant depends on temperature. (The *Krafft point* is the temperature at which the solubility of an ionic surfactant becomes equal to the critical micelle concentration. Below that temperature, micelles do not form. Some nonionic surfactants have a similar critical low temperature.)[157]

To mobilize DNAPL, an aqueous solution of the surfactant must be injected into the aquifer upgradient of (or into) the DNAPL residual. The distribution of the surfactant through the aquifer depends on the hydraulic conductivity and heterogeneity of the aquifer; surfactant use is most likely to succeed in a homogenous, relatively permeable aquifer. Field trials in sand and/or gravel aquifers have been more successful than those in low-permeability soils or fractured rock.[172]

The concentration of dissolved NAPL peaks relatively soon after the surfactant is distributed through an aquifer. As the remediation program continues, the dissolved-phase concentration of NAPL compounds tends to decrease. Laboratory studies have shown that the rate at which NAPL is solubilized depends on groundwater velocity and the NAPL–water interfacial area, among other factors. One series of experiments with surfactant flooding of columns of PCE-containing soil compared the dissolved PCE concentration per number of pore volume flushes at different hydraulic gradients. The results indicated that lower hydraulic gradients resulted in higher peak concentrations and a shorter tailing period (in terms of pore volumes).[160]

The mobilized NAPL must be captured in a groundwater extraction system. If capture is incomplete, the remediation system may worsen the environmental problem. Downward migration of mobilized DNAPL through a fractured aquitard is of particular concern.[173]

The hydraulic efficiency of a surfactant system can be increased — thereby decreasing the cost — by balancing several factors:[169]

- minimizing the volume of surfactant injected into the aquifer;
- minimizing the volume of groundwater pumped to the surface for treatment (while maximizing the capture of the groundwater treated with surfactant as noted above); and
- focusing surfactant injection on the likely area of NAPL residuals, rather than the entire dissolved-phase plume.

The use of vertical barriers has been suggested to direct clean (upgradient) groundwater around the area treated, minimizing the amount of groundwater to be treated and surfactant to be injected; vertical barriers have also been suggested to contain source areas and aid in the hydraulic control of surfactant-treated groundwater. Theoretical evaluation and bench-scale tests suggest that a vertical circulation well (e.g., the UVB well discussed in Section 3.5.2) may be an effective option for *in situ* treatment with surfactants.[169]

Finally, the extracted groundwater must be treated. Ideally, the treated water can be reinjected after adding additional surfactant; if the surfactant can pass through the groundwater treatment system, the surfactant addition (and cost) will be minimized. A surfactant can affect the treatment units. For example, surfactants can affect the adsorption of contaminants onto activated carbon, affect the volatility of a compound in an air stripper, or exert BOD. (For additional information on treatment and costs, see the discussions by Sabatini et al.[169] and Kueper et al.[174])

3.6.2.2.2   Cosolvents/Alcohol Flooding

3.6.2.2.2   Cosolvents/Alcohol Flooding — Alcohols such as methanol, ethanol, and propanol have a bipolar structure. As a result, alcohols have a potential to enhance the solubility of DNAPL or low-solubility contaminants in soil or groundwater similar to surfactants. However, alcohols do not form micelles.

Alcohols can be used in three ways:[175]

- Alcohols may be used in conjunction with surfactants as a *cosurfactant* or *cosolvent*. This application developed in the petroleum industry.
- *Cosolvent flooding* refers to the injection of a dilute aqueous solution of an alcohol, generally at a concentration between 1 to 5 vol%. Cosolvent flooding can increase the solubility of low-solubility contaminants and enhance their desorption from soil. It does not mobilize NAPL per se.
- *Alcohol flooding* is the injection of pure alcohol or a concentrated aqueous solution (70 to 90%) of alcohol. At high alcohol concentrations, the alcohol can partition into both the groundwater and NAPL, changing the viscosity, density, solubility, and interfacial tension of the NAPL. A sufficient change in the interfacial tension will mobilize NAPL.

Concentrated alcohol solutions can be toxic, flammable, and explosive. As a result, they must be handled carefully and high concentrations must be recovered from an aquifer. Low concentrations of alcohols (less than approximately 1%) which may remain in an aquifer are readily biodegradable.[175]

3.6.2.3   Steam Injection

Injecting steam or hot water into an aquifer can have two effects on DNAPL residual or pools:

- Increasing the temperature decreases the viscosity and potentially the density of DNAPL, and can volatilize DNAPL compounds. This increases the mobility of the DNAPL.
- Injecting a fluid (water or condensed steam) induces a hydraulic gradient which can mobilize DNAPL as described in Section 3.6.2.1. As with water flooding, horizontal and vertical hydraulic gradients must be controlled to prevent DNAPL from migrating further.

Injected steam displaces water and DNAPL to create a "steam zone". Contaminants — particularly those with boiling points less than that of water — volatilize with the increase in temperature. When the vapor-phase contaminants reach cooler soils beyond the influence of the steam injection wells, they can recondense to form a DNAPL *bank*. Steam injection wells and groundwater recovery wells must be sited and operated to capture the DNAPL, rather than allowing a DNAPL bank to move beyond the remediation area either vertically or laterally. The downward migration of mobilized DNAPL is of particular concern unless a relatively impermeable layer underlies the treatment area.[176] The example which follows describes one application of steam injection with practical detail.

Example 3.5:   This example illustrates the application of several of the technologies discussed in this chapter: DNAPL recovery, iron removal, biological treatment of groundwater, and reinjection of treated water. More importantly, it illustrates how even a carefully designed remediation system must be adapted in the field based on unanticipated operating results. This synopsis is derived from several sources.[177-180]

The Brodhead Creek site in Stroudsburg, Pennsylvania was used to manufacture natural gas from coal between 1888 and 1944. The site is underlain by coarse glacial gravels, which are underlain in turn by fine silty sand 20 to 30 ft below ground surface. Coal gasification generated several wastes, including coal tar which was initially disposed of in an open trench near Brodhead Creek. Later, coal tar was processed on site to make usable products. Residues from coal tar processing were discharged through a shallow injection well into the gravel layer.

When a seep of coal tar to Brodhead Creek was observed in 1980, a site investigation was performed to characterize the nature and extent of contamination. The investigation showed that tar had collected in the voids in the gravel, accumulating in a stratigraphic depression in the silty sand layer located near the former injection well. Investigators initially estimated the volume of free coal tar at 75,000 to 1.8 million gallons distributed over a 7.5-acre area. However, in calculating these volumes, investigators had not distinguished between coal tar residual and separate-phase coal tar. The estimated volume was soon revised downward to approximately 10,000 gallons of coal tar. Pumping-test data indicated that the transmissivity of water through the gravel was approximately 30,000 to 40,000 gal/d/ft; however, the transmissivity of coal tar was estimated at 50 gal/d/ft.

In response to these findings, a cement–bentonite slurry wall (648 × 17 ft) was installed along the creek and a coal tar pumping system was installed in 1982 to recover pumpable tar. The pumping system also withdrew groundwater to cause an upwelling of the static coal tar surface and enhance the recovery of the coal tar. The system recovered 8000 gallons of tar in 1 year. The tar was used for supplemental fuel at a chemical company.

The site was placed on the NPL in 1982. The remedial investigation (RI) occurred from 1987 to 1989. The site investigation indicated that coal tar had accumulated in two areas. In addition to the area defined in the initial investigation, tar had apparently accumulated behind the slurry wall.

The feasibility study (FS) was performed from 1989 to 1991. The EPA issued a Record of Decision (ROD) in 1991 which specified that free subsurface coal tar would be removed by enhanced product recovery and incinerated. The EPA and potentially responsible parties (PRPs) negotiated a Consent Decree for performance of the remedial design and remedial action. The PRPs signed the Consent Decree in December 1991. It was entered into court (and thus became an enforcable agreement) in September 1992. The Remedial Design was performed and reviewed by EPA in four steps: 30% design, 60% design, 90% design, and 100% design. EPA and the Pennsylvania Department of Environmental Resources approved the final design in May 1994.

The design detailed the construction, operation, and monitoring of the CROW™ system (i.e., Contained Recovery of Oily Waste). The process, which entails flushing an aquifer with steam or hot water, was originally developed in the 1970s to recover petroleum deposits from oil sands and deep shale deposits. Environmental applications of the technology were developed by the Western Research Institute.

Injection of steam or hot water reduces the density and viscosity of tar, allowing the tar to flow with the groundwater. Extraction wells are used to capture groundwater and the mobilized tar. Free tar is removed; however, a tar residual remains in the soil. Treatability tests for the Brodhead Creek Site indicated that: (1) the optimum flushing temperature was 156°F, and (2) 98% of the removable coal tar would be recovered after 19 pore volume flushes of water at that temperature.

The recovery system was to be installed into the larger of the two subsurface coal tar deposits at the site, which contained an estimated 6000 gallons of tar. Several water injection/extraction scenarios were modeled to optimize the design. The final design included two 14-in. diameter extraction wells and six 5-in. diameter injection wells. In order to control the hydraulic gradients, and thus the migration of tar and contaminated water, more water was extracted than was reinjected.

The system was designed to extract heated groundwater at 115 gpm. Tar was to be removed from the extracted liquid by adding an acid, to adjust the pH to 5.0 and break the tar–water emulsion, and then by gravity separation in three 20,000-gal tanks. The tanks to be used for gravity separation were uninsulated to allow the liquids to cool and enhance the separation. Tar which accumulated in the separation tanks was to be pumped to a 10,000-gal tank for storage until disposal. 100 gpm of water was to be pumped from the separation tanks to a water heater, heated to 180°F, and reinjected into the ground. The remaining 15 gpm was to be pumped to a treatment system.

The groundwater treatment system included a fluidized bed reactor (FBR) followed by activated carbon. The FBR, which was 15 ft high and 4.5 ft in diameter, contained granular activated carbon to act as the support medium for fixed-film biological growth. It also had nutrient and oxygen feed systems. The carbon in the FBR was suspended (fluidized) by the upward flow of water. The influent flow of 15 gpm was augmented with 175 gpm of water recycled from the effluent for a total flow of 190 gpm. Effluent from the FBR passed through a bag filter and then activated carbon to provide a polishing step at the request of EPA. Four 200-pound carbon adsorption units were installed, with the intent that they would provide sufficient treatment capacity for the life of the project. Treated water was discharged under a NPDES permit equivalent which limited the discharge of selected VOCs, PAHs, BOD, TSS, and pH.

Construction occurred between June and October 1994. Startup of the system began with the groundwater treatment system. The FBR was innoculated with bacterial sludge, and cold water was extracted from the aquifer and then injected into the aquifer. Once all systems were working, the heater was slowly turned on.

Almost immediately, the injection wells began to fail: three of the six wells could inject less than 1 gpm, rather than the design rate of 15 gpm. The wells were redeveloped, then examined with a down-hole camera. The water in the wells was quite turbid, growth was evident on the well screens, and the bottom of each well contained 6 to 12 in. of material. Analyses of this material indicated a large amount of iron, suggesting the growth of iron-oxidizing bacteria.

The wells were redeveloped using a chlorine-based detergent to kill the bacteria and remove particulates. "Juttering heads" were installed in each well to allow operators to periodically agitate each well at a low pH, then pump down the well to remove solids.

The activated carbon units also clogged very quickly and were inoperable within 5 weeks. Brownish–orange slimy material coated part of the carbon. That observation, and DO measurements that indicated a decrease in DO across the carbon units (from about 5 mg/L to about 2 mg/L), suggested the presence of iron-oxidizing bacteria. Further, the turbidity of the effluent from the FBR had increased, suggesting that biomass was leaving the reactor and entering the downstream units. A sand filter was installed after the FBR and before the bag filter to remove additional solids.

Meanwhile, the chemical concentrations in the effluent from the treatment system increased to the point where the EPA requested that the discharge to Brodhead Creek be stopped. Monitoring data suggested that PAHs attached to solid particles were passing through the treatment system. In addition, the water heater kept shutting down. As the water heated, iron oxidized and precipitated. The filter after the heater would quickly clog and block flow; as the water in the heater overheated as a result, the control system shut the heater down.

In hindsight, all of the operational problems — with the injection wells, activated carbon system, discharge limit exceedances, and water heater — ultimately were related to the levels of iron in the groundwater. The groundwater contained approximately 8 to 10 mg/L iron. When the problem became clear, about 3 months into operation, treatability tests were performed to develop an iron removal system. Several oxidants were tested: potassium permanganate, hydrogen peroxide, oxygen gas, and DO-rich water. Hydrogen peroxide was selected because it oxidized iron most effectively and quickly, and could be used at a neutral pH.

The treatment system was retrofitted for iron removal, using the existing tanks rather than new treatment units due to space limitations on site. Water from the initial settling tank was directed to a second tank for neutralization with sodium hydroxide and oxidation with hydrogen peroxide; a flocculent (polyacrylic acid) was also added to enhance iron removal. Oxidized iron was to settle out in that tank and a subsequent (third) tank. Water would then flow from the third tank through bag filters to either the FBR or the water heater. This water contained less than 2 mg/L iron.

Installation and operation of the iron removal system improved the overall system performance. However, the head losses from the bag filters and the limited capacity of the injection wells limited the injection flow rate from 100 gpm to 30 gpm.

The system was operating at steady state, extracting 40 gpm of 120°F water and rein-jecting approximately 25 gpm of 180°F water, by mid-July 1995. The biological ground-water treatment system effectively removed over 95% of the PAHs in the water (primarily two- and three-ring PAHs, at concentrations on the order of 10^2 μg/L in the influent).

When remediation ended in May 1996, 29 pore volumes of hot water had been flushed through the treatment zone. The system had recovered 1424 gal of tar, 6064 gal of water, and 3206 gal of inorganic solids (such as iron sludge).

Remediation cost a total of $2,914,000, including testing, design, construction, operation, and decommissioning. Of this total, research and development of the CROW™ process cost $846,000. Costs associated with the regulatory requirements for a Superfund site totaled an estimated $311,000. Costs associated directly with remediation totaled $1,757,000.

The project team offered several lessons learned from the operation of this system:

- Do not inject water into an aquifer less than 5 ft below ground surface, as the injected water may simply short-circuit and flood the ground surface.
- The capacity of injection wells will decrease over time, and injection wells will require routine maintenance.
- Design the system to remove iron.
- Model the groundwater extraction/reinjection system using a conservative (low) hydraulic conductivity.
- Install multiple filters to protect expensive treatment equipment, and size the pumps to allow for the head drop through the filters.

While the coal tar was being remediated, investigation of conditions in the deeper aquifer continued. In 1995, the EPA determined that no action needed to be taken beyond continued monitoring, since the shallow groundwater is not used for drinking water and the site conditions make excavation of further source material impracticable. EPA documented the decision not to proceed with groundwater remediation in a Technical Impracticability Waiver.

3.6.2.4 Oxidation

As described in Section 3.4.3.2, chemical oxidants have been used to treat a variety of contaminants in wastewater and extracted groundwater. Injection of oxidants to treat soil and groundwater contaminants *in situ* has been contemplated since the early 1980s.[181] In recent years, several researchers have examined the use of chemical oxidants to destroy NAPL residuals *in situ*. In general, chemical oxidation is most effective for contaminants which contain a carbon–carbon double bond (such as PCE and TCE), rather than single bonds (as found in TCA, for example).[182]

Researchers have been testing the use of potassium permanganate ($KMnO_4$).[183] This reagent can oxidize common chlorinated solvents, initially to carboxylic acids or diols and ultimately to mineralization.[182] For example,

$$TCE + 2\ KMnO_4 \rightarrow 2\ CO_2 + 2\ MnO_2(\downarrow) + 2\ KCl + HCl$$

$$PCE + 2\ KMnO_4 \rightarrow 2\ CO_2 + 2\ MnO_2(\downarrow) + 2\ KCl + Cl_2$$

Treatment requires the addition of reagent in excess of the amount predicted by the stoichiometry of the oxidation reactions, as potassium permanganate can react with organic aquifer materials. Treatment may also cause the pH to drop in unbuffered aquifers.[183]

Fenton's reagent has been used to treat groundwater and NAPL residuals *in situ*. As described in Section 3.4.3.2, Fenton's reagent is a mixture of hydrogen peroxide and a metal catalyst, such as iron, which generates a free radical that can oxidize organic contaminants. GEO-CARE, INC. has developed the Geo-Cleanse® process based on Fenton's chemistry. The vendor claims that the process has been used to successfully remediate TCE and methylene chloride residuals in clay and silty clay, as well as BTEX and phthalates at other sites.[182,184]

According to the patents,[185,186] an aqueous mixture of ferrous sulfate or other metallic catalysts, hydrogen peroxide (at 10 to 100%), and stabilizers is injected into an aquifer. The stabilizers act to slow the degradation of hydrogen peroxide. Surfactants may be added to enhance the availability of the organic contaminants and thereby increase the rate of reaction. Acids or bases may also be added to the solution in order to maintain the pH between 4 and 6.

The injection system is designed to inject fluids into the aquifer under pressure (5 psi or higher). The injection pressure helps to disperse the reagent through the aquifer, and increases the permeability of the aquifer by hydraulic fracturing (see Section 4.3.1.2.2). The injection wells may also be fitted with air spargers; air sparging is intended to enhance vertical circulation within the aquifer. The injection well network can be coupled with an extraction well network to provide hydraulic containment of the area undergoing treatment.

This process was tested at the Savannah River Site in 1997.[187] Separate-phase DNAPL found in monitoring wells contained 95% TCE, 5% PCE, and a small amount of PCBs. DNAPL had apparently accumulated above a semi-confining clay layer 152 ft BGS and some 20 ft below the water table. A 50- by 50-ft area containing approximately 64,000 ft^3 of contaminated soil was selected for testing. Three injection wells were installed on 17-ft centers in a circular pattern. GEO-CARE, INC. injected one batch of reagents per day over a 6-h period. Each batch treatment consisted of injecting catalyst solution (100 ppm ferrous sulfate which was pH adjusted with sulfuric acid), followed by injecting hydrogen peroxide and additional catalyst solution simultaneously in volumes varying from 500 to 1000 gallons. During treatment, observers noted gas bubbles generated by the reaction in monitoring wells. Table 3.4 summarizes the results achieved in the 6-d test.

Table 3.4 Pre- and Post-Treatment Data Oxidation of DNAPL by Fenton's Reagent Field Test at Savannah River Site

Parameter	Pre-Test	Post-Test
Average concentration of PCE in groundwater, mg/L	119.49	0.65
Average concentration of TCE in groundwater, mg/L	21.31	0.07
Estimated mass of PCE above clay, lbs	528.53	28.24
Estimated mass of TCE above clay, lbs	64.56	7.95
pH	5.71	2.44
Temperature, °C	19.2	34.7
Chloride concentration, mg/L	3.61	24.35

Note: Post-test results were measured immediately after treatment. Groundwater concentrations later rebounded; this result was attributed to contributions from outside the treatment area. Three months later, the pH had increased to 3.5.

From Jerome, K. M., Riha, B., and Looney, B. B., Final Report for Demonstration of In Situ Oxidation of DNAPL Using the Geo-Cleanse Technology, WSRC-TR-97-00283, U.S. Department of Energy, Office of Technology Development, September 19, 1997, 23 pp.

Treatment using potassium permanganate and treatment using Fenton's reagent differ in several respects:[182,188]

- The reaction rate for Fenton's reagent is much faster than for potassium permanganate.
- Fenton's reagent can be applied at higher doses than potassium permanganate due to the limited solubility of the latter.
- The rate of oxidation via potassium permanganate does not depend on the pH under typical field conditions. Oxidation via Fenton's reagent occurs under slightly acidic conditions.

In situ oxidation of NAPL residuals has limitations:

- To be effective, oxidants delivered into the subsurface must be thoroughly distributed throughout the NAPL/residual. The distribution of the reagent may be limited by the heterogeneity and/or permeability of the aquifer materials and the hydrophobic nature of the NAPL residual.
- Metallic oxides or hydroxides formed during the oxidation reaction can precipitate and lower the permeability of the soil.
- Most oxidation reactions are *exothermic* (that is, they give off heat).[182] As a result, oxidation of a large mass of contaminants beneath buildings or other structures, or near subsurface utilities, may generate significant heat and/or mobilize potentially toxic contaminant vapors. (Mobilization of vapors is of particular concern when this technology is applied to LNAPLs which can generate potentially explosive vapors.) The heat or pressure generated by the oxidation reaction can potentially damage nearby structures.
- High levels of organics will consume a large quantity of oxidants, increasing the cost of treatment. Carbonates and phosphates, which can scavenge free radicals, can also consume reagents.[182,188]
- Incomplete oxidation of certain compounds can produce toxic compounds.[182]

3.7 CONTROL OF AIR EMISSIONS FROM GROUNDWATER TREATMENT UNITS

Many of the groundwater and NAPL remediation technologies discussed in this chapter transfer contaminants to a vapor stream. Off-gases often require treatment to protect human health and the environment or to meet regulatory requirements. Following a general discussion of factors related to emission control are brief descriptions of the technologies commonly used to treat vapor emissions.

Many contaminants are explosive in the vapor phase, including light hydrocarbons such as methane, and hexane; aromatics such as benzene, light alcohols, and ketones; and many other compounds. Handling and treating vapor streams containing potentially explosive compounds requires special care. The range of concentrations at which such compounds can explode is bounded by the *lower explosive limit (LEL)* and *upper explosive limit (UEL)*. The LEL is the lowest concentration in air at which the vapor burns upon contact with an ignition source and the flame spreads through the flammable gas mixture.[189] In other words, at concentrations below the LEL, the concentration of the compound is not sufficient to sustain an explosion. Above the UEL, the concentration of air (oxygen) is not sufficient to sustain an explosion. Vapors containing potentially explosive contaminants are typically managed to remain below 25% LEL in off-gas streams, e.g., by dilution. Equipment and controls used in treatment systems handling potentially explosive compounds are made to be explosion-proof.

Air emissions are measured in units of ppmv and partial pressure (p_i). For a review of these units and the conversions between units, see Section 2.2.1.1.

3.7.1 Removal of Water or Other Liquids

Vapor emissions from groundwater treatment units are often laden with water vapor. The first step in treatment is often to remove liquids or reduce the relative humidity of vapor streams. (*Relative humidity* is the ratio of the partial pressure of the compound in the vapor phase to its vapor pressure.)[190] Commonly used techniques include the following:

- A *demister* is a knitted mesh, typically made of stainless steel or plastic fibers. A demister entrains small droplets of liquid as the vapor flows upward through the mesh.[191]
- A *knock-out tank* works very simply: by increasing the cross-sectional area in flow (relative to the influent pipe), the velocity of the vapor stream slows. As a result, water droplets and particulates fall out of the vapor stream by gravity.[191] A knock-out tank can be as simple as a 55-gal drum with a vapor inlet near the bottom of the drum, an outlet for the vapor at the top of the drum, and a tap for removing collected liquids from the bottom of the drum. Another common design for a knock-out tank brings the influent vapors in through a pipe inserted through the top of the tank and down through a demister. As the vapors expand, droplets form; the demister helps to remove the droplets as the vapors flow upward though the demister to exit the tank through a port located above the demister. Knock-out tanks are commonly used in dual-phase extraction systems and soil vapor extraction systems.[192]
- Either cooling or heating can be used to control the water content of a vapor stream. Cooling a vapor stream can cause water or organic liquids to condense. Condensation can be an effective form of treatment for off-gases containing >5,000 to 10,000 ppmv organics.[193] However, the organics which have a high enough vapor pressure to be removed in a remediation system which does not heat the soil or water will not condense readily in a heat exchanger unless the temperature is lowered significantly. Heating a vapor stream lowers the relative humidity, as discussed below in Section 3.7.2.

3.7.2 Vapor-Phase Carbon

Organic contaminants in a vapor stream can be adsorbed onto activated carbon and thereby removed from vapor emissions. This form of treatment is commonly used for relatively low concentrations, e.g., below 200 to 500 ppmv (and <25%

LEL).[194,195] At higher concentrations, thermal or catalytic oxidation becomes more cost effective than carbon use.

Vapor-phase carbon adsorbs a variety of organic compounds. In general, carbon has a higher adsorption capacity for heavier molecular weight compounds than for lighter molecules; unsaturated compounds and aromatic compounds tend to adsorb better than saturated or linear compounds. Very heavy compounds (i.e., molecular weight above 130 g/mole) are difficult to desorb from carbon. Adsorption capacity also increases with concentration.[196]

The adsorption capacity for a particular compound is represented by a Freundlich isotherm:[196]

$$W_e = kp_i^m \tag{3.17}$$

where: W_e = equilibrium adsorptivity, lb contaminant/lb carbon;
        p_i = partial pressure of contaminant in air stream; and
        k, m are empirical parameters, valid within a specified range of partial pressures for a given contaminant.

Table 3.5 provides isotherm parameters for some common contaminants. Note that these parameters depend on the temperature, the concentration range, and the type of carbon as well as the compound being adsorbed. The carbon usage estimated from these parameters is commonly multiplied by two to account for the fact that a carbon bed is taken off line before breakthrough (if only one unit is used, in contrast to the two units in series commonly used for water-phase treatment). If a vapor stream contains several contaminants, the carbon usage is conservatively estimated based on isotherm for the least-adsorptive contaminant. Lacking isotherm data, carbon usage can be grossly estimated as 10 lb contaminant per 100 lb carbon.[196]

Table 3.5 Adsorption Isotherms for Vapor-Phase Carbon

Compound	Temperature (°F)	Isotherm parameters k	m	Range of isotherm (psia)
Benzene	77	0.597	0.176	0.0001–0.05
Toluene	77	0.551	0.110	0.0001–0.05
m-Xylene	77	0.708	0.113	0.0001–0.001
	77	0.527	0.0703	0.001–0.05
Phenol	104	0.855	0.153	0.0001–0.03
Chlorobenzene	77	1.05	0.188	0.0001–0.01
Yinyl chloride	100	0.20	0.477	0.0001–0.05
Dichloroethane	77	0.976	0.281	0.0001–0.04
Trichloroethane	77	1.06	0.161	0.0001–0.04
Acetone	100	0.412	0.389	0.0001–0.05

Note: Data are for adsorption on Calgon type BPL carbon, 4 × 10 mesh.

Adapted from U.S. EPA, *Handbook — Control Technologies for Hazardous Air Pollutants,* EPA/625/6-91/014, Office of Research and Development, Washington, D.C., June 1991, 4:28–30.

Emissions from a vapor-phase carbon unit typically range from 50 to 150 ppmv, although concentrations can be as low as 20 to 25 ppmv.[196] The removal efficiency increases with the inlet concentration, e.g., 90% to 95% at 1000 ppmv, increasing to 99% at concentrations over 5000 ppmv.[193]

Efficient vapor-phase treatment requires that the gas stream have less than approximately 50% relative humidity if the contaminant concentration is less than 1000 ppmv. Emissions from air strippers are typically heated before they are passed through carbon beds to reduce the relative humidity. However, the adsorption capacity of carbon decreases at higher temperatures, becoming significant at about 130°F.[196] The increase in efficiency resulting from decreasing the relative humidity is balanced against the decrease in efficiency resulting from the higher temperature. Alternatively, vapor streams can be cooled to condense the water vapor, or diluted with clean air. The latter may require a larger-sized treatment unit, and as a result may not be cost effective.

The activated carbon used for vapor treatment has larger granules and a finer pore structure than the carbon used for liquid-phase treatment. The size of a vapor-phase carbon unit is based on the flow rate, the necessary contact time, and the projected carbon use rate. The materials of construction must be compatible with the compounds in the vapor stream; vapor-phase cannisters are commonly made of polypropylene, polyethylene, or steel (which may be lined or coated). Finally, the design must account for regeneration of the spent carbon. Smaller units are sent back to the manufacturer for reprocessing. Larger units may be designed for carbon replacement or in-place regeneration. Many vendors provide treatment units in various configurations; a few examples follow to indicate the general sizes of treatment units. A relatively small stream, up to 100 cfm, can be treated in a unit 24 in. in diameter and 36 in. high, roughly the size of a 55-gal drum, containing up to 200 lb carbon.[197,198] Larger units, designed to treat flows up to 1000 to 2000 cfm, contain roughly 2000 lb carbon. Such units are 4 to 5 ft in diameter, and 7 to 8 ft high.[199,200] Example 3.2 describes carbon units used to treat a very high (5500-cfm) flow rate.

When the carbon is exhausted, it can be regenerated in place by thermal oxidation or steam, or vacuumed out and replaced. About 3 to 5% of the organics adsorbed onto virgin carbon cannot be removed by regeneration.[196]

Ketones can cause operational problems in vapor-phase carbon treatment. Ketones polymerize on the carbon in an exothermic reaction. The polymers clog the pores on the carbon and reduce its effectiveness.[196] The heat released from the exothermic adsorption and polymerization can also cause carbon bed fires.[201]

3.7.3 Thermal Incineration

Thermal incineration is used to treat off-gases with relatively high vapor concentrations (ideally greater than 100 ppmv, and less than 25% of LEL).[193,194] A thermal incinerator is a refractory lined vessel fitted with a nozzle burner where vapors are combusted at a temperature of 1200 to 2000°F. Typically, thermal incinerators provide 1 sec or less of residence time.[202] The destruction efficiency generally ranges from 95% to greater than 99%.[193]

Supplemental fuel (natural gas or propane) is needed to maintain the required temperature when the vapor stream is relatively dilute. The cost of this supplemental fuel can make thermal incineration less cost effective than other methods for treating dilute vapor emissions.

Highly variable flow rates can cause operating problems. A dramatic increase in the flow rate decreases the residence time and mixing within the unit, causing less efficient combustion. High levels of sulfur or chlorine may also be cause for concern. Oxidation of these compounds produces sulfur dioxide and hydrogen chloride, respectively, which may have to be removed from the off-gases.

3.7.4 Catalytic Incineration

Catalytic incinerators are similar to thermal incinerators. However, a catalyst is used to enhance the rate of oxidation, and therefore catalytic incinerators can operate at temperature some 500°F lower than thermal incinerators. Fuel costs are resultingly lower. Catalytic incinerators are generally used to treat vapor streams containing 50 to 10,000 ppmv contaminants,[194] although at the lower end of this range carbon adsorption may be more cost effective. A catalytic incinerator can destroy some 90% of the compounds at an inlet concentration of 50 ppmv, increasing to 95%–98% destruction at concentrations over 100 ppmv. The highest destruction efficiencies (98%–99%) require larger catalyst volumes and/or higher temperatures than the typical design.[193]

The catalysts used to enhance oxidation are typically platinum or palladium. Chlorine, from chlorinated organic compounds, can poison these catalysts. Other catalysts may be used if chlorinated compounds are to be treated, including chrome/alumina, cobalt oxide, or copper oxide/manganese oxide. Compounds containing certain metals can also poison or inhibit catalysts. Finally, the high temperatures generated by oxidizing highly concentrated vapors can deactivate a catalyst. As a result, highly concentrated vapors (heat content greater than 15 BTU/scf) may require dilution before treatment.[195,203]

The catalyst bed is designed for maximum surface area. Configurations include a metal mesh-mat, ceramic honeycomb or other matrix structure, or pellets.[202,203]

The performance of a treatment unit depends on the temperature, residence time, type of catalyst, and type and concentration of contaminants. The vapor stream may be preheated using heat from the emission stream before entering a catalytic oxidizer in order to increase the treatment efficiency. In addition to measuring the concentrations of target compounds in the off-gas stream, the temperature rise and pressure drop across the catalyst bed are monitored to evaluate the condition of the system. The temperature of the gas stream rises across the catalyst bed as organic compounds oxidize. If the influent has remained the same, a decrease in the temperature differential indicates poorer performance. The pressure drops as the vapor stream flows through the catalyst bed; over time, this pressure drop decreases as the catalyst degrades and particles of the catalyst become entrained in the vapor stream. A catalyst generally requires replacement every 2 to 3 years.[195,203]

As discussed previously for thermal incinerators, high concentrations of sulfur or chlorine compounds may produce unacceptable concentrations of sulfur dioxide or hydrogen chloride in the off-gases (in addition to potentially poisoning the catalyst). A scrubber might be needed to remove those compounds.

3.7.5 Biofilters

While biofilters have been used for odor control in industrial applications since the 1950s, they have not been commonly used for emissions control on groundwater treatment systems. Interest in biofilters has grown in the 1990s because of their relatively low cost compared to other alternatives.[195]

A biofilter is essentially a thin-film bioreactor. Biomass grows on a support medium such as soil, peat, oyster shells, pelletized activated carbon, or ceramic media. Nutrients are supplied as needed. As the vapor stream flows through the treatment unit, contaminants adsorb and biodegrade in the biofilm. Vapor-phase contaminants at concentrations less than 1000 ppmv can reportedly be treated in a contact time of 90 sec or less.[194] As biomass builds up, it may slough off (ceramic packing) or clog the medium (media such as peat), requiring that the medium be replaced.[195,204]

Problems

3.1 As noted in Example 3.2, remediation of the Verona Well Field was projected to require 5 years. Operation of the pump-and-treat system has continued for over 12 years. Based on the information provided in the example and with the benefit of hindsight, why was the initial projection wrong?

3.2 Slug tests in a water table aquifer indicate a hydraulic conductivity of approximately 75 gal/d/ft². A layer of clay underlays this sand and gravel aquifer at a depth of 12 ft. The depth to the water table is approximately 2 ft. Assume a storativity of 0.3.

 [a] Estimate the radius of influence of a 6-in.-diameter pumping well, assuming that it reaches steady state in 16 h.
 [b] Estimate the flow rate corresponding to that radius of influence.

3.3 Consider a site where levels of benzene in groundwater are being monitored to determine the effectiveness of groundwater extraction. The aquifer comprises silty sand with occasional lenses of sandy silt. Table 3.6 summarizes the analytical results from a representative well. The remediation goal is to restore the aquifer to a maximum concentration of 5 μg/L.

 [a] Based on these data, will the remediation system achieve the remediation goal within 10 years? Why or why not?
 [b] If the answer to part [a] was *yes*, project how long it will take to achieve the remediation goal (expressed as number of months since start-up of the system). If the answer to part [a] was *no*, briefly describe two options to consider for management of the problem.
 [c] Estimate and plot the total mass of benzene removed from the aquifer in the first 5 years of operation. Compare the mass removal in the first year to the total mass removed in the second through fifth years of operation by expressing each as a fraction of the total mass removed in 5 years. What does this evaluation suggest about the relative efficiency of mass removal as remediation continues into the long term?

Table 3.6   Problem 3.3 Groundwater
Monitoring Data from
Remediation Program

Month	Benzene concentration (μg/L)
0	1600
0.5	1620
0.75	1379
1	1273
2	1174
3	700
5	340
6	330
7	250
8	190
9	160
10	170
11	190
12	300
13	310
14	280
15	300
16	250
17	220
18	270
19	245

3.4 The objective of this problem is to develop a conceptual remediation solution and to identify critical site characteristics or concerns.

Contamination at a Superfund site resulted from historical wastewater treatment practices. Wastewater containing hexavalent chromium (abbreviated Cr^{6+} or $Cr(VI)$) was treated in a series of wastewater treatment impoundments, as sketched in Figure 3.19 and described in Table 3.7. As a result of historical practices, the sludge in the impoundments, the underlying soil, and the groundwater were contaminated with both $Cr(VI)$ and with trivalent chromium (abbreviated Cr^{3+} or $Cr(III)$).

Sludge remains in the impoundments awaiting cleanup. The sludge contains both hexavalent and trivalent chromium. Table 3.7 summarizes the data obtained from analyzing the sludge.

The impoundments were lined with geomembrane liners; however, over the years the liners developed numerous holes in the seams and in the liners themselves that allowed water and sludge from the impoundments to leach downward and contaminate the underlying soil and groundwater. Nonetheless, the liners still provide some containment; as a result, the impoundments are full of rainwater in addition to residual sludge from historical use of the impoundments. The liners were placed on a layer of crushed stone (0.5 to 1.0 ft thick), which covered the native sandy soil.

Soil beneath the impoundments is thought to be contaminated with both $Cr(VI)$ and $Cr(III)$. However, due to the geometry of the impoundments and the desire to protect the remaining integrity of the liners, the soil was not sampled. Based on visual observations, at least a portion of the soil beneath the impoundments is contaminated with chromium at levels similar to those found in the sludge in the impoundments.

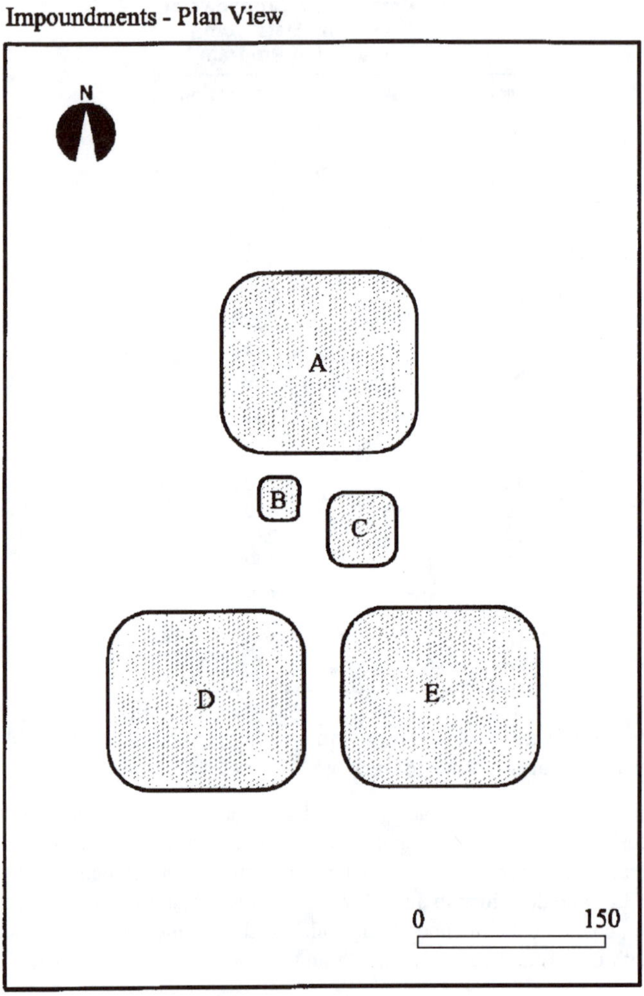

Impoundments - Plan View

Figure 3.19 Superfund site, Problem 3.4.

The state has a policy that unsaturated soil should be cleaned up to 100 mg/kg total chromium (maximum).

Groundwater at and near the site is also contaminated. In the vicinity of the impoundments, the depth to the groundwater table is approximately 14 ft. An aquifer comprising fine sand approximately 35 ft thick underlies the site. Groundwater flows to the southwest at a rate of approximately 400 ft/year. During a pumping test, a well pumped at approximately 100 gpm had a radius of influence of approximately 400 ft.

A plume of groundwater contamination approximately 700 ft wide emanates from the site. The plume extends approximately 4200 ft to the southwest. (see Figure 3.19). This plume contaminated several private drinking water wells, which were shut down in response to the contamination. Concentrations of Cr(VI) in unfiltered samples of the groundwater range up to 87 mg/L; concentrations of Cr(III) range up to 7.3 mg/L.

Extent of Plume

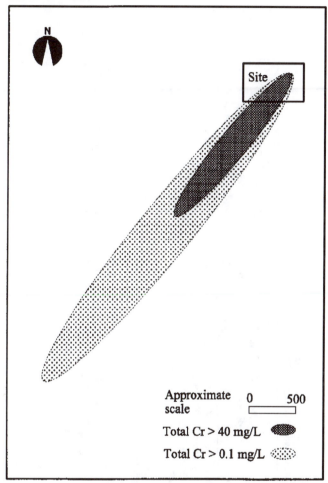

Figure 3.19   *Continued.*

Groundwater must be remediated before the plume spreads to contaminate more downgradient wells.

The clean-up goal for groundwater is 0.1 mg/L total chromium. Preliminary limits on the discharge of extracted groundwater to a local surface water body are 10 μg/L Cr(VI) and 200 μg/L Cr(III).

Provide a conceptual design for the groundwater remediation system, and the rationale for proposing that program. Include in your conceptual design, as appropriate to the system:

- a brief description of the system;
- plan view of the layout of the remediation system;
- cross section showing *in situ* components, if any;

Table 3.7 Problem 3.4 Former Wastewater Treatment Impoundments, Superfund Site

Impoundment	Function	Nominal dimensions (ft)	Maximum depth (ft)	Volume (cy)	Sludge TCLP leachate[a] (mg/L Cr)	Sludge Total Cr (mg/kg)
A	Flow equalization	145 × 165	12	2300	4.5 to 23.0	18,000 to 47,000
B	Adjust pH to 2 to 3 with H2SO4, then reduce Cr+6 to Cr+3 with FeSO4	35 × 35	13	[b]	N/A[c]	N/A
C	Adjust pH to 9 with NaOH to precipitate Cr(OH)3	58 × 58	5	[b]	N/A	N/A
D	Sludge settling	145 × 165	12	5300	0.023 to 0.36	9,300 to 42,000
E	Sludge settling	145 × 165	13	2000	0.22 to 0.48	10,000 to 40,000

[a] Maximum concentration for the toxicity characteristic is 5.0 mg/L.
[b] Lagoons contain a combined total of approximately 600 cubic yards of sludge.
[c] Not analyzed.

- estimate of the total quantity of water to be pumped, if any;
- block-flow diagram of the treatment system;
- wastes generated from the remediation system;
- list of key assumptions; and
- additional data or testing, if any, needed to refine the design.

3.5 An air stripper is used to treat groundwater contaminated with VOCs. The influent contains 1000 µg/L benzene, 900 µg/L toluene, and 850 µg/L xylenes. The column is 4 ft in diameter and the packing height is 20 ft. Water flows through the column at 200 gpm; the blower supplies 2700 cubic feet per minute (cfm) of air. The air-to-water ratio is 100. The air stripper removes nearly 100% of the contamination from the groundwater.

   [a] Estimate the pounds of liquid-phase carbon required per year to treat the groundwater if carbon was used to treat the groundwater *instead* of air stripping. What factors affect the accuracy of this estimate?

   [b] Estimate the pounds of vapor-phase carbon that would be required to treat the air emissions from the air stripper in 1 year.

   [c] What conclusion can you draw from these calculations regarding the relative efficiency of vapor-phase carbon and liquid-phase carbon?

3.6 Develop a conceptual design for remediation of groundwater at the ASR site by:

   [a] a pump and treat system; and

   [b] *in situ* treatment of the shallow aquifer.

Use the following assumptions, as needed:

- Clean-up levels for the groundwater are 0.005 mg/L TCE, 0.005 mg/L PCE, 0.07 mg/L 1,2-DCE, 0.002 mg/L VC, and 0.0005 mg/L PCBs.
- Iron filings have a hydraulic conductivity and porosity of approximately 5×10^{-5} cm/sec and 0.4, respectively.
- Preliminary limits for discharging extracted groundwater to surface water (approximately 600 ft off site) are 3.5 mg/L TCE, 1.2 mg/L PCE, and 0.0019 mg/L PCBs. Reinjected groundwater must meet the site clean-up goals for groundwater before reinjection.

Include the components of a conceptual design outlined in Problem 3.4. In addition, provide a screening-level capital cost estimate for each system, based on Table 3.8, and a list of major operation and maintenance cost items.

3.7 Would you use water flooding to remediate the ASR site? Surfactants? In each case, why or why not?

3.8 Additional data have been collected for the HypoChem facility described in Problem 2.6:

- A new well, MW-5, was installed 100 ft due east of MW-2. The well was screened from 5 to 10 ft below ground surface. A sample of groundwater from MW-5 contained 0.083 mg/L toluene.
- Samples from selected wells were analyzed for inorganic parameters. Table 3.9 summarizes the data.
- The field team performed slug tests in three wells, MW-1, MW-3S, and MW-5. These tests indicated an average hydraulic conductivity of 1.9×10^{-3} cm/sec.

Table 3.8 Problem 3.6 Unit Costs

Item		Unit	$/unit
Mobilization/site setup		ea	6,000
Extraction/	Vertical well installation (no soil disposal)	ft	65
injection	Horizontal drain in trench lined with HDPE	sf	20
wells, piping	Well pump/controls	ea	3,500
	Transfer piping, installed in trench	lf	20
	Soil incineration, TTD[a]	ton	1,000
	Soil disposal, haz waste, TTD	ton	200
Containment	Interlocking steel sheet piling, installed	sf	30
Treatment	Treatability testing	ea	15,000
	Chemical precipitation/clarification, package unit, up to 50 gpm	ea	45,000
	Air stripper, 4 tray, 5-HP blower, 420 cfm, 1–50 gpm water	ea	13,000
	Activated carbon, up to 50 gpm (1,000 lb carbon)	ea	4,800
	UV/oxidation, installed, 50 gpm unit	ea	85,000
	Cartridge filter, SS housing w/replaceable filter, 50 gpm	ea	800
	Tank, 15,000 gal	ea	25,000
	Plate and frame filter press	ea	49,000
	Transfer pump	ea	1,000
	Blower, 250 cfm, 1½ HP, 12 in. pressure	ea	700
	Vapor-phase carbon, 375#, 40–250 scfm	ea	1,200
	Catalytic oxidation unit, 250 scfm	ea	40,000
	Knock-out tank	ea	800
Miscellaneous	Iron filings, delivered	ton	700
	Pea gravel, delivered	CY	24
	Prefabricated remediation building	ea	20,000
	Excavation	CY	15

Note: Use these unit costs only to solve Problem 3.6. Do not use these unit costs in professional practice. Assume costs are installed costs, unless noted.

[a] TTD transportation, taxes, disposal.

Table 3.9 Problem 3.8 Inorganic Parameters in Groundwater, HypoChem Facility

Well	pH	Fe (mg/L)	Mn (mg/L)	Ca (mg/L)	Mg (mg/L)	TSS (mg/L)
MW-1	7.1	8.7	2.4	10.7	3.2	4.3
MW-3S	6.5	23.6	17.8	8.6	5.1	5.1
MW-5	6.8	11.9	3.7	12.6	8.1	6.8

State regulations require that groundwater be restored to drinking water quality, i.e., 1 mg/L. The state also requires protection of the ecosystem.

[a] Develop a conceptual design for a groundwater remediation system.

[b] What additional data are necessary to design a comprehensive approach to site remediation?

3.9 What parameters would you monitor to evaluate the natural attenuation of groundwater containing PCE?

Table 3.10   Problem 3.10

Remediation option	Contaminant					
	BTEX	PAH	PCB	Metals	Chlorinated ethenes	TPH
Air stripping						
Carbon adsorption						
Bioremediation						
Precipitation						
Chemical oxidation						
Ion exchange						

3.10 Indicate on the matrix in Table 3.10 whether or not groundwater containing each of the listed contaminants can be remediated by the technologies indicated.

Please use the following abbreviations in your answer:

Y   **Yes**, the technology can generally be applied to the contaminant.
N   **No**, the technology should not be applied to the contaminant.
M   The technology **may** be applied under certain limited circumstances; *briefly indicate the circumstances in your answer.*

Note that, for the sake of simplicity in answering this problem, you should consider the application of the technology *regardless of pretreatment requirements.*

REFERENCES

1. U.S. EPA, *Final Guidance: Presumptive Response Strategy and Ex-Situ Treatment Technologies for Contaminated Groundwater at CERCLA Sites*, EPA/540/R-96/023, Office of Emergency and Remedial Response, Washington, D.C., Pre-publication copy, October 1996, 2.
2. U.S. EPA, Preamble to the National Contingency Plan, Prepublication copy, February 7, 1990, response to comments on Section 300.430(f)(5)(iii)(A), 250.
3. U.S. EPA, *Guidance for Evaluating Technical Impracticability of Ground Water Restoration*, OSWER Directive 9234.2-25, EPA/540-R-93-073, September 1993.
4. U.S. EPA, *Groundwater Currents: Developments in Innovative Groundwater Treatment*, Office of Solid Waste and Emergency Response, Issue No. 16, September 1996, 1.
5. U.S. EPA, *Handbook — Remedial Action at Waste Disposal Sites (Revised)*, EPA/625/6-85/006, Office of Emergency and Remedial Response, Washington, D.C., October 1985, 5:84–94.
6. Powers. J. P., *Construction Dewatering — New Methods and Applications*, 2nd ed., John Wiley & Sons, New York, 1992, 361.
7. Cherry, J. A., Feenstra, S., and MacCay, D. M., Concepts for the Remediation of Sites Contaminated with Dense Nonaqueous Phase Liquids (DNAPLs), in *Dense Chlorinated Solvents and other DNAPLS in Groundwater*, Pankow, J. F. and Cherry, J. A., Eds., Waterloo Press, Portland, OR, 1996, 486.
8. U.S. EPA, *Handbook — Remedial Action at Waste Disposal Sites (Revised)*, EPA/625/6-85/006, Office of Emergency and Remedial Response, Washington, D.C., October 1985, 5:111.
9. U.S. EPA, *Handbook — Remedial Action at Waste Disposal Sites (Revised)*, EPA/625/6-85/006, Office of Emergency and Remedial Response, Washington, D.C., October 1985, 5:23–26.

10. U.S. EPA, *Handbook — Ground Water, Volume II: Methodology*, EPA/625/6-90/016b, Office of Research and Development, Washington, D.C., July 1991, 2–9.
11. Powers. J. P., *Construction Dewatering — New Methods and Applications*, 2nd ed., John Wiley & Sons, New York, 1992, 288–308.
12. Powers, J. P., *Construction Dewatering — New Methods and Applications*, 2nd ed., John Wiley & Sons, New York, 1992, 208–218.
13. Powers, J. P., *Construction Dewatering — New Methods and Applications*, 2nd ed., John Wiley & Sons, New York, 1992, 224–233.
14. Driscoll, F. G., *Groundwater and Wells*, 2nd ed., Johnson Division, St. Paul, MN, 1986, 454–456.
15. Powers. J. P., *Construction Dewatering — New Methods and Applications*, 2nd ed., John Wiley & Sons, New York, 1992, 73, 78, 94–106.
16. U.S. EPA, *Handbook — Remedial Action at Waste Disposal Sites (Revised)*, EPA/625/6-85/006, Office of Emergency and Remedial Response, Washington, D.C., October 1985, 5:12–22.
17. Powers, J. P., *Construction Dewatering — New Methods and Applications*, 2nd ed., John Wiley & Sons, New York, 1992, 158.
18. Powers, J. P., *Construction Dewatering — New Methods and Applications*, 2nd ed., John Wiley & Sons, New York, 1992, 114.
19. Powers, J. P., *Construction Dewatering — New Methods and Applications*, 2nd ed., John Wiley & Sons, New York, 1992, 129.
20. Cohen, R. M., Vincent, A. H., Mercer, J. W., Faust, C. R., and Spalding, C. P., *Methods for Monitoring Pump-and-Treat Performance*, EPA/600/R-94/123, U.S. EPA, Office of Resarch and Develoment, Ada, OK, June 1994, 31–35.
21. Powers, J. P., *Construction Dewatering — New Methods and Applications*, 2nd ed., John Wiley & Sons, New York, 1992, 65–71, 199–201.
22. U.S. EPA, *Handbook — Remedial Action at Waste Disposal Sites (Revised)*, EPA/625/6-85/006, Office of Emergency and Remedial Response, Washington, D.C., October 1985, 5:30–36.
23. Powers, J. P., *Construction Dewatering — New Methods and Applications*, 2nd ed., John Wiley & Sons, New York, 1992, 3–19.
24. U.S. EPA, *Handbook — Remedial Action at Waste Disposal Sites (Revised)*, EPA/625/6-85/006, Office of Emergency and Remedial Response, Washington, D.C., October 1985, 5:9.
25. Powers, J. P., *Construction Dewatering — New Methods and Applications*, 2nd ed., John Wiley & Sons, New York, 1992, 430.
26. U.S. EPA, *Handbook — Remedial Action at Waste Disposal Sites (Revised)*, EPA/625/6-85/006, Office of Emergency and Remedial Response, Washington, D.C., October 1985, 5:46–73.
27. Homma, F., A viscous fluid model for demonstration of groundwater flow to parallel drains, Bulletin 10, International Institute for Land Reclamation and Improvement, Wageningen, The Netherlands, 1983, 15–16.
28. Powers, J. P., *Construction Dewatering — New Methods and Applications*, 2nd ed., John Wiley & Sons, New York, 1992, 101–102.
29. Cohen, R. M., Vincent, A. H., Mercer, J. W., Faust, C. R., and Spalding, C. P., *Methods for Monitoring Pump-and-Treat Performance*, EPA/600/R-94/123, U.S. EPA, Office of Research and Development, Ada, OK, June 1994, 13–73.
30. Environmental Sciences and Engineering, *Technological Limits of Groundwater Remediation: a Statistical Evaluation Method*, API Publication Number 4510, American Petroleum Institute, June 1991, 1–5.

31. Mercer, J. W., Skipp, D. C., and Griffen, D., *Basics of Pump-and-Treat Ground-Water Remediation Technology*, EPA-600/8-90/003, U.S. EPA, Office of Research and Development, Robert S. Kerr Environmental Research Laboratory, Ada, OK, 1990, 16.

32. U.S. EPA, *Evaluaton of Ground-Water Extraction Remedies: Phase II, Volume 1 — Summary Report*, Publication 9355.4-05, Office of Emergency and Remedial Response, Washington, D.C., February 1992.

33. National Research Counsel, *Alternatives for Ground Water Cleanup*, Prepublication Copy, National Academy Press, Washington, D.C., June 1994, 83–88.

34. U.S. EPA, *Final Guidance: Presumptive Response Strategy and Ex-Situ Treatment Technologies for Contaminated Groundwater at CERCLA Sites*, EPA/540/R-96/023, Office of Emergency and Remedial Response, Washington, D.C., Pre-Publication Copy, October 1996, A-8.

35. Powers. J. P., *Construction Dewatering — New Methods and Applications*, 2nd ed., John Wiley & Sons, New York, 388.

36. American Public Health Association, American Public Works Association, and Water Environment Federation, *Standard Methods for the Examination of Water and Wastewater*, 18th ed., Greenberg, A. E., Clesceri, L. S., Eaton, A. D., and Franson, M. A. H., Eds., American Public Health Association, Washington, D.C., 1992.

37. *Recommended Standards for Water Works, Great Lakes Upper Mississippi River Board of State Public Health & Environmental Managers*, Health Research Inc., Albany, New York, 1987.

38. Nyer, E. K., *Groundwater Treatment Technology*, Van Nostrand Reinhold, 1985, 117.

39. American Society of Civil Engineers and American Water Works Association, *Water Treatment Plant Design*, 2nd ed., McGraw-Hill, New York, 1990, 326–328.

40. U.S. EPA, *Handbook — Remedial Action at Waste Disposal Sites (Revised)*, EPA/625/6-85/006, Office of Emergency and Remedial Response, Washington, D.C., October 1985, 10:23–25.

41. Viessman, W. and Hammer, J., *Water Supply and Pollution Control*, 5th ed., Harper Collins College Publishers, New York, 1993, 651–652.

42. Weber, W. J., *Physicochemical Processes for Water Quality Control*, John Wiley & Sons, New York, 1972, 128–130.

43. Nyer, E. K., *Groundwater Treatment Technology*, Van Nostrand Reinhold, 1985, 120–123.

44. Viessman, W. and Hammer, J., *Water Supply and Pollution Control*, 5th ed., Harper Collins College Publishers, New York, 1993, 697–699.

45. Weber, W. J., *Physicochemical Processes for Water Quality Control*, John Wiley & Sons, New York, 1972, 163–164.

46. Johnson Division, Microfloc® Products, Gravity Filters — Complete Filtration Systems, sales sheet by the Johnson Division, 1987, St. Paul, MN.

47. U.S. EPA, *Superfund Innovative Technology Evaluation Program, Technology Profiles*, 9th ed., EPA/540/R-97/502, Office of Research and Development, Washington, D.C., December 1996, 54–55, 66–67.

48. U.S. EPA, *Engineering Bulletin: Air Stripping of Aqueous Solutions*, EPA/540/2-91/022, Center for Environmental Research Information, Cincinnati, OH, October 1991, 2.

49. Nyer, E. K., *Groundwater Treatment Technology*, Van Nostrand Reinhold Company, New York, 1985, 49–50.

50. U.S. EPA, *Handbook — Remedial Action at Waste Disposal Sites (Revised)*, EPA/625/6-85/006, Office of Emergency and Remedial Response, Washington, D.C., October 1985, 10:50.

51. Treybal, R. E., *Mass-Transfer Operations*, 3rd ed., McGraw-Hill, New York, 1980, 200–201.
52. Treybal, R. E., *Mass-Transfer Operations,* 3rd ed., McGraw-Hill, New York, 1980, 300–313.
53. Fair, J. R., Steinmeyer, D. E., and Crocker, B. B., Liquid-Gas Systems, in *Perry's Chemical Engineers' Handbook*, 6th ed., Perry, R. H., Green, D. W., and Maloney, J. O., Eds., McGraw-Hill, New York, 1984, 18:23.
54. U.S. EPA, *Air Stripping of Contaminated Water Sources — Air Emissions and Controls*, EPA-450/3-87-017, Control Technology Center, Research Triangle Park, NC, August 1987, 3:9, 4:7, 6:9–12.
55. U.S. EPA, *Fact Sheet — Verona Well Field*, as of March 1996, downloaded May 14, 1997 from http://www.epa.gov.
56. Treybal, R. E., *Mass-Transfer Operations*, 3rd ed., McGraw-Hill, New York, 1980, 158–185.
57. Treybal, R. E., *Mass-Transfer Operations*, 3rd ed., McGraw-Hill, New York, 1980, 140–145.
58. Marks, P. J., Wujcik, W. J., and Loncar, A. F., *Remediation Technologies Screening Matrix and Reference Guide*, 2nd ed., NTIS PB95-104782, U.S. Army Environmental Center, Aberdeen Proving Ground, MD, October 1994, 4:178.
59. Greg Leininger of Calgon Carbon Corporation, personal communication, August 28, 1991.
50. North East Environmental Products, VOC and Radon Removal from Water, Shallow Tray Aeration Systems, West Lebanon, NH, undated.
61. American Society of Civil Engineers/American Water Works Association, *Water Treatment Plant Design*, 2nd ed., McGraw-Hill, New York, 1990, 288, 295, 302.
62. American Society of Civil Engineers/American Water Works Association, *Water Treatment Plant Design*, 2nd ed., McGraw-Hill, New York, 1990, chap. 11.
63. Baker, M. N., *The Quest for Pure Water: the History of Water Purification from the Earliest Records to the Twentieth Center*, The American Water Works Association, New York, 1948, 1.
64. Corapcioglu, M. O. and Huang, C. P., The adsorption of heavy metals onto hydrous activated carbon, *Water Res.*, 21, 9, 1031–1044, 1987.
65. Huang, C. P., *The Removal of Heavy Metals by Activated Carbon Process from Water and Wastewater in the Absence of Complex Formation*, Seminar on Adsorption, U.S. EPA, Office of Exploratory Research, EPA/600/X-85/122, June 1985.
66. U.S. EPA, *Capsule Report: Aqueous Mercury Treatment*, EPA/625/R-97/004, Office of Research and Development, Washington, D.C., July 1997, 3:1–3.
67. Bernardin, F. E., Cyanide detoxification using adsorption and catalytic oxidation on granular activated carbon, *Water Pollut. Control Fed.*, 45, 2, 221–231, 1973.
68. Lehr, J. H., Granular-Activated Carbon (GAC): Everyone Knows of It, Few Understand It, *Groundwater Monitoring Review*, Fall 1991, 5–8.
69. Tchobanoglous, G. and Burton, F. L., *Wastewater Engineering: Treatment, Disposal, and Reuse/Metcalf & Eddy, Inc.*, 3rd ed. revised, McGraw-Hill, New York, 1991, 315–317.
70. Carbtrol Corporation, Sales sheet, Carbtrol® Water Purification Canister, 200 lb Activated Carbon, L-1, March 23, 1995.
71. Calgon Carbon Corporation, Sales sheet, DISPOSORB, July 1989.
72. Carbtrol Corporation, Sales sheet, Carbtrol® Water Purification Absorbers, Model L-5, June 18, 1996.

73. U.S. EPA, *Engineering Bulletin: Granular Activated Carbon Treatment*, EPA/540/2-91/024, Office of Emergency and Remedial Response, Washington, D.C., October 1991, 2–4.

74. U.S. EPA, *Treatability Database, Version 5.0*, Risk Reduction Engineering Laboratory, Cincinnati, OH, April 1994.

75. Kim Friedman, Calgon Carbon Corporation, personal communication, October 15, 1991.

76. U.S. EPA, *Emerging Technology Summary: Demonstration of* Ambersorb® 563 Adsorbent *Technology*, EPA/540/SR-95/516, National Risk Management Research Laboratory, Cincinnati, OH, 5 pp.

77. Carbtrol Corporation, Sales sheet, Clay/Anthracite Canister, Granular Petroleum Adsorbent CA-1, July 16, 1992.

78. Calgon Carbon Corporation, Klensorb 100™ Application Bulletin, undated.

79. Hairston, D. H., Zealous Zeolites, *Chem. Eng.*, July 1996, 57–60.

80. U.S. EPA, *Dynaphore, Inc. Forager™ Sponge Technology: Innovative Technology Evaluation Report*, EPA/540/R-94/522, Office of Research and Development, Washington, D.C., June 1995, 76 pp.

81. U.S. EPA, *Emerging Technology Summary: Removal and Recovery of Metal Ions from Groundwater*, EPA/540/55-90/005, RREL, Cincinnati, OH, 1990.

82. Miller, S. A., Ambler, C. M., Bennett, R. C., Dahlstrom, D. A., Darji, J. D., Emmett, R. C., Gray, J. B., Gurnham, C. F., Jacobs, L. J., Klepper, R. P., Michalson, A. W., Oldshue, J. Y., Silverblatt, C. E., Smith, J. C., and Todd, D. B., Liquid solid systems, *Perry's Chemical Engineers' Handbook*, 6th ed., Perry, R. H., Green, D. W., and Maloney, J. O., Eds., McGraw-Hill, New York, 1984, 19:41.

83. Nyer, E. K., *Groundwater Treatment Technology*, Van Nostrand Reinhold, New York, 1985, 129.

84. Marks, P. J., Wujcik, W. J., and Loncar, A. F., *Remediation Technologies Screening Matrix and Reference Guide*, 2nd ed., NTIS PB95-104782, U.S. Army Environmental Center, Aberdeen Proving Ground, MD, October 1994, 4:185.

85. Henry, J. D., Corder, W., Winston, W. S., Hoglund, R. L., Lemlich, R. L., Li, N. N., Moyers, C. G., Newman, J., Pohl, H. A., Pollock, K., Prudich, M. E., Spiegler, K. S., and Von Halle, E., Novel separation processes, *Perry's Chemical Engineers' Handbook*, 6th ed., Perry, R. H., Green, D. W., and Maloney, J. O., Eds., McGraw-Hill, New York, 1984, 17:22–27.

86. Tchobanoglous, G. and Burton, F. L., *Wastewater Engineering: Treatment, Disposal and Reuse*, 3rd ed., Metcalf & Eddy, Inc., McGraw-Hill, New York, 1991, 403.

87. Grady, C. P. L. and Lim, H. C., *Biological Wastewater Treatment: Theory and Applications*, Marcel Dekker, New York, 1980, 9–12.

88. Nyer, E. K., *Groundwater Treatment Technology*, Van Nostrand Reinhold, New York, 1985, 84–85.

89. Tchobanoglous, G. and Burton, F. L., *Wastewater Engineering: Treatment, Disposal and Reuse*, 3rd ed., Metcalf & Eddy, Inc., McGraw-Hill, New York, 1991, 423–424.

90. Tchobanoglous, G. and Burton, F. L., *Wastewater Engineering: Treatment, Disposal and Reuse*, 3rd ed., Metcalf & Eddy, Inc., McGraw-Hill, New York, 1991, 368–373.

91. Grady, C. P. L. and Lim, H. C., *Biological Wastewater Treatment: Theory and Applications*, Marcel Dekker, New York, 1980, 321.

92. Grady, C. P. L. and Lim, H. C., *Biological Wastewater Treatment: Theory and Applications*, Marcel Dekker, New York, 1980, 299.

93. U.S. EPA, *Site Technology Capsule: ZenoGem™ Wastewater Treatment Process*, EPA/540/R-95/503a, Office of Research and Development, Cincinnati, OH, August 1995, 11 pp.

94. U.S. EPA, *Emerging Technology Summary: Pilot-Scale Demonstration of a Two-Stage Methanotrophic Bioreactor for Biodegradation of Trichloroethylene in Groundwater*, EPA/540/R-93/505, Office of Research and Development, Cincinnati, OH, August 1995, 5 pp.

95. U.S. EPA, *Biological Treatment of Wood Preserving Site Groundwater by BioTrol, Inc., Applications Analysis Report*, EPA/540/A5-91/001, Office of Research and Development, Cincinnati, OH, September 1991, 39 pp.

96. Allied Signal, *Technology Profile — Immobilized Cell Bioreactor (ICB)*, Allied Signal Inc., Morristown, NJ, undated, 2 pp.

97. Allied Signal, *Advanced Environmental Solutions — Immobilized Cell Bioreactor*, Allied Signal Inc., Morristown, NJ, undated, 4 pp.

98. U.S. EPA, *Emerging Technology Summary: Handbook for Constructed Wetlands Receiving Acid Mine Drainage*, EPA/540/SR-93/523, Office of Research and Development, Cincinnati, OH, September 1995.

99. U.S. EPA, *Ultrox International Ultraviolet Radiation/Oxidation Technology — Application Analysis Report*, EPA/540/A5-89/012, Office of Research and Development, Washington, D.C., September 1990, 11–15.

100. Smith, P. W., Rayox® — a Second Generation Enhanced Oxidation Process for the Destruction of Waterborne Organic Contaminants, presented at: Haz Mat South Conference, Atlanta, GA, 1991.

101. U.S. EPA, *Perox-pure™ Chemical Oxidation Technology, Peroxidation Systems Inc. — Application Analysis Report*, EPA/540/AR-93/501, Office of Research and Development, Washington, D.C., July 1993, 7–16.

102. Solarchem Environmental Systems, *The UV/Oxidation Handbook*, Solarchem Environmental Systems, Las Vegas, NV, 1994, 5:2–3.

103. Barberi, M., Minero, C., Pelizzetti, E., Borgarello, E., and Serpone, N., Chemical Degradation of Chlorophenols with Fenton's Reagent, *Chemosphere*, 16(10–12), 2225–2237, 1987.

104. GEO-CARE, INCORPORATED, THE GEOCLEANSE PROCESS®, An *In Situ* Oxidative Contaminant Reduction System for Remediation of Soil and Goundwater, Ramsey, NJ, 1997.

105. Monsen, R. M. and Sherman, B. D., *Pretreatment of a Creosoting Plant Wastewater with Hydrogen Peroxide*, presented at: Industrial Waste Symposia of the Water Pollution Control Federation 60th Annual Conference, Philadelphia, PA, October 1987.

106. Aronstein, B. N., Lawal, R. A., and Maka, A., Chemical Degradation of Cyanides by Fenton's Reagent in Aqueous and Soil-Containing Systems, *Environ. Toxicol. Chem.*, 13(11), 1719–1726, 1994.

107. Pardieck, D. L., Bouwer, E. J., and Stone, A. T., Hydrogen peroxide use to increase oxidant capacity for *in situ* bioremediation of contaminated soils and aquifers: a review, *J. Contaminant Hydrol.*, 9, 221–242, 1992.

108. Johnson, T. L. and Tratnyek, P. G., A Column Study of Geochemical Factors Affecting Reductive Dechlorination of Chlorinated Solvents by Zero-Valent Iron, in *Proceedings of the 33rd Hanford Symposium on Health & the Environment — In Situ Remediation: Scientific Basis for Current and Future Technologies*, Vol. 2, Batelle Press, Pasco, WA, 1994, 931–947.

109. Gillham, R. W., Blowes, E. W., Ptacek, C. J., and O'Hannesin, S. F., Use of Zero-Valent Metals in *In Situ* Remediation of Contaminated Ground Water, in *Proceedings of the 33rd Hanford Symposium on Health & the Environment — In Situ Remediation: Scientific Basis for Current and Future Technologies*, Vol. 2, Batelle Press, Pasco, WA, 1994, 913–930.

110. Gillham, R. W. and O'Hannesin, S. F., Enhanced Degradation of Halogenated Aliphatics by Zero–Valent Iron, *Ground Water*, 32(6), 958–967, 1994.

111. Chuang, F-W., Larson, R. A., and Wessman, M. S., Zero-Valent Iron-Promoted Dechlorination of Polychlorinated Biphenyls, *Environ. Sci. Technol.*, 29(9), 2460–2463, 1995.

112. U.S. EPA, *Superfund Innovative Technology Evaluation Program — Technology Profiles*, 9th ed., EPA/540/R-97/502, Office of Research and Development, Washington, D.C., December 1996, 64–65.

113. U.S. EPA, *SITE Demonstration Bulletin: Metal-Enhanced Abiotic Degradation Technology, EnviroMetal Technologies, Inc.*, EPA/540/MR-95/510, Center for Environmental Research Information, Cincinnati, OH, May 1995.

114. Schenck, R. C., President, CERCONA Inc., Dayton Ohio, personal communication, February 1996.

115. Kaplan, D. I., Cantrell, K. J., and Wietsma, T. W., Formation of a Barrier to Groundwater Contaminants by the Injection of Zero-Valent Iron Colloids: Suspension Properties, in *Proceedings of the 33rd Hanford Symposium on Health & the Environment — In Situ Remediation: Scientific Basis for Current and Future Technologies*, Vol. 2, Batelle Press, Pasco, WA, 1994, 821–837.

116. Orth, W. S. and Gillham, R. W., Dechlorination of Trichloroethene in Aqueous Solution Using Fe°, *Environ. Sci. Technol.*, 30(1), 66–71, 1996.

117. Johnson, T. L., Scherer, M. M., and Tratnyek, P. G., Kinetics of halogenated organic compound degradation by iron metal, *Environ. Sci. Technol.*, 30(8), 264–2460, 1996.

118. U.S. EPA, *In Situ Remediation Technology Status Report: Treatment Walls*, EPA 542-K-94-004, Office of Solid Waste and Emergency Response, Washington, D.C., April 1995, 26 pp.

119. U.S. EPA, *Use of Monitored Natural Attenuation at Superfund, RCRA Corrective Action, and Underground Storage Tank Sites, Interim Final*, OSWER Directive 9200.4-17, Office of Solid Waste and Emergency Response, Washington, D.C., November 1997, 31 pp.

120. Freeze, R. A. and Cherry, J. A., *Groundwater*, Prentice-Hall, Englewood Cliffs, NJ, 1979, 95–118.

121. U.S. EPA, *Handbook — Ground Water, Volume II: Methodology*, EPA/625/6-90/016b, Office of Research and Development, Washington, D.C., July 1991, 56–57.

122. Sewell, G. W. and Gibson, S. A., Microbial Ecology of Adaptation and Response in the Subsurface, in Proceedings of the Symposium on Natural Attenuation of Chlorinated Organics in Ground Water, Dallas, TX, September 11–13, 1996, EPA/540/R—96/509, September 1996, 14–16.

123. Freeze, R. A. and Cherry, J. A., *Groundwater*, Prentice-Hall, Englewood Cliffs, NJ, 1979, 245.

124. Weidemeier, T. H., Swanson, M. A., Moutoux, D. E., Wilson, J. T., Kampbell, D. H., Hansen, J. E., and Haas, P., Overview of the Technical Protocol for Natural Attenuation of Chlorinated Aliphatic Hydrocarbons in Ground Water under Development for the U.S. Air Force Center for Environmental Excellence, in Proceedings of the Symposium on Natural Attenuation of Chlorinated Organics in Ground Water, Dallas, TX, September 11–13, 1996, EPA/540/R-96/509, 35–59.

125. McCarty, P. L., Biotic and Abiotic Transformations of Chlorinated Solvents in Ground Water, in Proceedings of the Symposium on Natural Attenuation of Chlorinated Organics in Ground Water, Dallas, TX, September 11–13, 1996, EPA/540/R-96/509, September 1996, 5–9.

126. Gossett, J. M. and Zinder, S. H., Microbiological Aspects Relevant to Natural Attenuation of Chlorinated Ethenes, in Proceedings of the Symposium on Natural Attenuation of Chlorinated Organics in Ground Water, Dallas, TX, September 11–13, 1996, EPA/540/R-96/509, September 1996, 10–13.

127. U.S. EPA, *How to Evaluate Alternative Cleanup Technologies for Underground Storage Tank Sites: a Guide for Corrective Action Plan Reviewers*, EPA/510-B-94-003, September 1996, VII:6–31.

128. U.S. EPA, *How to Evaluate Alternative Cleanup Technologies for Underground Storage Tank Sites: a Guide for Corrective Action Plan Reviewers*, EPA/510-B-94-003, September 1996, XVII:9–10.

129. Marley, M. C. and Bruell, C. J., *In Situ Air Sparging: Evaluation of Petroleum Industry Sites and Considerations for Applicability, Design and Operation*, API Publication Number 4609, American Petroleum Institute, Washington, D.C., March 1995, 2:2–11.

130. Marley, M. C. and Bruell, C. J., *In Situ Air Sparging: Evaluation of Petroleum Industry Sites and Considerations for Applicability, Design and Operation*, API Publication Number 4609, American Petroleum Institute, Washington, D.C., March 1995, 5:2.

131. Marley, M. C. and Bruell, C. J., *In Situ Air Sparging: Evaluation of Petroleum Industry Sites and Considerations for Applicability, Design and Operation*, API Publication Number 4609, American Petroleum Institute, Washington, D.C., March 1995, 4:17–18.

132. Kueper, B. H., Water flooding and air sparging, Presented at: Short Course — DNAPL Site Characterization & Remediation, December 2–5, Boston, MA, University Consortium Solvents-in-Groundwater Research Program, course notes Waterloo Educational Services, Guelph, Ontario.

133. Hinchee, R. E., Air sparging state of the art, in *Air Sparging for Site Remediation*, Hinchee, R. E., Ed., Lewis Publishers, Boca Raton, FL, 1994, 1–12.

134. Cherry, J. A., Feenstra, S., and Mackay, D. M., Concepts for the Remediation of Sites Contaminated with Dense Non-Aqueous Phase Liquids (DNAPLs), in *Dense Chlorinated Solvents and other DNAPLs in Groundwater*, Pankow, J. F. and Cherry, J. A., Eds., Waterloo Press, Portland, OR, 1996, 491–492.

135. Pankow, J. F., Johnson, R. L., and Cherry, J. A., Air Sparging in Gate Wells in Cutoff Walls and Trenches for Control of Plumes of Volatile Organic Compounds (VOCs), *Ground Water*, 31(4), 654–663, 1993.

136. Marley, M. C. and Bruell, C. J., *In Situ Air Sparging: Evaluation of Petroleum Industry Sites and Considerations for Applicability, Design and Operation*, API Publication Number 4609, American Petroleum Institute, Washington, D.C., March 1995, 6:6–8.

137. Driscoll, F. G., *Groundwater and Wells*, 2nd ed., Johnson Division, St. Paul, MN, 1986, 811.

138. U.S. EPA, *Superfund Innovative Technology Evaluation Program — Technology Profiles*, 9th ed., EPA/540/R-97/502, Office of Research and Development, Washington, D.C., December 1996, 192–193.

139. SBP Technologies, Inc., UVB *In Situ* Technology Installations, Sales Brochure, 1996.

140. U.S. EPA, *SITE Technology Capsule-Unterdruck-Verdampfer-Brunnen Technology (UVB) Vacuum Vaporizing Well*, EPA/540R-95/500a, Office of Research and Development, Cincinnati, OH, July 1995, 2–6.

141. Herrling, B., Buermann, W., and Stamm, J., *In situ* remediation of volatile contaminants by a new system of "Vacuum-Vaporizer-Wells", in *Subsurface Contamination by Immiscible Fluids*, Weyer, K. U., Ed., A. A. Balkema, Rotterdam, 1991 (prepublication copy).

142. U.S. EPA, *Handbook — Remedial Actions at Waste Disposal Sites (Revised)*, EPA/625/6-85/006, Office of Emergency and Remedial Response, Washington, D.C., October 1985, 9:12–17.

143. U.S. EPA, *How to Evaluate Alternative Cleanup Technologies for Underground Storage Tank Sites: a Guide for Corrective Action Plan Reviewers*, EPA/510-B-94-003, September 1996, VII:14.

144. Leeson, A. and Hinchee, R., *Manual — Bioventing Principles and Practice, Volume 1: Bioventing Principles*, EPA/540/R-95/534a, U.S. EPA, Office of Research and Development, Washington, D.C., September 1995, 23–24.

145. Leeson, A. and Hinchee, R., *Manual — Bioventing Principles and Practice: Volume 1: Bioventing Principles*, EPA/540/R-95/534a, U.S. EPA, Office of Research and Development, Washington, D.C., September 1995, 3–4.

146. Pardieck, D. L., Bouwer, E. J., and Stone, A. T., Hydrogen peroxide use to increase oxidant capacity for *in situ* bioremediation of contaminated soils and aquifers: a review. *Contaminant Hydrol.*, 9, 221–242, 1992.

147. U.S. EPA, *How to Evaluate Alternative Cleanup Technologies for Underground Storage Tank Sites: a Guide for Corrective Action Plan Reviewers*, EPA/510-B-94-003, September 1996, VIII:11-15, 34.

148. Leeson, A. and Hinchee, R., *Manual — Bioventing Principles and Practice, Volume 1: Bioventing Principles*, EPA/540/R-95/534a, U.S. EPA, Office of Research and Development, Washington, D.C., September 1995, 18–21.

149. U.S. EPA, *Superfund Innovative Technology Evaluation Program, Technology Profiles*, 9th ed., EPA/540/R-97/502, Office of Research and Development, Washington, D.C., December 1996, 246–247, 336–337.

150. Murray, W., Dooley, M., Belcher, D., Odell, K., Johnson, J., Dogon, A., and Hardin, J., *Bioremediation of a Chlorinated Solvent Plume under an Operating Manufacturing Plant*, Presented at: 1997 Technical Seminar: Renewal, Rehabilitation, and Upgrades in Civil and Environmental Engineering, Maine Section ASCE, March 19, 1997, Lewiston, ME.

151. Edwards, R. W., Duster, D., Faile, M., Gallant, W., Gibeau, E., Myller, B., Nevling, K., and O'Grady, B., Preliminary Performance Results From a Zero-Valent Metal Reactive Wall for the Passive Treatment of Chlorinated Organic Compounds in Groundwater, copy provided by Richard Muza, U.S. EPA, Region VIII, Denver, CO, August 1997.

152. Muza, R., *Reactive Wall Demonstration Project, Lowry Air Force Base, Colorado*, presentation materials provided by R. Muza, U.S. EPA, Region VIII, Denver, CO, August 1997.

153. Newell, C. J., Acree, S. D., Ross, R. R., and Huling, S. G., *Ground Water Issue: Light Nonaqueous Phase Liquids*, EPA/540/S-95/500, U.S. EPA, Office of Solid Waste and Emergency Response, Washington, D.C., 1995, 14–23.

154. QED Environmental Systems Inc.™, *Tough Pumps: Landfill and Cleanup Pumping Equipment Catalog*, Ann Arbor, MI, June 1997, 14–15.

155. U.S. EPA, *How to Evaluate Alternative Cleanup Technologies for Underground Storage Tank Sites: a Guide for Corrective Action Plan Reviewers*, EPA/510-B-94-003, September 1996, XI:22–27.

156. U.S. Army Corps of Engineers, *Engineering and Design, Soil Vapor Extraction and Bioventing*, EM 1110-1-4001, Washington, D.C., 30 November 1995, 4:4.

157. West, C. A. and Harwell, J. H., Surfactants and subsurface remediation, *Environ. Sci. Technol.*, 26(12), 2324–2330, 1992.

158. Kueper, B. H., Water flooding and air sparging, Presented at: Short Course — DNAPL Site Characterization & Remediation, December 2–5, Boston, MA, University Consortium Solvents-in-Groundwater Research Program, course notes. Waterloo Educational Services, Guelph, Ontario.

159. Kueper, B., Pitts, M., Wyatt, K., Simkin, T., and Sale, T., *Technology Practices Manual for Surfactants and Cosolvents*, AATDF Report TR-97-2, The Advanced Applied Technology Demonstration Facility Program, Rice University, Houston, TX, prepared for the U.S. Department of Defense, February 1997, 3:7.

160. Kueper, B., Pitts, M., Wyatt, K., Simkin, T., and Sale, T., *Technology Practices Manual for Surfactants and Cosolvents*, AATDF Report TR-97-2, The Advanced Applied Technology Demonstration Facility Program, Rice University, Houston, TX, prepared for the U.S. Department of Defense, February 1997, 4:2–12.

161. U.S. EPA, Dense Nonaqueous Phase Liquids — a Workshop Summary, Dallas, TX, April 16–18, 1991, EPA/600/R-92/030, Robert S. Kerr Environmental Research Laboratory, Ada, OK, 10–11.

162. Longino, B. L. and Kueper, B. H., The use of upward gradients to arrest downward migration in the presence of solubilizing surfactants, *Can. Geotech. J.*, 32(2), 296–308, 1995.

163. Kueper, B. H., Chown, J. C., and McWhorter, D. B., The Use of Upward Hydraulic Gradients to arrest Downward DNAPL Migration in Fractures, Presented at: GEOENVIRONMENT 2000, Characterization, Contamination, Remediation and Performance in Environmental Geotechnics, an ASCE Specialty Conference, February 22–24, 1995, New Orleans, LA.

164. Kueper, B., Pitts, M., Wyatt, K., Simkin, T., and Sale, T., *Technology Practices Manual for Surfactants and Cosolvents*, AATDF Report TR-97-2, The Advanced Applied Technology Demonstration Facility Program, Rice University, Houston, TX, prepared for the U.S. Department of Defense, February 1997, 4:31.

165. Vigon, B. W. and Rubin, A. J., Practical considerations in the surfactant-aided mobilization of contaminants in aquifers, *J. Water Pollut. Control Fed.*, 61(7), 1233–1240, 1989.

166. Fountain, J. and Waddell-Sheets, C., A Pilot Scale Test of Surfactant Pump and Treat, presented at the 86th Annual Meeting and Exhibition, Air & Waste Management Association, Denver, CO, June 13–18, 1993.

167. Wunderlich, R. W., Fountain, J. C., and Jackson, R. E., *In Situ* Remediation of Aquifers Contaminated with Dense Nonaqueous Phase Liquids by Chemically Enhanced Solubilization, *J. Soil Contamination*, 1(4), 361–378, 1992.

168. Pitts, M. J., Wyatt, K., Sale, T. C., and Piontek, K. R., Utilization of Chemically-Enhanced Oil Recovery Technology to Remove Hazardous Oily Waste from Alluvium, SPE25153, Presented at: SPE International Symposium on Oilfield Chemistry, New Orleans LA, March 2–5, 1993.

169. Sabatini, D. A., Knox, R. C., and Harwell, J. H., *Surfactant-Enhanced DNAPL Remediation: Surfactant Selection, Hydraulic Efficiency, and Economic Factors*, EPA/600/S-96/002, U.S. EPA, Risk Reduction Research Laboratory, Ada, OK, August 1996, 15 pp.

170. Kimball, S. L., The use of surfactants to enhance pump-and-treat processes for *in situ* soil remediation, in *Remediation of Hazardous Waste Contaminated Soils,* Wise, D. L. and Trantolo, D. J., Eds., Marcel Dekker, New York, 1994, chap. 4.

171. Brown, M. J. and Burris, D. R., Enhanced Organic Contaminant Sorption on Soil Treated with Cationic Surfactants, *Ground Water*, 34(4), 734–744, 1996.

172. Kueper, B., Pitts, M., Wyatt, K., Simkin, T., and Sale, T., *Technology Practices Manual for Surfactants and Cosolvents*, AATDF Report TR-97-2, The Advanced Applied Technology Demonstration Facility Program, Rice University, Houston, TX, prepared for the U.S. Department of Defense, February 1997, 2:4.

173. Kueper, B. H., Surfactant and alcohol flooding for DNAPL removal, Presented at: Short Course — DNAPL Site Characterization & Remediation, December 2–5, Boston, MA, University Consortium Solvents-in-Groundwater Research Program, course notes. Waterloo Educational Services, Guelph, Ontario.

174. Kueper, B., Pitts, M., Wyatt, K., Simkin, T., and Sale, T., *Technology Practices Manual for Surfactants and Cosolvents*, AATDF Report TR-97-2, The Advanced Applied Technology Demonstration Facility Program, Rice University, Houston, TX, prepared for the U.S. Department of Defense, February 1997, chap. 6.

175. Kueper, B., Pitts, M., Wyatt, K., Simkin, T., and Sale, T., *Technology Practices Manual for Surfactants and Cosolvents*, AATDF Report TR-97-2, The Advanced Applied Technology Demonstration Facility Program, Rice University, Houston, TX, prepared for the U.S. Department of Defense, February 1997, 4:22–25.

176. Kueper, B. H., Steam flooding, Presented at: Short Course — DNAPL Site Characterization & Remediation, December 2–5, Boston, MA, University Consortium Solvents-in-Groundwater Research Program, course notes. Waterloo Educational Services, Guelph, Ontario.

177. Moeller, J. W., Gerrish, J. A., and Johnson, L. A., *Implementation of a Full-Scale Enhanced Recovery System for Coal Tar at a Former MGP Superfund Site*, in *Contaminated Soils*, Vol. 1, Proceedings of the 10th Annual Conference on Contaminated Soils, University of Massachusetts, Amherst, MA, October 1995, Calabrese, E. J., Kostecki, P. T., and Bonazountas, M., Eds., Amherst Scientific Publishers, Amherst, MA, 1996, 59–82.

178. U.S. EPA, Brodhead Creek Stroudsberg, Monroe County, PA, Fact Sheet, undated — downloaded from http://www.epa.gov on May 22, 1997.

179. Environmental Research and Technology, Inc. and Koppers Company, Inc., *Handbook on Manufactured Gas Plant Sites*, prepared for Utility Solid Waste Activities Group, Superfund Committee, Washington, D.C., September 1984, 4:63–71, 7:64–73.

180. Leuschner, A. P., Moeller, M. W., Gerrish, J. A., and Johnson, L. A., Case study: MGP site remediation using enhanced DNAPL recovery, in *Contaminated Soils,* Vol. 2, Proceedings of the 11th Annual Conference on Contaminated Soils, University of Massachusetts, Amherst, MA, October 1996, Calabrese, E. J., Kostecki, P. T., and Bonazountas, M., Eds., Amherst Scientific Publishers, Amherst, MA, 1997, 607–620.

181. U.S. EPA, *Handbook — Remedial Action at Waste Disposal Sites (Revised)*, EPA/625/6-85/006, Office of Emergency and Remedial Response, Washington, D.C., October 1985, 8:52–54.

182. Haselow, J. S., Syaler, G., and Wilson, J., *In Situ Treatment Technologies for Dense Non-Aqueous Phase Liquids*, Report prepared for Battelle Pacific Northwest Laboratories, June 1996, 14–21.

183. Cherry, J. A., *In situ* destruction of solvent DNAPL: chemical and biochemical examples, Presented at Short Course — DNAPL Site Characterization & Remediation, December 2–5, Boston, MA, University Consortium Solvents-in-Groundwater Research Program, course notes. Waterloo Educational Services, Guelph, Ontario.

184. GEO-CARE, INC., The Geo-Cleanse® Process — an *In Situ* Oxidative Contaminant Reduction System for Remediation of Soil and Groundwater, Ramsey, NJ, 1997.

185. Wilson, J. T., Remediation apparatus and method for organic contamination in soil and groundwater, U.S. Patent 5,525,008, June 11, 1996.

186. Wilson, J. T., Remediation apparatus and method for organic contamination in soil and groundwater, U.S. Patent 5,611,642, March 18, 1997.

187. Jerome, K. M., Riha, B., and Looney, B. B., *Final Report for Demonstration of In Situ Oxidation of DNAPL Using the Geo—Cleanse Technology*, WSRC-TR-97-00283, U.S. Department of Energy, Office of Technology Development, September 19, 1997, 23 pp.

188. Vella, P. A. and Veronda, B., *Oxidation of Trichloroethylene: a Comparison of Potassium Permanganate and Fenton's Reagent*, presented at the Third International Symposium on Chemical Oxidation, Technology for the Nineties, Vanderbilt University, Nashville, TN, 1994.

189. U.S. EPA, *Handbook — Control Technologies for Hazardous Air Pollutants*, EPA/625/6-91/014, Office of Research and Development, Washington, D.C., June 1991, 4:3.

190. Himmelblau, D. M., *Basic Principles and Calculations in Chemical Engineering*, 3rd ed., Prentice-Hall, Englewood Cliffs, NJ, 1974, 207.

191. Fair, J. R., Steinmeyer, D. E., Penney, W. R., and Crocker, B. B., Liquid-gas systems, in *Perry's Chemical Engineers' Handbook*, 6th ed., Green, D. W. and Maloney, J. O., Eds., McGraw-Hill, New York, 18:73–79.

192. U.S. EPA, *Soil Vapor Extraction Technology — Reference Handbook*, EPA/540/2-91/003, February 1991, 109.

193. U.S. EPA, *Handbook — Control Technologies for Hazardous Air Pollutants*, EPA/625/6-91/014, Office of Research and Development, Washington, D.C., June 1991, 3:2–3; 4:55–64.

194. U.S. EPA, *Manual: Bioventing Principles and Practice*, Vol. II, Bioventing Design, EPA/625/XXX/001, Office of Research and Development, Washington, D.C., September 1995, 74–75.

195. U.S. Army Corps of Engineers, *Engineering and Design — Soil Vapor Extraction and Bioventing*, EM 1110-4001, Washington, D.C., 30 November 1995, 5:49–51.

196. U.S. EPA, *Handbook — Control Technologies for Hazardous Air Pollutants*, EPA/625/6-91/014, Office of Research and Development, Washington, D.C., June 1991, 4:28–30.

197. Carbtrol® Corporation, Air Purification Canisters, 140–200 lb Activated Carbon, Sales sheet AT-116/#1, May 6, 1994.

198. Calgon Carbon Corporation, High FLow VentSorb® Emission Control Units, Bulletin 27-174d, February 1990.

199. Carbtrol® Corporation, Air Purification Adsorbers, 1,000–3,000 lb Activated Carbon, Models G-4, G-6, G-9, Sales sheet AT-411/#1, June 20, 1996.

200. Carbtrol® Corporation, Air Purification Adsorbers, 1,600 lb Activated Carbon G-7, 3,000 lb. Activated Carbon G-8, Sales sheet AT-415/#2, June 27, 1996.

201. Marks, P. J., Wujcik, W. J., and Loncar, A. F., *Remediation Technologies Screening Matrix and Reference Guide*, 2nd ed., NTIS PB95-104782, U.S. Army Environmental Center, Aberdeen Proving Ground, MD, October 1994, 4:224.

202. van der Vaart, D. R., Vatvuk, W. M., and Wehe, A. H., Thermal and catalytic incinerators for the control of VOCs, *J. Air Waste Manage.*, 41(1), 92–98, 1991.

203. U.S. EPA, *Handbook-Control Technologies for Hazardous Air Pollutants*, EPA/625/6-91/014, Office of Research and Development, Washington, D.C., June 1991, 4:10–13.

204. U.S. EPA, *Emerging Technology Bulletin: Bioscrubber, Aluminum Company of America*, EPA/540/F-93/507, July 1993.

CHAPTER 4

Soil Remediation

Soils, sludges, or sediments are typically remediated to:

- reduce the potential risk to human health or the ecosystem;
- reduce a source of groundwater contamination (so-called *source control*), or, less frequently, to surface water; and/or
- meet land disposal restrictions [40 CFR 268] before disposal in a landfill.

Options for remediation include containment, landfilling, treatment in place, or excavation and treatment. Remediation technologies serve to immobilize contaminants, separate them from the soil, or destroy them. Some technologies accomplish two or more of these functions, depending on the type of contaminants present.

Although Superfund sites represent only a fraction of the hazardous waste sites in the U.S., the Superfund program maintains the most comprehensive data bases on the use of remediation technologies. Those data indicate the trends in the types of technologies used to remediate soils. Table 4.1 lists technologies commonly used for soil remediation under Superfund and shows how the use of those technologies has changed over time. Early in the Superfund program, most soils were capped, excavated, and landfilled, or treated by solidification/stabilization or incineration. The number and type of available treatment technologies grew rapidly in response to the number of sites requiring remediation and regulatory mandates under CERCLA and RCRA which emphasized treatment over containment. Soil vapor extraction, thermal desorption, and bioremediation are now commonly used technologies under Superfund and other federal programs.[1]

This chapter discusses excavation and materials handling (Section 4.1), immobilization technologies (Section 4.2), separation technologies (Section 4.3), destruction technologies (Section 4.4), and air-pollution control technologies commonly used in soil treatment systems (Section 4.5).

4.1 SOIL EXCAVATION AND MATERIALS HANDLING

The logistics of excavating soil can drive the costs and schedule of remediation. Those logistics can be one of the reasons for choosing *in situ* (in-place) remediation

Table 4.1   Trends in Soil Remediation Based on U.S. EPA Records
of Decision

Category	Type of remediation	Percent of RODs in that year[a]			Total number of RODs 1982–1995
		1985	1990	1995	
Immobilization	Containment/disposal	60	28	41	383
	Solidification/stabilization	4	19	4	206
Separation	Soil washing	0	4	0	9
	Soil flushing	2	1	0	16
	Soil vapor extraction	4	13	8	139
	Thermal desorption	2	6	6	50
Destruction	Incineration	12	18	9	168
	Bioremediation	0	5	7	69
Other treatment/not specified		5	5	19	30
Other actions (e.g., institutional controls, monitoring, relocation)		12	2	6	34
Total number of source control ROD (not including "no action")		57	125	117	1104

[a] Totals may sum to more than 100% due to rounding.

From U.S. EPA, *Innovative Treatment Technologies: Annual Status Report,* 8th ed., EPA-542-R-96-010, Number 8, Office of Solid Waste and Emergency Response, Washington, D.C., November 1996, 6–11, A-1.

over excavation and disposal or treatment. Large excavation jobs require careful planning and flexible field management to respond to unexpected conditions.

4.1.1   Planning

A large soil excavation at a hazardous waste site requires extensive planning. Design drawings show the extent of excavation and the final grades to be achieved after remediation is complete. Technical specifications describe the materials and methods that the contractor must use. Excavation plans for hazardous waste sites also commonly include dewatering plans, if saturated soil must be excavated; health and safety plans, as required by OSHA; and sampling plans, to ensure that the work meets clean-up goals. Factors such as the need for dewatering, health and safety precautions, and the adequacy of existing data must be considered even in the preliminary stages of a feasibility study or conceptual design. Small excavation jobs require less rigorous planning, although the engineer in charge must consider the same factors.

The first step in planning an excavation is to define the area and volume to be excavated based on the available data and the clean-up levels. Volume and cost estimates must account for the measures taken to prevent an excavation from caving in. When contaminated soil deeper than 5 ft below the ground surface must be excavated, federal regulations on worker protection [29 CFR 1926, Subpart P] require that vertical side walls must be shored to prevent collapse, the sides must be benched (excavated in "stair steps"), or the side slopes must be graded at a secure angle. Those slopes are typically as steep as possible to minimize the volume of clean soil which must be excavated. Table 4.2 indicates the maximum slopes for various types of soil.

Table 4.2   General Guidelines for Soil Excavation

Soil type[2]	Maximum allowable slope, simple slope excavation to 20 ft.[a] (horizontal:vertical)[2]	Soil	Bank density[b] (ton/bcy)[3]	Swell factor[b] (%)[3]
Type C: granular soils including gravel, sand, loamy sand; or submerged soil	1½:1	Sand (dry)	1.44	11
		Gravel (dry)	1.51	15
Type B: granular cohesionless soils including angular gravel (similar to crushed rock), silt, silt loam, sandy loam	1:1	Silt	1.62	36
		Earth, loam (damp)	1.685	40
Type A: cohesive soils such as clay, silty clay, clay loam	¾:1	Clay (damp)	1.675	40

[a] The design of side slopes depends on several engineering and regulatory considerations not reflected in this summary table. See federal regulations at 29 CFR 1926.652, and Appendix A and Appendix B to Subpart P.

[b] These values are approximate. Reported values for these parameters vary. Church[3] notes that average values of soil density and swell factor for a specific type of earthen material are subject to a variation of ±10% and 33%, respectively. (This variation is a multiplier, not a percentage to be added or subtracted from the value of the swell factor.) For values for other types of earth materials, see, for example, the *Excavation Handbook*.[3]

The volume of soil is calculated in *cubic yards*, abbreviated *cy* or simply *yards*. The *bank measure*, also called *bank yards* or abbreviated *bcy*, is the in-place volume of soil and is calculated from the geometry of the excavation. Soil expands or *swells* upon excavation to a volume designated by *loose yards*, sometimes abbreviated *lcy*. Table 4.2 lists the swell factors for common types of soil. The bank measure is increased by the swell factor to calculate the loose cubic yards. Cost estimates must factor in the swell so as not to underestimate the volume of soil requiring disposal or treatment and the associated cost.

Example 4.1:   Soil in a former tank farm is contaminated over an area roughly 50 ft by 50 ft to the depth of the water table, approximately 10 ft. (The contamination did not spread laterally from the surface before it reached the water table.) The ground surface is level.

Sketch an excavation plan. Calculate the total volume to be excavated and the fraction of that volume which will be uncontaminated soil. Assume that the walls of the excavation will be sloped at 1.5:1.

Figure 4.1 shows the dimensions of the excavation. The volume of soil to be excavated is commonly estimated using software associated with Computer-Aided Design packages used to draft excavation plans. Alternatively, the volume can be estimated "by hand" using one of several methods.[4] The calculations can become quite complicated when the existing ground surface is uneven and/or the depth and lateral extent of contaminated soil to be excavated varies irregularly. In the *subdivision method*, the total volume to be excavated is divided into a series of smaller geometric solids. The volume of each of the smaller solids is calculated, and then the volumes summed to estimate the total volume to be excavated. The volume shown in Figure 4.1 can be

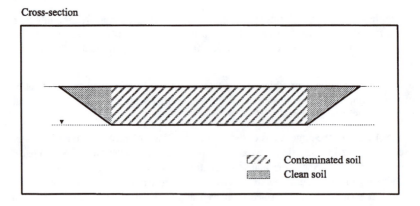

Plan view

Top of excavation

Bottom of excavation

Cross-section

Contaminated soil
Clean soil

0   10'

Figure 4.1   Example 4.1: dimensions of excavation.

subdivided into a rectangular prism, four wedges (along the walls), and four pyrimids (at the corners):

Rectangular prism:	$50' \times 50' \times 10' =$	25,000 cf
Wedges:	$4 \times \{(50' \times 15' \times 10')/2\} =$	15,000 cf
Pyrimids:	$4 \times \{\{Area \times Height\}/3\} =$	
	$4 \times ((15' \times 15' \times 10')/3) =$	3,000 cf
	TOTAL:	40,000 cf

The total volume to be excavated according to this plan is 1481 bcy, say 1500 bcy. Of that volume, approximately 45% is uncontaminated.

In uncohesive soil, the uncontaminated soil can be difficult to segregate from contaminated soil. Disposal or treatment of the uncontaminated soil with the remainder as hazardous waste would add significantly to the cost of the project. In some cases, it is more economical to shore the excavation than to slope the sides. Shoring adds to the cost of excavation, but can limit the volume of soil for eventual treatment or disposal and the associated cost.

Usually excavations at hazardous waste sites extend to — not below — the water table. Soil beneath the water table is often contaminated as a result of groundwater contamination, and is not itself source material that must be removed. However, decision makers sometimes choose to excavate saturated soil for several reasons, e.g.,

- Excavation sometimes extends to the seasonal groundwater low to remove the maximum volume of unsaturated soil at a site and/or to remove light nonaqueous phase liquid (LNAPL) residual smeared by water table fluctuations. If remediation occurs when the groundwater level is not at the seasonal low level, the excavation must be dewatered.
- Occasionally, excavation will extend further into an aquifer in an attempt to remove dense nonaqueous phase liquid (DNAPL) residual. (However, as described previously, DNAPL residuals can be quite difficult to pinpoint. As a result, extensive and costly excavation may remove only a small fraction of the DNAPL residual in an aquifer.)

When an excavation must extend below the water table, design plans must include dewatering plans. Contractors use a variety of dewatering techniques, commonly including the use of well points. Extracted groundwater is usually contaminated and must be treated. As a result, extensive dewatering can be quite costly.

Project planning includes the preparation of a Health and Safety Plan (HASP). The HASP typically includes the type of air monitoring which will be performed during the excavation, identifies site-specific hazards, specifies the level of personal protective equipment that must be worn as defined under OSHA [29 CFR 1910], and outlines the procedures to be taken in case of emergency.

Regulators usually require some level of *post-excavation sampling* or *confirmation sampling*. Samplers collect soil from the bottom and side walls of an excavation for analysis to determine whether or not the remaining soil meets the clean-up levels. If relatively few data were available to plan the work, samples may be collected as the excavation progresses to guide the work. Such samples are analyzed in an on-site mobile laboratory or on a fast-turnaround schedule in an off-site laboratory to minimize the standby time of the excavation crews. Post-excavation sampling, confirmation sampling, and potentially, in-progress sampling can add considerable cost to an excavation project.

Several land features can constrain excavation or subject the work to specific regulatory requirements. As a result, excavation costs increase. The design team must consider the existence and cost implications of:

- Buried utilities. Subsurface utility lines and process piping often cross areas that must be excavated at industrial or urban sites. Excavation equipment must work around these lines, or, in some cases, subsurface lines must be relocated to allow for excavation of contaminated soil.
- Aerial utilities. Above-ground electrical lines and pipelines can restrict the type of equipment used or the work area available for some types of equipment.
- Physical structures. Contamination sometimes extends up to or beneath the foundations of a building. Excavations close to a building require shoring to prevent damage to the building.
- Water bodies. Federal and state regulations limit excavation and backfilling in wetlands and flood plains. Regulations contain specific requirements for minimizing and mitigating damage to sensitive areas and may require permits to do the work.
- Public impacts. In residential areas, the public has site-specific concerns related to construction noise, truck traffic, dust, property damage, site security, and safety. In some cases, excavation plans allow for relocation of affected residents during the work.
- Plant operations. At active industrial facilities, remediation must be planned to minimize disruption to the production schedules of the facility. Workers at the facility may have specific health and safety concerns regarding site remediation. The facility may have its own health and safety requirements for contractors coming on to the site.

4.1.2 Mobilization, Excavation, and Production

Work begins with *mobilization* of people and equipment to the site. On large projects, workers set up facilities for decontamination of people and equipment, sanitary facilities, and an office trailer. Workers may build a security fence to restrict access to the excavation area. The work areas — including the excavation, soil stockpile areas, and health and safety zones — must be laid out. The area to be excavated may require clearing of vegetation or pavement. Some projects require construction of access or haul roads.

The type of construction equipment used on a site depends on the nature of the work, the work area, and the size of the excavation. Some of the equipment commonly used to excavate and move soil about a site is described briefly below. When excavation is a necessary part of a soil remediation plan, feasibility study planning should take into consideration the probable types of equipment to be used so as to ensure the viability of the concept and the magnitude of the cost. Only in rare occasions is a piece of equipment specifically identified in the technical specifications for a remediation project: the design engineer typically leaves the choice of equipment to the contractor.

A bulldozer, or dozer, pushes dirt with a hydraulically controlled blade. Dozers are used to move earth, grade soil, and excavate shallow layers of soil. Dozers are either track- or rubber-tire mounted. Tracks or crawlers are "self-laying steel tracks of variable cleat design and width which provide good ground contact and excellent flotation and traction capabilities"; as a result, crawler-mounted dozers are better suited than wheel-mounted equipment for uneven or unstable ground.[5] Wheel-mounted dozers, however, are more maneuverable than crawler-mounted units on level grade. Depending on the size of the dozer, the maximum production rate, or amount of material that a dozer can move over an average distance of 100 ft in an hour, is between 100 and 2700 lcy per hour. The actual production rate for a particular piece of equipment depends on the level of health and safety protection required, the operator's skill, the type of soil, the grade or slope of soil to be excavated, and the visibility on site, among other factors.[6] It is estimated by a method similar to the calculation shown below for an excavator.

A backhoe, or hydraulic excavator, is usually used for trenching or to otherwise excavate soil below grade. Soil is excavated using a toothed bucket attached to a boom or dipper stick. The choice of equipment depends on the size of the excavation, the terrain, and the equipment availability. Excavator buckets usually range in size from 0.5 to 4 cy. The size of the bucket is matched to the size of the job. For example, an excavator with a 1-cy bucket would typically be used to excavate a soil volume of no more than 2000 bcy, and an excavator with a 2-cy bucket would typically be used to excavate 4000 to 6000 bcy.[7] The length of the dipper stick determines the maximum depth to which a given unit can dig. Excavators can be track or wheel mounted, and, as described for dozers, the type of mounting determines their mobility.

The production rate of a backhoe depends on the size of the excavator and its *bucket capacity* (BC), the bucket fill factor, the cycle time, and the level of health and safety precautions. The *bucket factor* (BF) characterizes the amount of soil that will actually be carried in the bucket. It is the ratio of the bank measurement of soil that will actually be in the bucket to the dipper capacity or BC, expressed as a percentage or fraction of a full bucket.[8] The bucket fill factor depends on the type of soil: wet or moist soil will tend to stick together and will heap in the bucket, unlike loose, dry soil. For soil, the BF generally ranges from 0.80 to 1.10, and for blasted rock, typically 0.40 to 0.75.[9] The *cycle time* (CT) is the amount of time needed to load the bucket, swing the backhoe arm to the side of the excavation, dump the bucket load, and swing back to the excavation. This factor depends on the job conditions and on the size and type of equipment used, but typically varies from 17 to 28 sec.[9]

At an uncontaminated site, the production rate is estimated as follows:[8]

$$P = \frac{50 \text{ min}}{CT} \cdot BF \cdot BC \qquad (4.1)$$

where P is the production rate, in bcy/hour, and the other variables are as defined above. The 50-min factor accounts for the normal maximum working time per hour. This factor decreases for excavation around drums or subsurface utilities, or other difficult working conditions.

Front-end loaders are tractors equipped with buckets that can be used for shallow excavations, lifting, hauling, and dumping materials. Like dozers and excavators, front-end loaders can be crawler or wheel mounted. The production rate is calculated similar to the production rate for an excavator.[5,7,8]

Example 4.2   For the project described in Example 4.1, calculate the number of hours required to excavate the soil. Assume that the work will be done with a 1-cy hydraulic excavator with an 18-sec cycle time; the BF is 1.00; and that the workers will use Level C health and safety precautions.

From Equation 4.1,

$$P = \frac{50 \text{ min/hour}}{(18 \text{ sec})/60 \text{ sec/min}} \cdot 1.00 \cdot 1 \text{ cy} = 167 \text{ bcy/hour}$$

Productivity must be adjusted to account for the effect of taking health and safety precautions. From Section 2.7.4, Level C precautions can reduce equipment productivity to roughly 75% of levels for routine production work. The adjusted productivity is then 130 bcy/hour.

From Example 4.1, approximately 1500 bcy of soil must be excavated. Excavation will require approximately:

$$Duration = \frac{Volume}{Productivity} = \frac{1500 \text{ bcy}}{130 \text{ bcy/hour}} = 12 \text{ hours}$$

Excavating dry soil generates dust and causes volatile organic compounds (VOCs) to volatilize. These air emissions may cause health or odor problems. As a result, workers monitor the levels of dust and/or target compounds in the air in the work area. The area may be sprayed with a mist of water or a vapor-suppression foam to control emissions. Alternatively, temporary enclosures (sprung structures) are built over work areas to contain emissions.[10,11]

Excavated soil may be stockpiled if it is to be treated or held before loading for transportation off site. Alternatively, excavated soil can be loaded directly onto trucks for transportation off site. Minimizing handling — such as moving from the excavation to a truck, rather than to a stockpile and then to a truck — helps to minimize the cost of the work.

After contaminated soil is removed and soil samples are collected from the excavation to confirm that the remaining soil meets the clean-up levels, the excavation is often surveyed to document the limits of the work. Common fill is placed in the excavation and compacted. The surface of the fill is graded to provide the final site contours. Depending on the planned site use, the area may be paved or covered with a layer of topsoil and revegetated. At the end of the job, people and equipment *demobilize* from the site.

4.1.3 Materials Handling

Excavated soil is commonly pretreated to allow for shipping or for treatment. Sorbent materials may be added to take up excess water, or the soil may be sized to allow for limitations on treatment equipment.

Soil to be transported off site for disposal should not contain "free liquids" as determined using the paint filter test. As a result, wet soils or sludges are sometimes mixed with absorbant material such as wood fiber, ash, or kiln dust. The ratio of sorbent material to soil depends on the sorbent and soil conditions, but can be in the range from 1:1 to 2.5:1 sorbent to waste.[12] While the sorbent itself is usually not very expensive, the cost of transporting and disposing of the saturated sorbent material mixed with soil can be quite costly.

Most equipment used to treat soil has size limitations. As a result, soil must be screened before treatment to remove oversized particles. A *grizzly*, or bar screen, can separate materials over 2 in. in diameter. Revolving screens and shaking screens are used to separate out materials over ½ in. in size.[13] The oversized rocks and debris may be landfilled, or in some cases, crushed or shredded so that they meet the size limitations for further treatment.

4.1.4 Transportation Off Site

Bulk soils may be shipped off site for disposal or treatment by truck, or, when available, by rail. Depending on their size, dump trailers can carry 20 to 36 cy soil; 20-cy trailers are commonly used. Roll-off boxes can carry 20 to 30 cy. Loading can be volume limited or weight limited (by road weight limits). For example, a 20-cy dump trailer can carry only 18 tons of soil per load on certain roadways.

Shipping costs are commonly billed per *loaded mile*, or mile between the site and the treatment or disposal facility. Charges can also include *demurrage*, or an hourly cost for time spent idling at the site waiting for the truck to be filled or paperwork to be filed. Ancillary charges may also be incurred, including: a fee for waste acceptance at the treatment or disposal facility; a charge for a truck liner; or a surcharge to manage the presence of free liquid in the shipment once it reaches the facility.

Federal and state regulations govern the shipment of waste soil. Depending upon the waste, it may be regulated as hazardous waste under RCRA or analogous state regulations. RCRA regulations require that the waste be shipped under a manifest. The generator must fill out a manifest form for each truckload of waste. The manifest provides information about the physical and chemical nature of the waste, its origin, and destination. The Hazardous Materials Transportation Act also regulates the shipment of hazardous material. The regulations promulgated under that Act [49 CFR 100-199] specify the requirements and limitations for packaging and shipping waste. (See Section 2.6 for a further description of applicable regulations.)

4.2 IMMOBILIZATION TECHNOLOGIES

Contaminants in soils, sludges, or sediments can be immobilized by capping or landfilling the material. Alternatively, such materials can be treated to immobilize contaminants within the matrix. Immobilization technologies include asphalt batching, stabilization/solidification, and vitrification.

4.2.1 Disposal/Landfilling

Off-site disposal can be the most cost-effective option for certain wastes. It is still the easiest, cheapest option in most cases for handling small amounts of soil or waste. The cost of landfilling soils or solid wastes depends on the waste classification, treatment required to meet land disposal restrictions (LDRs), if any, and proximity to a disposal facility. The decision to dispose of soil must balance these cost factors against the potential liabilities from transportation and disposal of contaminated material.

Not all of the waste from a hazardous waste site is hazardous waste. "Waste" soil or sludge can be classified as hazardous waste or non-hazardous waste under federal regulations. State regulations can be more stringent than federal regulations; in addition, certain states have so-called *special waste* classifications.

A waste is a hazardous waste under federal RCRA regulations when it is (1) listed in the regulations [40 CFR 261.30-261.33]; or (2) a characteristic hazardous waste, determined by chemical testing [40 CFR 261.20-261.24]. (See Section 2.5 for additional information on waste classification.)

A disposal facility requires a waste *profile* before it will accept waste for disposal. The profile includes information on the source of the waste, its physical state and chemical composition, and the volume and type of shipping. The generator must also certify that the waste meets — or does not meet — regulatory criteria for waste classification under RCRA, DOT, The Toxic Substances Control Act, the Nuclear Regulatory Commission, and other regulations. The facility may also request a sample of the waste.

Not all hazardous wastes can be put in a landfill. So-called LDRs promulgated under RCRA [40 CFR 268] require that most hazardous wastes meet certain concentration limits or are treated using a specified technology before disposal. Certain hazardous wastes cannot be placed in a landfill under any circumstances (see Section 2.6.3.2).

Example 4.3   This example describes a site where contaminated soils were to be managed by a combination of excavation, treatment, and disposal; containment; and institutional controls on land use. It illustrates how risk-management decisions can direct the course of remediation. Finally, this example shows the length of time that can elapse between recognition and remediation of a problem. The site history which follows is based on information available as of late 1997.[14-19]

For 50 years, beginning in 1928, a brass and bronze foundry operated in the eastern section of Portsmouth, VA. The foundry melted used railroad car journal bearings and recast the molten metal into sand molds to make new bearings. The waste sand from the molds, which contained high levels of lead and other metals, were apparently disposed of in open areas around the foundry. Some 3500 tons of lead-bearing furnace sands were disposed of on a 1-acre parcel adjacent to the foundry. Over time, much of this sand was apparently used as fill for nearby residential and commercial development. As a result, soils were contaminated with lead and other heavy metals from foundry sands. Some soils also contained polynuclear aromatic hydrocarbons (PAHs) and polychlorinated biphenyls (PCBs).

The Portsmouth Redevelopment and Housing Authority built Washington Park, a 160-unit low-income housing project, near the foundry site in 1964. The Authority was not aware that the building site contained contaminated fill. In 1974, the City of Portsmouth sold 17 parcels of land in a two-block area south of the foundry for residential development. The homes built on this land came to be known as the Effingham residences.

In 1982, local doctors observed that children in the area had elevated levels of lead in their blood. In 1984 and 1986, the U.S. EPA collected soil samples from the area. Those samples contained lead concentrations between 450 and 12,800 mg/kg. As a result of those data, and under the terms of a Consent Agreement and Order with the U.S. EPA, the Abex Corporation removed 6 to 8 in. of soil from some residential yards and a playground. Abex also covered two disposal areas with asphalt and fenced the areas to limit access.

The EPA proposed the Abex Corporation site for inclusion in the National Priorities List in 1988, and the site was listed in 1990. The Superfund site includes the original 2-acre foundry property, the 1-acre disposal area, and adjacent properties contaminated by the former foundry operations. These properties include the Washington Park development, the Effingham residences, a drug rehabilitation center, and several vacant lots.

Between 1991 and 1994, five different clean-up plans were proposed for the site. The plans differed in three respects: the extent and location of soil excavation, the area to be capped, and the use of institutional controls. The evolution of the final remediation plans is summarized below.

Abex began site investigations in 1989 and completed the investigation of soils near the foundry site in 1992. The EPA proposed a remedial action plan in the spring of 1992. Although a 30-day public comment period on a proposed remedial action is typical, the EPA continued discussions with the public for some 4 months. In September 1992, the U.S. EPA determined that soils which exceeded specified clean-up levels within a 700-ft radius of the foundry property should be excavated and landfilled off site, for a total estimated cost of $28,891,243. The foundry buildings and asphalt and concrete pavement were to be demolished to allow for soil excavation. The Record of Decision (ROD) specified two clean-up levels: 500 mg/kg lead for surface soils (0- to 1-ft depth), and 1000 mg/kg lead for subsurface soils (1-ft depth to the water table, some 3 to 4 ft below ground surface [BGS]). The EPA expected that some of the soil which exceeded these clean-up levels would fail the toxicity characteristic leaching procedure (TCLP) test and be classified as a characteristic hazardous waste. The ROD specified that such soil would be treated to reduce TCLP-leachable lead using stabilization before disposal off site.

The remedy included removal of contaminated soil from residential yards and replacement with clean soils. EPA planned to shore or support the homes to protect them from structural damage while soil was excavated from beneath the foundations. The residents were to be relocated during construction. The local land use was to remain residential, and no institutional controls would be used to limit land use after remediation.

In 1993, the City of Portsmouth and other parties responsible for site remediation notified the EPA that two blocks of residential property at the site were to be rezoned for commercial/light industrial use. The City indicated that future development of the remaining industrial property would be controlled. The Washington Park development would remain as low-income housing.

Based on the proposed land use restrictions, the EPA changed the clean-up plan in 1994. The revised plan included the purchase and demolition of 22 private homes in the rezoned area and construction of a municipal facility. As a result of the land use change, less soil would be excavated and removed from the site than previously planned. Some soil would be excavated, stabilized, and transported off site for disposal. That soil would be replaced with clean fill. The concrete pad and asphalt paving at the former foundry would remain as a cap over some soils. The estimated cost of the revised remedy was $31,507,670.

In December 1995, the Abex Corportion, the City of Portsmouth, and the Portsmouth Redevelopment and Housing Authority signed a Consent Order with EPA. In that Order, they agreed to design and perform the cleanup of the site. By January 1997, the initial design plans were complete. The foundry buildings were demolished by May 1997.

Meanwhile, Abex continued to prepare design plans for soil remediation. EPA required three design submittals: conceptual, pre-final, and final design. Site investigations continued into the extent of contamination in groundwater, surface water, sediments, and soils outside the 700-ft radius. Potential ecological impacts will be considered.

4.2.2 Capping

Caps are used to cover waste or contaminated soil. Cap designs range in complexity from simple asphalt pavement to engineered multilayer caps, sometimes called "RCRA caps". A cap is used to:

- prevent direct contact with contaminated material, and/or
- minimize infiltration of rainwater or snowmelt into the contaminated material, in order to minimize contaminant leaching to groundwater.

A cap may also be used to:

- eliminate the contamination of surface-water runoff which would otherwise contact with contaminated soil, sediment, or waste, and/or
- prevent generation of contaminated dust or volatilization of contaminants.

Caps are commonly used to contain waste in landfills. Many old landfills are now Superfund sites; the U.S. EPA has a policy that capping is usually an appropriate component of remedial actions for such sites.[20] Capping is not appropriate for many sites. It is rarely used to contain small volumes of highly toxic or leachable waste. It will not effectively contain, for example, waste in the saturated zone, or liquids in drums. Finally, capping is not appropriate where the long-term use of the land cannot be restricted or the cap will not be maintained.

Certain state regulations describe standards for caps used for site remediation (e.g., the Massachusetts Contingency Plan at 310 CMR 40.0996(4)(c)). At a federal Superfund site, RCRA requirements for landfill caps are "applicable or relevent and appropriate requirements".

RCRA regulations [40 CFR 264.310] provide performance standards, not construction specifications. They require that the final cover over a landfill must be designed and constructed to:

1. provide long-term minimization of migration of liquids through the closed landfill;
2. function with minimum maintenance;
3. promote drainage and minimize erosion or abrasion of the cover;
4. accommodate settling and subsidence so that the integrity of the cover is maintained; and
5. have a permeability less than or equal to the permeability of any bottom liner system or natural subsoils present.

The cover must be maintained and groundwater conditions monitored.

4.2.2.1 Construction of a Multilayer Cap

Design engineers typically meet the RCRA performance standards using three functional layers: bottom, or low permeability, middle or drainage, and top or erosion control.[21,22] The choice of materials used in each of these functional layers depends on the local availability and price of alternative materials. Examples of materials used to construct these layers are described below and shown in Figure 4.2.

A cap may also include, in addition to the three principle functional layers,

- a gas vent layer. A cap commonly includes gas vents through the cap to allow gases generated by decomposition of organic wastes to escape from beneath the cap. A layer of coarse-grained porous material may be placed between the waste and the low-permeability layer to convey gas to the vents.
- a *biobarrier*, such as a 2- to 3-ft layer of cobbles, may be placed immediately above the low-permeability layer. A biobarrier prevents animals from burrowing into the cap.
- geotextiles (as described further below), to reinforce and/or separate other cap materials.

4.2.2.1.1 Bottom or Low-Permeability Layer — Construction of a cap begins with grading and compacting the waste or soil to be capped. This step provides a structural base for the load of the cap materials and helps to create the contours required to enhance runoff of precipitation. Creating the desired grade may require borrow material which adds to the expense of a cap.

The next step is to construct the low-permeability layer. This layer may comprise a geomembrane and/or a layer of compacted clay.

Low-permeability layer: geomembrane

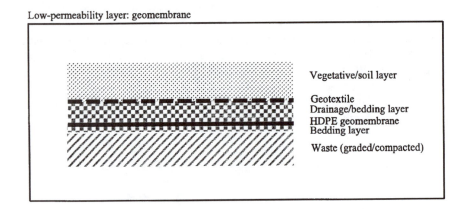

Low-permeability layer: compacted clay

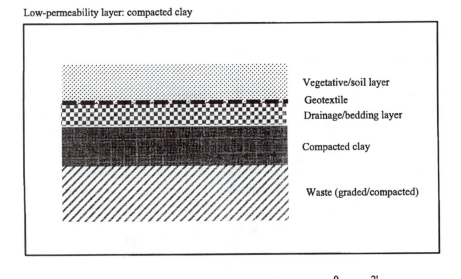

Figure 4.2   Schematic of multilayer cap.

A *geomembrane*, also known as a *flexible membrane liner*, is an essentially impermeable membrane commonly made of high-density polyethylene (HDPE) or other polymeric material. A geomembrane typically has a hydraulic conductivity on the order of 10^{-11} to 10^{-13} cm/sec.[23] Variations include geomembrane/bentonite (clay) composites and geomembrane/geotextile composites. Design engineers specify various physical properties for geomembranes based on American Society for Testing and Materials (ASTM) or other standard tests. These properties include thickness, weight, tensile strength, puncture resistance, and other characteristics.[24] Thickness is measured in *mils*: 20 mils is equivalent to 0.5 mm. Thinner geomembranes are quite flexible; thicker geomembranes (60 mil and higher) have a thickness and flexibility similar to the linoleum used to cover floors. Geomembranes can be purchased in standard sizes (e.g., 20- to 140-mil thickness in 22½-ft widths) or they can be custom fabricated.

If a geomembrane is used as the low-permeability material, a bedding layer such as 6 in. of clean sand is placed over the soil/waste to provide a smooth layer free of materials that could puncture the membrane. The contractor unrolls sheets of the geomembrane over the area to be capped and weights them down with sandbags. A specially trained and certified crew joins the seams between the panels using a solvent, adhesive, heat, or other methods. The seams are tested to ensure minimal leakage. The geomembrane must be placed at a minimum 5% slope to allow for runoff. The maximum allowable slope is typically established at 3H:1V; however, the slope may be further restricted by the slope stability.

Alternatively, or in addition to a geomembrane, a layer of clay is used as the low-permeability layer. This layer might comprise, for example, a 24-in. thickness of compacted clay at a hydraulic conductivity of 1×10^{-7} cm/sec or less and a minimum slope of 3%. The raw material is carefully specified according to liquid and plastic limits, percent fines, percent gravel, and water content so that the clay layer can meet the performance standard. The clay is placed and compacted in *lifts*, or layers. For example, three lifts may be used to achieve a total thickness of 24 in. The compacted clay is tested to ensure that the cap meets the design specifications.

4.2.2.1.2 Middle or Drainage Layer — A drainage layer covers the low-permeability material. This layer is designed to convey rainwater and snowmelt off the cap. It may comprise, for example, a 12-in. layer of sand with a hydraulic conductivity of 1×10^{-2} cm/sec or greater. The thickness is usually established through modelling, commonly using the EPA's HELP model. The U.S. EPA developed the *HELP model* to evaluate the potential long-term performance of alternative landfill designs and to enable the designer to assess the sensitivity of design parameters.[25] The model performs a water balance which accounts for precipitation, surface runoff, evapotranspiration, lateral drainage from each liner/drain system, percolation or leakage through each liner and from the bottom of the profile, moisture storage, snow accumulation, and the depth of saturation on the surface of liners.

Geosynthetic drainage materials may also be used for the drainage layer. A *geonet* is formed by extruding HDPE or medium-density polyethylene (MDPE) into a net of intersecting ribs. Geonets are characterized by their thickness, tensile strength, compressive strength, shear strength, hydraulic conductivity, and "creep", or deformation under pressure.[26]

A *geotextile* layer often covers the drainage layer to prevent the fine particles from overlying soils from clogging the drainage layer. Geotextiles are fabrics made of synthetic fibers. These fibers are either woven or matted together to make a flexible, porous fabric.[27] Geotextiles are used to reinforce geomembranes and to separate functional layers.

4.2.2.1.3 Top or Erosion-Control Layer — The top layer of the cap is soil which supports vegetation and enhances runoff. The thickness of this layer may be increased beyond that simply needed to support vegetation in order to protect the low-permeability layer from frost or the drainage layer from freezing. Grass and other shallow-rooted vegetation is planted to prevent erosion.

4.2.2.2 Maintenance and Monitoring

Design plans typically include plans to monitor and maintain a landfill or area containing contaminated soil after it is capped. Regulatory agencies require potentially responsible parties (PRPs) or facility owner/operators to maintain, monitor, and report on those activities for a period of typically 30 to 50 years.

Monitoring activities usually include:[28]

- routine inspection of the area, to determine whether or not the cap has been damaged by vandalism, erosion, or subsidence, and whether or not the surface-water management system has been compromised;
- collection and analysis of groundwater samples from monitoring wells around the cap, to determine groundwater elevations and whether or not and to what extent contaminants attenuate or continue to leach from the soil/waste; and
- collection and analysis of emissions from gas vents, to determine whether or not hazardous substances are being emitted.

The use of a capped area must be restricted and the cap must be permanently maintained to ensure long-term containment of contaminated soil or waste. Maintenance includes mowing the vegetation to prevent deep-rooted plants from growing and puncturing the low-permeability layer. Eroded areas must be repaired and reseeded.

The low-permeability layer may periodically require repair or replacement due to breaches caused by settlement or vandalism. The low-permeability layer can rupture if the waste subsides considerably over time; however, low-permeability layers are designed to minimize the damage due to subsidence. Certain chemicals, ultraviolet light, and oxidation, among other factors, can degrade geomembranes. However, burying the geomembrane between layers of clean materials in a cap, which is typically part of the design, minimizes such degradation.[29]

4.2.3 Asphalt Batching

Soil contaminated with total petroleum hydrocarbons (TPH), typically from a leaking underground fuel storage tank, can be incorporated into asphalt. The soil is substituted for a portion of the stone aggregate ordinarily used to make asphalt concrete. There are two types of asphalt-batching processes: hot-mix asphalt batching and cold-mix asphalt batching.

4.2.3.1 Hot-Mix Process

In the *hot-mix* process, the aggregate (soil) is dried at temperatures ranging from 300 to 600°F to remove moisture. The drier is akin to the rotary driers used in thermal desorption (Section 4.3.2), and typically includes air-pollution control systems to remove particulates and organics from the off-gases. After drying, the soil is then mixed with hot liquid asphalt. The product *cures* or hardens as it cools.[30-32]

When petroleum-contaminated soils are mixed with the clean aggregate and dried, volatile compounds volatilize and, depending on the drier temperature and retention time, oxidize. Most asphalt-batching plants have extensive facilities to control air emissions. Volatilization of VOCs during the drying step is critical. Large quantities of the lighter petroleum hydrocarbons act as solvents to soften the final asphalt product, resulting in an inferior product.

Volatilization and thermal destruction are the primary methods of remediation of lighter hydrocarbons in the hot-mix asphalt process. Solidification/stabilization is the secondary method of remediation. Heavier hydrocarbons remaining in the soil after the drying step are encapsulated within the asphalt/aggregate matrix.

In many states, commercial hot-mix asphalt plants are permitted to accept TPH-contaminated soils as a recycling operation. A commercial plant is typically responsible for the ultimate use of the asphalt.

In the *cold-mix* asphalt-batching process, aggregate is mixed with an asphalt emulsion in a *pug mill*. (A pug mill is a large self-contained unit with a horizontal-blade mixing apparatus. Pug mills are used in common construction operations, for example, to mix cement.) The asphalt emulsion contains particles of asphalt suspended in an aqueous solution. Emulsifying agents (surfactants) are used to keep the asphalt in suspension. The proportions are approximately, by weight, 50 to 65% heated liquified asphalt, 35 to 50% water, and 1 to 2% emulsifying agent. Mixing the emulsion with soil causes the emulsion to *break*, or separate into water and asphalt phases. The asphalt droplets coat the aggregate particles with a thin film.[30]

When the asphalt emulsion/aggregate mixture is applied to the area to be paved and compacted, the water is expelled under pressure. The asphalt and aggregate mixture cures, developing into a bituminous concrete, as the remaining water evaporates. High humidity, low temperature, and rainfall hinder curing.[30] The final volume of paving material is roughly the same as the in-place volume of contaminated soil.

Cold-mix asphalt batching occurs, as the name implies, at ambient temperature. The ambient temperature has two effects. First, the product cannot be made or applied when the ambient temperature is below 32°F. Second, unlike the hot-mix process, volatile compounds remain in the mixture. Large proportions of VOCs can soften the final product. Leaching of these VOCs from the final product is also a concern.

In many states, mobile cold-mix asphalt plants can be brought on site to make asphalt concrete from contaminated soils. The product must then be used on site. Cold-mix asphalt can be used as base materials for private parking lots and roads. It does not have the strength of a hot-mix product.

Cold-mix asphalt batching may be subject to state regulations. Those regulations may limit, for example, the concentrations of contaminants in the feed, the materials used for the asphalt, the properties of the product, and the conditions under which the asphalt can be placed.

Other factors may also limit the application of asphalt batching. Typically,

- only non-hazardous waste containing TPH may be treated;
- hot-mix plants may have upper limits on the level of TPH based on a state-issued recycling permit. Higher prices may be charged for highly contaminated soils. The plant permit limits may preclude acceptance of highly contaminated soils. Such limits vary, but can be on the order of 30,000 to 60,000 ppm TPH;
- hot-mix plants in cold climates may not operate in the winter, since the use of the product in construction slows during the winter;
- a high proportion of fines in soil can result in unacceptable product. Fines cannot be readily and uniformly coated by the asphalt binder, and tend to separate into uncoated pockets in the bituminous concrete. Soils with a high proportion of clays are unacceptable;
- stones or debris larger than 2 to 3 in. in diameter must be screened out of the soil before asphalt batching;[32]
- long-term liability may remain; and
- the volume of soil that can be treated may be limited by the use for the product.

4.2.4   Solidification/Stabilization

Solidification/stabilization processes treat waste by:

- improving the handling and physical characteristics of the waste, as in the sorption of free liquids;
- decreasing the surface area of the waste mass across which transfer or loss of contaminants can occur; and/or
- limiting the solubility of hazardous constituents.

The terms "solidification" and "stabilization" are often used interchangeably. The term *fixation* refers to either solidification or stabilization.

Solidification implies that treatment produces a solid block of waste material which has high structural integrity (sometimes referred to as a *monolith*). The contaminants do not necessarily react with reagents, but are mechanically locked within the solidified matrix, an effect called *microencapsulation*.[33] *Thermoplastic microencapsulation* refers to processes in which waste particulates are blended with melted asphalt or materials such as polyethylene, polypropylene, wax, or elemental sulfur. Treatment of soil by mixing with asphalt was discussed previously. Other matrices, such as polyethylene, are typically used for small-scale applications where cost is not necessarily the limiting factor.

Stabilization techniques limit the solubility or mobility of the contaminants. Stabilization may or may not change or improve the physical characteristics of the waste. Stabilization usually involves adding materials which ensure that the hazardous constituents are maintained in their least mobile or toxic form. Examples include the addition of lime or sulfide to a metal sludge to precipitate the metal ions, or the addition of an absorbent to an organic waste.[32]

Stabilization/solidification processes are usually used to treat inorganic wastes, commonly soils and sludges containing metals. Stabilization/solidification is the "Best Developed Available Technology" (BDAT) for many metal-containing wastes, and is required before landfilling such wastes under RCRA LDR [40 CFR 268]. Treatment of organic wastes, such as oily sludges containing PCBs or PAHs, has met with variable success. VOCs tend to volatilize during the mixing of soil with stabilization/solidification agents. An exothermic stabilization/solidification reaction increases volatilization. VOCs are generally not immobilized.[34]

4.2.4.1 Mechanics of Treatment

Treatment may require that the soil or sludge be dewatered or that larger-sized stones or debris be removed or crushed. The equipment used to mix reagents with the waste often has limits on the maximum particle size. Depending on the equipment and objectives of treatment, the upper limit typically ranges from 1 to 3 in., but may be higher.

The waste may also be stabilized using chemical additives to reduce the solubility of contaminants before it is solidified. Materials containing hexavalent chromium may be pretreated with ferrous sulfate to reduce chromium to the less soluble, less toxic trivalent form. Arsenic can be immobilized by oxidizing As(III) to the less toxic, less mobile As(IV) and then treating the As(IV) with ferrous sulfate to form insoluble $FeAsO_4$.[35] Wastes containing lead may be pretreated with trisodium phospate, to immobilize the metal as lead phosphate.[35]

Soil or sludge may be treated in containers such as drums, in a mobile processing unit, or in place (sludge in lagoons, or soil *in situ*).

Figure 4.3 shows a block flow diagram of one type of mobile treatment unit used to treat high-solids waste. This treatment unit operates in a batch mode, using a pug mill to mix the soil and reagents. The treated waste is discharged to a containment area to cure. Ultimately, most treated wastes are capped or landfilled to prevent physical or chemical disturbance.

Sludge in lagoons or shallow soils can be mixed with solidification/stabilization reagents using common construction equipment such as a backhoe. However, thorough mixing and immobilization of the contaminants can be difficult to achieve.

Several vendors have equipment designed to treat soils at depth. These systems typically include an above-ground reagent mixing and delivery system and one or more augers.

As the augers bore into the soil, a slurry of solidification/stabilization reagents in water is injected into the soil through the auger. The rotation of the auger mixes the reagent into the soil column. Treatment of a large area is achieved by overlapping bore holes.[36,37] As with any *in situ* treatment method, process control and results can be difficult to monitor.

The augers range in diameter from 2.5 to 12 ft. Smaller-diameter augers are used in finer-grained soils to ensure adequate mixing, and in deeper soil applications. These augers typically enable treatment to a depth of up to 40 ft, although treatment of deeper soils is possible.[36]

The cost of *in situ* solidification/stabilization using an auger system typically exceeds the cost of excavation and solidification/stabilization when the target soils are less than 8 to 10 ft deep. The cost differential results in part from the high cost of mobilizing the specialized equipment for *in situ* treatment.[36]

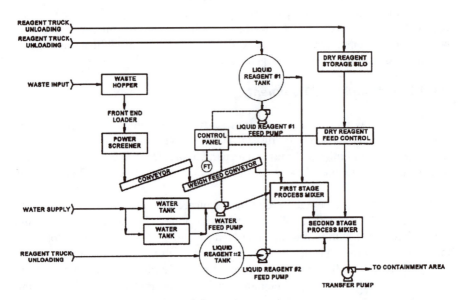

Figure 4.3   Process flow diagram, solidification and stabilization, Chemfix Technologies, Inc. (From U.S. EPA, *Superfund Innovative Technology Evaluation Program Technology Profiles*, 9th ed., EPA/540/R-97/502, Office of Research and Development, Washington, D.C., December 1996, 46–47.)

4.2.4.2 Reagents

Stabilization/solidification processes generally fall into two categories: (1) pozzolan or silica based, and (2) cement based. Many processes use a combination of silica and cement. Vendors typically include proprietary additives. Each of the two broad categories of processes is described below.

4.2.4.2.1 Pozzolan- or Silicate-Based Processes — *Pozzolanic materials* set to a solid mass when mixed with hydrated lime ($Ca(OH)_2$). These materials contain silicic acid and often contain appreciable levels of aluminum oxide. Pozzolanic materials include diatomaceous earth, some fly ash from powdered coal furnaces, blast-furnace slag, and cement kiln dust. Standard ASTM testing systems and specifications exist for pozzolanic materials.

In a pozzolan-based solidification process, the waste is mixed with fly ash or another silicate material, then lime or cement. Polyvalent metals — such as the calcium in lime, metals in portland cement, or the waste itself — initiate the precipitation and/or gelation of silicates.[38] The heavy metals in waste become part of the calcium silicate and aluminate colloidal structures, or adsorb to the surface of the pozzolanic structure.[39]

Common mixes include lime/fly ash and lime/cement kiln dust.[40] The amount and ratios of the silicate material and lime varies with the waste. For example, the

U.S. EPA reports on the amounts of lime and a mixture of lime, fly ash, and bentonite required to treat a liter of several types of waste:[41]

- metal hydroxide sludge: 2.9 kg lime/L sludge; 1.1 kg lime, fly ash, and bentonite/L sludge;
- sulfuric acid plating waste: 3.0 kg lime/L waste; 2.3 kg lime, fly ash, and bentonite/L waste;
- oily metal sludge: 0.6 kg lime/L sludge; 0.54 kg lime, fly ash, and bentonite/L sludge.

These mixtures are provided only to illustrate that solidification can require a relatively high proportion of reagents, increasing the volume and weight of waste which must ultimately be contained or landfilled. Waste-specific testing is necessary to determine the appropriate reagents and amounts before treating the waste at a particular site.

Other materials may be added in addition to lime, kiln dust, or fly ash. For example, bentonite is added to make the reagent slurry easier to pump and decrease the permeability of the treated waste.[36] Selected clays may be added to absorb liquids and bind specific anions or cations. Carbon or zeolites may also be added to bind specific contaminants. Finally, emulsifiers and surfactants may be added to allow the incorporation of organic liquids such as oil residues.[38] The doses of reagents and additives must be determined by bench- and/or pilot-scale testing.

The solid that is produced varies from a clay-like material to a hard, dry solid similar to concrete in appearance.[38] The treated material typically has a permeability between 10^{-6} and 10^{-7} cm/sec, and an unconfined compressive strength between 30 and 200 psi[37] or more.

In general,

- lime/fly ash treatment is relatively inexpensive;
- wastes solidified with lime/fly ash are generally not as physically durable as pozzolan–portland cement-treated wastes which have formed a monolith;[38,40]
- the use of lime creates alkaline (basic) conditions and as a result, generally reduces metals leaching;
- the treated waste contains water that is not chemically bound, and may lose some water after treatment;[38] and
- pozzolan-based processes are potentially more applicable to organic wastes than cement-based processes, although oil and grease can coat the waste particles and interfere with the formation of bonds between calcium silicate and aluminum hydrates.[38,40]

4.2.4.2.2 Cement-Based Processes — Cement-based systems are the second type of solidification/stabilization processes. Solidification can be achieved by mixing waste with portland cement (usually type I or II). Additives may include, for example,

- pozzolanic material, to aid processing and assist in containment of metals through the formation of silicate gels;[42]
- clays, to absorb liquid and bind specific compounds;
- lime (CaO), to raise the pH and the reaction temperature, and thereby improve setting characteristics;

or other additives.

The amount and ratios of cement and additives varies with the waste. Treatment might require, for example, 30% portland cement, 2% sodium silicate.[40] Reagent doses must be determined by bench- and/or pilot-scale testing. Depending on the additives and amounts, the end product may be a monolithic solid or have a crumbly, soil-like consistency.[38] Monolithic solids typically have a permeability between 10^{-5} and 10^{-9} cm/sec, and an unconfined compressive strength between 20 and 1000 psi.[40]

In general,

- treatment by a cement-based process is relatively inexpensive;
- portland cement alone will not immobilize organics;[38]
- most metals precipitate as insoluble hydroxides or carbonates at the pH of the cement;[38]
- strong acids can leach metals from the solid,[38] and destroy the concrete matrix after setting has occurred;
- the waste must contain less than 150 mg/kg sulfates if type I portland cement is to be used; type II and type V cements can tolerate higher levels of sulfates;[38,40]
- without pretreatment, certain waste constituents can hinder setting and the permanent stability of cement: sodium salts of arsenate, borate, phosphate, iodate, and sulfide; salts of magnesium, tin, zinc, copper, and lead; organic matter, some silts and clays, and coal or lignite;[38] and
- the treated waste comprises a porous matrix; contaminants which are not chemically bound can leach from the matrix, although additives can be used to decrease leaching.

4.2.4.3 Results of Treatment

A variety of tests are used to gauge the potential long-term stability of treated waste. Three tests indicate the potential to leach contaminants. In the TCLP test, a sample of treated waste is ground and then subjected to an 18-h extraction with an acetic acid solution; the leachate is then analyzed for contaminants [40 CFR 268]. The multiple extraction procedure tests the potential for treated waste to leach under more extreme conditions. In that test, a sample of treated waste material is ground and subjected to 24-h extraction with acetic acid solution, followed by nine sequential extractions with acidic rain-simulated leachate. Other leaching tests are also used to gauge the effectiveness of solidification/stabilization.[43] Finally, the permeability (hydraulic conductivity) of a sample may be measured to assess the resistance of the material to water flow.

Three additional tests are used to measure the physical integrity of treated waste. Unconfined compressive strength (ASTM D 1633) is measured to assess the structural integrity or durability of a monolith. The U.S. EPA guideline for unconfined compressive strength is 50 psi, based on the pressure at the bottom of a landfill containing waste at a bulk density of 70 lb/ft^3.[40] Unconfined compressive strength does not indicate the potential for contaminant mobility. Two tests are used to gauge the potential effects of weathering: wet/dry resistance (ASTM D 4843) and freeze/thaw resistance (ASTM D 4842).

As noted previously, solidification/stabilization is the BDAT for many wastes containing metals. EPA reports indicate that the results for immobilization of organics are very waste and process specific. Results obtained under EPA-sponsored tests or remediation work include the following:[44]

- Excavated soil containing copper, lead, and PCBs was treated using a pozzolan-based process offered by Chemfix Technologies. The total lead concentration in the untreated waste approached 14%. The concentrations of lead and copper in TCLP leachate from the treated waste were 94% to 99% less than in the leachate from untreated waste. The volume of waste material increased by 20% to 50%.
- Geo-Con treated PCB-contaminated soil *in situ* using a proprietary cement–organo clay mixture. Geo-Con added the reagents and mixed them with soil using an auger system. The total PCB concentrations decreased, apparently as a result of mixing soils contaminated at different levels. TCLP leachate from untreated soils contained PCBs at levels up to 13 μg/L; the concentrations in TCLP leachate from treated samples were less than 1 μg/L. The EPA concluded that the process appeared to immobilize PCB, but because of the very low values measured, the conclusions could not be confirmed.
- Soils at the Selma Pressure Treating Superfund Site in Selma, CA, were remediated at full scale using an organic stabilization and chemical solidification technology offered by STC Remediation. The soil contained pentachlorophenol, arsenic, chromium, and copper. Treatment reduced the total extractable concentrations of pentachlorophenol by up to 97% (from 5 to less than 0.3 mg/L). Arsenic and copper were immobilized, and chromium levels in TCLP leachate remained within regulatory limits. Long-term monitoring of the waste 18 and 36 months after treatment indicated comparable results.

Successful treatment of a waste depends on the physical and chemical characteristics of the waste:

- Consistent treatment can be difficult to achieve with a highly variable waste stream.
- A high proportion of fines can impede solidification.
- Depending on the process, a high level of organics can interfere with treatment.
- Volatilization of organics due to the mixing and heat of reaction may be of concern.
- Treatment increases the volume and weight of the waste, particularly for cement-based treatment. The volume and weight can increase by up to 100%.[34,38] Since treated waste must ultimately be contained, treatment can significantly increase disposal costs, e.g., by 20% to 55%.[42]
- Solidification/stabilization can require large volumes of reagents. Transportation costs for these materials can be quite high, particularly if reagents must be transported for long distances.[34]
- Materials handling can be problematic. For example, cements can harden in equipment.

4.2.5 Vitrification

Vitrification is a thermal treatment process that melts soil or sludge into a glass-like material. This material comprises fused inorganic oxides, notably silica oxides, but not organics. Nonvolatile metals are immobilized within the glass. Volatile metals — such as lead, cadmium, and zinc — volatilize. These metals must be captured in the off-gas treatment system. Organic compounds pyrolyze or combust.

Vitrification requires an extraordinary amount of energy: some 800 to 1000 kWh per ton of soil processed *in situ*.[45] Because of the energy required to vitrify waste, this type of treatment is used to treat relatively small quantities of wastes which are difficult to treat by other methods. Vitrification can be used to treat radioactive wastes, metal sludges, asbestos-containing waste, or soil or ash contaminated with metals. While soil or waste containing organic contaminants can be treated using vitrification, vitrification technologies are usually directed toward inorganic contamination.

Most soils comprise primarily metal oxides such as SiO_2, Al_2O_3, and Fe_2O_3. These alkali metals can conduct electricity in the molten state and can produce a vitrified product upon cooling. Some geologic materials, such as limestone or dolomitic soils, do not contain sufficient alkali metals to vitrify without adding fluxants (silicate materials). Some waste materials, such as ash and certain process sludges, contain sufficient alkali metals to be readily vitrified.[46]

As of late 1996, 16 vendors reportedly offered vitrification processes, 8 at full scale.[47] Soil can be vitrified *in situ* or excavated and vitrified in one of several types of process equipment. These two options, *in situ* vitrification (ISV) and vitrification of excavated material, are each discussed below.

4.2.5.1 Vitrification of Excavated Soil

A variety of unit processes are available to convert excavated soil or waste into a glass-like product. These include a slagging incinerator, plasma torch melting, cyclone furnace vitrification, and adaptations of traditional glass-making techniques, among other processes. The inert product can be landfilled, or, depending on the volume, stability, and market conditions, the slag can potentially be recycled in various construction materials.[47,48]

Vitrification differs from incineration in that it requires a higher temperature and residence time than incineration. A flux material such as lime may be added to enhance slagging. With these operational changes, a solids incinerator can produce slag instead of ash from contaminated soils or sludges.[48,49]

A *rotary kiln*, which is commonly used as an incinerator, can be operated in a slagging mode. A rotary kiln is a refractory-lined cylinder mounted on a slight incline. The wastes are injected into the high end of the kiln and pass downward through the combustion zone as the kiln slowly rotates, mixing the waste. The slag flows from the lower end of the kiln. Rapid cooling of the slag — in water, for example — increases the immobilization of heavy metals in the slag,[48] although quench cooling can produce a more easily fractured solid than air cooling.[50] The gases pass through a secondary combustion chamber, then through air pollution control units for particulate and acid gas removal.

When a rotary kiln operates in an ashing mode, retention times in the kiln can vary from several minutes to an hour or more, and the combustion temperature is typically 1200 to 1800°F (see Section 4.4.3). In a slagging kiln, the retention time is typically 60 to 100 min and the temperature typically 2200 to 2600°F.[51]

Construction and operation of a slagging incinerator costs more than for an ashing incinerator. Successful operation in the slagging mode is not easy. It requires careful control of the waste feed. Slag which is too viscous can plug the kiln; slag which is too thin can expose and erode the refractory lining of the kiln or reduce hold-up time. These characteristics of the slag can be controlled somewhat by monitoring and controlling the nature of the feed.[48]

In a *plasma centrifugal furnace*, the energy from an arc generated by a plasma torch melts metal-bearing solids. Contaminated soil is fed into the top of a rotating reactor well or tub. The plasma arc heats the material to temperatures on the order of 3000°F. Organic contaminants volatilize and combust; any combustible gases remaining are incinerated in an afterburner. Off-gases pass through a series of emissions-control devices: a quench tank, where a heat exchanger cools the gas stream; a venturi scrubber, to remove particulates; a packed-bed scrubber, to remove acids; and a demister, to remove water. The operators periodically slow the rotation of the tub and release part of the melt from the bottom of the tub.[47,52]

A *cyclone furnace* has also been used to vitrify soil. Soil — dry or in a slurry — is fed into a cyclone furnace similar to a coal-fired utility boiler. Natural gas and preheated combustion air enter tangentially into the burner. This gas stream induces turbulence within the furnace. The force of the gas stream pushes soil particles to the walls of the furnace, where they melt into a slag layer. The slag exits the furnace at the base of the cyclone at a temperature of 2400°F. The slag is quenched in a water-cooled tank. Off-gases are cooled to approximately 200°F, then pass through a baghouse to remove particulates. The fly ash collected in the baghouse must be landfilled or may be recycled through the unit. A scrubber removes acid gases.[53]

Finally, soil or sludge can be treated using a variation of the commercial glass-making process. The waste is mixed with glass-forming additives (high silica material), then fed to a vitrification chamber. The amount of glass-forming additives that must be added varies with the type of waste to be treated and must be determined by testing; one vendor has noted that the mixture can contain up to 95% waste, 5% glass.[54]

In the vitrification chamber, the mixture is heated to temperatures on the order of 1600°F. The vitrification chamber is refractory lined and water cooled. Vapors from the process are collected and treated. Off-gases are then treated to remove acidic components and particulates. The molten discharge cools to a glass-like product, similar to that produced by ISV. One vendor reports that treatment results in a volume reduction of up to 50%, and a weight reduction on the order of 15%. Another vendor notes that this form of treatment can be limited by the concentrations of certain compounds in the waste, including organics, sulfur, and volatile heavy metals.[47]

4.2.5.2 In Situ Vitrification

Battelle's Pacific Northwest Laboratory developed ISV in the early 1980s for the Department of Energy (DOE) as a means to treat radioactive wastes.[55] The technology is available through a single vendor, the Geosafe Corporation.

Three process trailers contain power control and off-gas treatment equipment. A fourth trailer brings the off-gas hood to the site. Additional trucks bring ancillary equipment to the site. Setting up the treatment system requires approximately 3 weeks.[56]

ISV begins with the insertion of four 12-in.-diameter graphite electrodes into the soil in a square array. This array can vitrify soil in an area with a maximum width of some 35 to 40 ft. Larger areas are treated with multiple arrays of electrodes. The electrodes, which are initially inserted into shallow soils, are moved downward as treatment progresses to allow treatment to a maximum depth of approximately 20 ft.

The contractor places a conductive mixture of flaked graphite and glass frit just below the soil surface between the electrodes to act as a conductive starter path. An electrical potential is then applied to the electrodes, establishing an electrical current in the starter path. The current generates heat along the starter path, raising the temperature of the surrounding soil to a temperature between 1600 and 2000°C (2900 to 3600°F). Geosafe notes that ISV can operate at a higher temperature than melter- or furnace-based technologies because of equipment limits on the latter.[46]

Oxidation eventually consumes the graphite starter path and the current is transferred to the molten soil, or the *melt*. As the melt moves downwards and outwards at a rate of 1 to 2 in. per hour, it incorporates nonvolatile elements and destroys organic components by pyrolysis. The pyrolysis byproducts migrate to the surface of the vitrified zone. When they reach the oxygen environment at the surface, these compounds combust. An off-gas hood over the treatment zone collects volatilized metals, the combustion products of organic contaminants, and steam generated from the water in the soil. The off-gas stream must be treated to remove contaminants. Soil can be treated at a rate of approximately 4 to 6 ton/h.[45] Soil can be treated to a maximum depth of approximately 20 ft in relatively homogenous soils.

The melted soil eventually cools into a glassy solid. Since the void spaces initially in the soil are destroyed, ISV results in a volume reduction of 30% to 50%.[57] For typical soils, the final density of ISV glass is approximately 2.5 to 2.8 g/cm^3, which is more dense than concrete. The tensile and compressive strength of unfractured ISV product is reportedly an order of magnitude greater than unreinforced concrete.[58]

Currently there is little full-scale experience with this process. Commercial operations began in the early 1990s. By 1995, the U.S. EPA had selected ISV for a dozen waste sites, but six of these decisions had been reversed. By late 1996, remediation was occurring at two sites and complete at two sites.[59] At the Wasatch Chemical Superfund Site in Utah, Geosafe treated about 6000 tons of soil and debris containing dioxin, pentachlorophenol, herbicides, and pesticides. The technology was also demonstrated at the former Parsons Chemical Site in Michigan, where about 330 cy of soils contained pesticides, mercury, arsenic, chromium, lead, and low levels of dioxins.[45]

ISV has several limitations:

- Volatile metals — lead, cadmium, arsenic, and mercury — can volatilize from the soil during treatment, particularly if located in shallow soils. These metals must be collected in the off-gas treatment system.
- The water content of saturated soils can make ISV costs prohibitive, due to the energy consumed by vaporizing water. Saturated soils can potentially be vitrified if the soils are of low permeability (1×10^{-4} cm/sec or less)[60] or some means are taken to limit recharge.[50]

- The technology is not recommended for sites where metals in the soil exceed 25% by weight or where inorganic debris exceeds 20% of the soil by volume.[60,61]
- Loosely packed rubbish, coal, or other combustible solids can catch fire and may need to be removed from soil before treatment.[62]
- Combustion byproducts from combustible organics in the waste may overwhelm the off-gas control system if the organics exceed some 5% to 10% (by weight) of the soil or waste.[50,60]
- Buried drums containing vaporizable solids (e.g., organic liquids) cannot be readily treated.[62]

4.3 SEPARATION TECHNOLOGIES

Contaminants can be separated from soils based on their volatility at ambient or elevated temperatures, their solubility in water or another solvent, or their electrical charge. Separation technologies produce a stream of concentrated contaminants which must be further treated to destroy or immobilize the contaminants.

4.3.1 Soil Vapor Extraction

Soil vapor extraction, abbreviated SVE and sometimes called *soil venting, in situ vaporization*, and *enhanced volatilization*, has been used since the early 1970s to remove volatile contaminants from soil.[63] SVE systems typically use a series of extraction wells screened in the unsaturated zone to collect soil vapors containing volatile contaminants. A blower mounted at the surface induces a vacuum on the wells and stimulates air flow through the soil to the wells. Contaminants adsorbed to the soil and in the liquid in soil pores volatilize to the air in the soil pores. The vacuum on the extraction wells draws the contaminant-laden soil gas into the SVE system. Extracted vapors are usually treated at the surface to remove or destroy the contaminants.

4.3.1.1 Physical Principles

The effectiveness of a SVE system depends primarily on the volatility of the contaminant and the permeability and homogeneity of the soil. As discussed further below, SVE is most successful for highly volatile compounds and homogenous, highly permeable soils.

SVE is often used to remediate sites contaminated from spills or leaks of NAPLs comprising VOCs. Recall that NAPLs spilled onto soil move downward and outward through seams and lenses of the most permeable soil materials in response to gravity. As the liquid moves through the soil, it must displace air and water in the soil pores. LNAPLs migrate to the water table and begin to spread laterally. DNAPLs can sink through an aquifer until the mass is exhausted or until the liquid meets a low-permeability layer. The contamination distributes through the soil in several phases: in the unsaturated zone, contamination can remain as NAPL residuals in soil pores or rock fractures; contaminant molecules adsorbed to soil grains; contaminant molecules dissolved in the pore water; and in the vapor phase in soil gas. At an undisturbed site, contamination reaches near equilibrium among these phases.

SVE draws air through the soil pores, preferentially through the most permeable materials. This air flow removes contaminant mass by *advection*, i.e., transport with the flow of the air resulting from a pressure gradient.[64,65]

As the vapor-phase contamination is removed, contaminants in the liquid phase and adsorbed to the solid phase continue to volatilize into the soil gases. The rate at which contaminants volatilize depends upon their concentration, volatility, water solubility, and tendency to sorb to the soils. The rate of desorption of contaminants from soils and the rate of volatilization from liquid phases often limit the rate at which contaminants are removed by SVE.[66]

In heterogeneous soils, advection does not remove contaminant molecules from within the pores of fine-grained material because the air flow through those materials is limited. Contaminant molecules trapped in the pores of fine-grained soils must diffuse to the more permeable zones in order to be swept along in the vapor stream and captured by the SVE system. *Diffusion* refers to the migration of contaminant molecules from an area of relatively high concentration to an area of relatively low concentration as a result of random molecular motion.[67] The rate of diffusion is typically much lower than the rate of advection, and can be the limiting factor in remediation by SVE.[66]

SVE is used to remove both nonchlorinated and chlorinated VOCs from unsaturated soil. It is also used to treat lighter semivolatile organic compounds such as phenol and naphthalene.[68] It is commonly used to remediate soil contaminated by leaking fuel-storage tanks.

The potential effectiveness of SVE can be gauged by three parameters which indicate the volatility of a contaminant: boiling point, Henry's law constant, and vapor pressure. In general, SVE can be used to remove compounds with a boiling point greater than 250 to 300°C, or a Henry's law constant greater than 100 atm.[69] SVE should work best on contaminants with a vapor pressure greater than 0.5 mm Hg, and will not readily remove compounds with a vapor pressure less than 1×10^{-7} mmHg. However, the less volatile compounds may biodegrade in a bioventing system (see Section 4.4.1.1.3). Since the vapor pressure of a compound depends on the temperature, heating the soil will enhance SVE of less volatile compounds.[70] See the discussion of *in situ* thermal desorption in Section 4.3.2.2, and the related discussion of steam injection for NAPL recovery in Section 3.6.2.3. See also, for example, References 71, 72, 73.

The potential effectiveness of SVE also depends on the nature of the soil. Air flows preferentially through permeable soils such as sands and gravels, or through seams of such soils in silts and clays. For SVE, permeability is characterized by *air permeability* rather than intrinsic permeability (Equation 3.15) or hydraulic conductivity. Air permeability reflects the resistance to flow through the air-filled pore space rather than through the entire pore space. Water in the soil pores reduces the air permeability, and as a result, the air permeability is usually lower than intrinsic permeability. As the degree of water saturation increases, the relative permeability of the soil to air decreases nonlinearly, and effective SVE becomes more difficult. The relationship between air permeability and soil moisture content is location and site specific.[74]

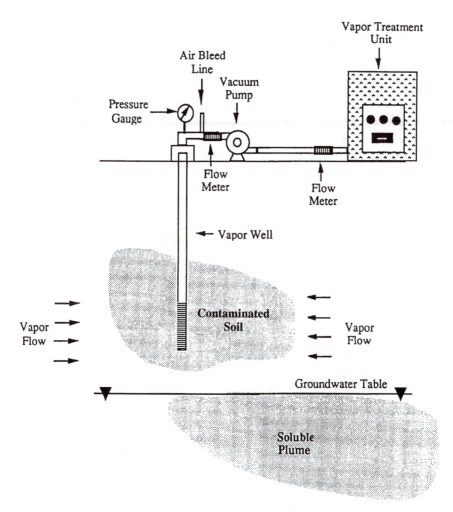

Figure 4.4   Schematic of soil vapor extraction system. (From Johnson, P. C., *Hyperventilate Users Manual, A Software Guidance System Created for Vapor Extraction Operations*, EPA 500-C-B-92-001, U.S. EPA Office of Solid Waste and Emergency Response, Washington, D.C., March 1992.)

4.3.1.2   Design of a Soil Vapor Extraction System

A SVE system comprises a series of extraction wells and associated piping; a blower or vacuum pump to induce the air flow; and a treatment system designed to remove contaminants from the extracted vapors. Figure 4.4 shows a typical system. Each of the components is described briefly below, as are design variations sometimes used to enhance air flow through soil.

4.3.1.2.1 Soil Vapor Extraction Wells — SVE systems typically include vertical wells similar to groundwater extraction wells (Section 3.2.1.1) or vacuum extraction wells used for free-product recovery (Section 3.6.1.1.3). SVE wells commonly have a PVC casing 4 to 6 in. in diameter, although larger and smaller wells are sometimes used. The well is typically screened over the stratum requiring treatment.[69,75]

SVE systems commonly include a network of extraction wells. These wells are linked through a *manifold* to the *header*, or pipe which leads to the blower and treatment system. Particularly for smaller sites, the well spacing, applied vacuum, and air flow rate may be chosen simply based on experience at other sites. At larger or more complex sites, pilot-test data are used to design a full-scale system, often using one of the many available models (e.g., see References 76, 77). As with groundwater models, the user must consider the underlying assumptions and limitations of the model when applying the model results.

A design team may determine the number and spacing of the wells from the radius of influence (ROI) of each individual well, similar to the design of a groundwater extraction well network. The ROI of an extraction well is the area over which an extraction well can affect the air pressure. It depends on the air permeability of the soil, the applied vacuum, and the air flow rate.[78] The ROI of a single well is typically between 5 ft (in fine-grained soils) to 100 ft (in coarse-grained soils).[69] All else being equal, the ROI typically increases with the depth of the screened interval.[78]

The ROI of an extraction well indicates the area over which the change in pressure can contain contaminant vapors. The change in pressure may not cause a significant vapor flow or effect remediation, nor can a design based on a ROI be tied directly to remediation goals.[79] Instead of designing a system solely based on the ROI, the number and spacing of vapor extraction wells can be based on a theoretical performance requirement. Two examples of such design bases follow.

Johnson and Ettinger[79] suggest that the minimum number of wells (N_{wells}) can be calculated from the minimum air flow required to remove a mass of contamination (M_{cont}, kg) over a desired remediation time (T_R, seconds). The calculation is based on the maximum amount of contamination that can volatilize at equilibrium:

$$N_{wells} = \frac{\alpha M_{cont}}{Q_{well} T_R} \qquad (4.2)$$

where α [m^3 vapor/kg contaminant] is the minimum volume of air per unit contaminant mass required to achieve remediation under ideal flow conditions and Q_{well} is the estimated flow rate to a single well. For single-component systems under ideal flow conditions, $1/\alpha$ is the saturated vapor concentration.[79]

An alternative approach is to design the extraction system based on the air velocity that will theoretically result in an air exchange rate (pore volumes per time) that will achieve the clean-up goal in the desired period of time:

$$N_{wells} = \frac{nV}{Q_{well} t} \qquad (4.3)$$

where n is the soil porosity [unitless], V is the volume of soil to be treated, and t is the desired pore volume exchange time [hours].[69] The time required for pore volume exchange, t, can be related to the total time required for site remediation, T_R, by the number of pore volume exchanges required to achieve the remediation goal. Some experts suggest that as few as 200 to 400 pore volume exchanges suffice, while others indicate that 2000 to 5000 may be necessary. The higher estimates may reflect mass-transfer limitations due to diffusion or desorption. Estimating the number of pore volume exchanges for a specific site requires pilot testing or experience at a similar site.[80]

Equation 4.3 incorporates many simplifying assumptions. For a more complete discussion of this approach, see References 81, 82.

The preceding discussion focused on vertical wells. However, not all SVE systems use vertical wells. Common variations include the use of horizontal wells or trenches to treat contaminants *in situ,* and treatment of excavated soil in an above-ground soil pile.[68]

Horizontal wells and vapor extraction trenches are used at sites with shallow groundwater and/or soil contamination near the surface (e.g., less than 10 ft deep).[69] Trenches are constructed similar to groundwater extraction trenches, except that a SVE trench is sealed to prevent short-circuiting of air flow from the surface to the piping.[83]

Soils can be excavated and treated by SVE in a soil pile. Certain RCRA regulations can pertain to the construction of a soil pile: waste piles [40 CFR 264.250] or containment buildings [40 CFR 264.1100]. The area where soils are to be treated is usually prepared by constructing a low-permeability liner to control leaching and run-on/run-off of precipitation. A network of slotted pipe placed on the liner is connected to a blower to induce air flow. Finally, workers pile soil over the piping to a typical depth of 1.2 to 3 m. The pile may be covered or constructed within a building to protect the pile from the weather. The construction crew must avoid compacting the soil for optimum treatment.[84]

4.3.1.2.2 Optimizing Air Flow Through the Soil — Several techniques are used to optimize the flow of air through soil. These include surface sealing, air injection wells, and, less commonly, soil fracturing to increase the soil permeability.

The ground surface is often sealed using pavement, a geomembrane made of HDPE or low-density polyethylene, or compacted clay. Many industrial sites are already paved. However, if the pavement was constructed with a porous subgrade, it may not effectively seal the surface. Surface sealing limits the vertical short-circuiting of air — or flow of air from the surface directly downward to the well screen — and increases the effective ROI of a SVE well as a result. Surface sealing also limits infiltration and the corresponding decrease in air permeability which results from increased soil moisture.[69,84]

Many designs incorporate either passive or active air injection wells. A well left open to the atmosphere allows the SVE system to draw air into the subsurface at the location of that well and bounds the ROI of an extraction well. Air can also be actively injected through a well using a blower or compressor, similar to an air sparging well. Alternatively, steam may be injected to increase the vapor pressure of a target contaminant. A designer or operator may site an active air injection well in a stagnant zone at the edge of the ROI of a single well, or at the confluence of the radii of influence of several wells. The injection well may be screened across the entire depth of contamination, or screened at a particular depth to enhance the air flow through a difficult-to-treat soil stratum. Active air injection adds to the capital cost of a SVE system. Because active air injection increases the flow of air and dilutes the contamination, it also increases the costs for operating the vapor treatment system.[85]

Hydraulic fracturing and *pneumatic fracturing* create fractures in soil or rock to increase air permeability and thus advective contaminant transport. Fracturing can be used to enhance the performance of a SVE system when the performance is not diffusion limited. Soil fracturing also has the potential to enhance other *in situ* soil and groundwater remediation technologies.

Hydraulic fracturing has been successfully used to enhance oil recovery from low-permeability rock since at least the 1950s. Researchers at the University of Cincinnati and the U.S. EPA began testing hydraulic fracturing as an enhancement to SVE in the late 1980s.[86,87]

Injection of a sand slurry into soil at high pressure can create sand-filled soil fractures up to 1 in. thick and 30 ft in radius from the well, at target depths ranging from 5 to 40 ft BGS. Hydraulic fracturing begins with drilling a 6- to 8-in. borehole using a drill rig with a hollow-stem auger. A *fracturing lance*, which consists of an outer casing and an inner rod which both have hardened conical points, is driven through the borehole to the desired depth. The lance is then retracted to expose soil at the bottom of the casing. Next, steel tubing is inserted into the casing. Water pumped at high pressure (ca. 3500 psi) through the tubing cuts a disc-shaped notch into the soil at the bottom of the casing. Finally, a hydraulic fracture is created by pumping a sand slurry into the soil through the notch at 10 to 25 gpm. Fractures growing from that nucleus extend some 30 ft from the borehole.[88]

Hydraulic fractures can be created in rock and in relatively uniform silty clays that are overconsolidated (permeability less than 10^{-7} cm/sec). Hydraulic fracturing is not effective in normally consolidated clays. One set of field trials in silty clay soils indicated that hydraulic fracturing can increase the flow of air through soil by up to an order of magnitude, and the flow of water by a factor of 25 to 40%.[88] Note that this increase indicates flow through the fractures; diffusion through the unfractured soil to the fractures may still limit the rate of remediation.

Pneumatic fracturing, developed for use at hazardous waste sites by Accutech and the Hazardous Substance Management Research Center at the New Jersey Institute of Technology, uses compressed air to fracture soil or rock such as shale. Compressed air (up to 500 psig) is injected into the formation in short bursts (<1 min) through uncased boreholes, causing the formation to fracture at weak points. Specific depth intervals are targeted using a proprietary system which isolates sections of the borehole using inflatable rubber "packers".[89] Unlike the sand-filled fractures created by hydraulic fracturing, pneumatic fractures may close as time passes.[86]

4.3.1.2.3 Blowers and Vacuum Pumps — A blower or vacuum pump is used to impose a vacuum on the SVE wells, creating a pressure drop in the soils in the unsaturated zone and inducing air flow toward the well. The type and size of the unit are based on the desired vacuum (pressure head drop) and air flow rate, and the pressure losses due to friction in the extraction and treatment system.[90] The vacuum at the wellhead is usually between 3 and 100 in. of water, typically 20 to 40 in. of water vacuum. Tighter soils require the highest vacuums. The typical air flow rate is between 10 and 100 cubic feet per minute (cfm) per well.[69,91] Depending on the contaminants treated, the blower may need to be explosion proof.

Commonly used units include regenerative blowers, rotary-lobe blowers, and liquid-ring vacuum pumps:[90,92]

- A *regenerative blower* is a type of centrifugal blower. In general, a centrifugal blower imparts energy to an air stream via a rapidly rotating impeller or propeller. A regenerative rotary blower has a short-bladed turbine impeller. Leeson and Hinchee[92] provide a concise description of its operation: "As the regenerative blower impeller rotates, centrifugal acceleration moves the air from the base of the blade to the blade tip. The fast-moving air leaving the blade tip flows around the housing contour and back down to the base of the next blade, where the flow pattern repeats. This repeated acceleration allows a regenerative blower to produce higher differential pressure than a conventional single-stage radial flow design. The regenerative blowers can also produce higher vacuum at the suction port compared with a pure radial design..."[92] Regenerative blowers are typically used for applications requiring a relatively low vacuum (less than 80 in. of water).
- A *rotary lobe blower* is a type of positive displacement blower. A twin-lobe blower contains two figure-eight-shaped impellers mounted on parallel shafts within a housing. These impellers rotate in opposite directions. As each lobe rotates past the blower inlet, it captures a volume of intake air. As the lobe continues to rotate toward the outlet, the captured air is compressed, then expelled from the blower. Rotary-lobe blowers are usually used for applications requiring a medium-range vacuum (20 to 160 in. of water).
- A *liquid ring vacuum pump* is also a positive displacement mechanism. A rotary-vaned impeller mounted off-center in a cylindrical pump casing transfers both liquid and gas through the pump. The rotation of the impeller causes the liquid to form a ring around the inside of the casing. Again, Leeson and Hinchee[92] provide a concise description: "Because the impeller is off center, the cavity formed between two impeller vanes and the water seal changes size as the vanes move around the pump housing. Air enters the pump where the cavity formed by the vanes and the water seal is large and discharges where the cavity is small, thus increasing the pressure of the pumped gas."[92] A water-sealed pump generates contaminated wastewater from the liquid ring which must be treated. Liquid ring vacuum pumps can generate a vacuum level close to zero.

SVE systems typically operate 24 h per day. The noise from the blower can be an issue with the community if the site is a residential area.[93] The blower can be equipped with a silencer if noise is an issue.

4.3.1.2.4 *Vapor Treatment System*

— Extracted soil vapors are treated to remove water and organic contaminants. Treatment units typically include a knock-out tank with a demister, and activated carbon. A treatment system could also include a catalytic oxidation unit, thermal oxidation unit, or biofilter. See Section 3.7 for information on these treatment units.

The level of volatile compounds in the off-gas from a SVE system typically drops rapidly. As a result, the most cost-effective approach to emissions control can be to lease an oxidizer unit for emissions control during the initial period of operation, then switch to activated carbon as the concentrations in the extracted vapors drop.

4.3.1.3 Effectiveness

Operators monitor several parameters to gauge the effectiveness of SVE systems. These include the air pressure at the extraction wells and monitoring wells; the vapor flow rate at each well and at the blower; and the concentrations of target compounds in the soil gas, the extracted vapor stream, and in the emissions from the treatment unit. Other parameters may be monitored to characterize the performance of the vapor treatment system.[91,94]

The concentrations of volatile organic contaminants in the extracted vapor typically drop rapidly during the initial period of operation, often followed by a slower decrease in concentration. This pattern resembles the asymptotic behavior described in Section 3.2.4 for groundwater extraction systems. Soil vapor concentrations may rebound after an SVE system is shut down.

SVE systems can effect treatment rapidly. Based on work at 103 Superfund sites, treatment typically requires 1 month to 5 years or more.[68]

Example 4.4:   The EPA selected and implemented SVE as a source-control measure at the Thomas Solvent Raymond Road (TSRR) area in Battle Creek, MI. This project was one of the earlier applications of SVE. The EPA described the project in a cost and performance report which formed the basis for this summary.[95]

Between 1963 and 1984, a facility at the TSRR site stored, blended, repackaged, and distributed solvents. As a result of spills and tank leaks, the facility contaminated soil and groundwater with VOCs, primarily tetrachloroethylene (PCE) and 1,1,1-trichloroethane. The site was one of three source areas which contaminated the Verona Well Field (described in Example 3.2).

The remedial investigation (RI) completed in 1983, indicated that soils in the vadose zone at TSRR contained some 1700 lb of VOC over an area of approximately 1 acre. The surficial soil at the site is predominantly fine sand. This layer of sand, some 15 ft thick, overlies sandstone layers some 100 to 120 ft thick. The sandstone, in turn, overlies shale.

Data collected over the course of the project characterize the sand layer as having an estimated porosity of 30 to 40% and a hydraulic conductivity of 0.0025 cm/sec. With a moisture content of approximately 5%, the vadose zone has an estimated air permeability of 10 cm/sec. The depth to groundwater is typically 14 to 16 ft BGS, although this was lowered by groundwater pumping to 16 to 25 BGS.

In 1985 the EPA determined that SVE should be used to remove the VOCs from the vadose zone. The project team initially expected that 98% of the estimated mass of VOCs could be removed in 2 years. The goal for source control evolved over the course of the project. The 1985 ROD did not specifiy clean-up goals for the SVE system. The contract documents subsequently prepared for construction and operation and maintenance specified two goals: (1) no sample of treated soil should exceed 10 mg/kg VOCs,

Table 4.3   Clean-Up Levels and Soil Data for Thomas Solvent Raymond Road

Compound	Maximum concentration, unsaturated soil, mg/kg	Clean-up level for soil, mg/kg	Range of concentrations in treated soil, mg/kg
PCE	1800	0.010	ND to 0.711
1,1,1-TCA	270	4	ND to 0.004
TCE	550	0.06	ND to 0.047
Toluene	730	16	ND to 0.073

and (2) less than 15% of the samples of treated soil should contain VOCs at concentrations greater than 1 mg/kg VOCs. In 1991, after SVE had begun, the EPA signed a second ROD. That ROD specified clean-up levels for VOCs in soil and groundwater, including the values shown in Table 4.3. The clean-up level for PCE subsequently changed again, to 0.014 mg/kg, after the State of Michigan enacted a law (Act 307) which included soil clean-up standards.

The EPA began a SVE pilot test at the site in November 1987 using four SVE wells, operated at air flow rates of 60 to 165 standard cubic feet per minute (scfm) per well and a vacuum at the wellhead of 3 to 4 in. Hg. The project team began by operating each well independently to determine the ROI; the relationship between the vacuum and vapor flow rate; evaluate the effect of underground storage tanks (UST) on the vacuum pressure distribution; and to identify VOC loading rates from each well as a function of the vacuum pressure and air flow rate. The tests showed that the ROI was greater than 50 ft. After a total operation time of 69 h over a period of 15 d, the pilot system recovered approximately 3000 lb of contaminants, nearly twice the initially estimated mass over the entire TSRR site.

In 1988, contractors collected additional soil samples before construction of the full-scale system in order to better characterize the mass of VOCs in the vadose zone. Those data indicated that the vadose zone contained 13,000 to 16,500 lb of VOCs. The initial estimate of 1,700 lb had been based on a soil sampling technique used early in the Superfund program which was later found to produce VOC results lower than the actual values.

Construction of the full-scale SVE system began in early 1988. The system included 23 extraction wells, each constructed with a 2- or 4-in. PVC casing and screened from approximately 5 ft BGS to 3 ft below the water table. Each well was fitted with a throttling valve to control the air flow, a sample port, and a vacuum pressure gauge. A surface collection manifold connected the extraction wells and conveyed the vapors to the treatment system. Vapors initially entered a centrifugal air/water separator. The two vacuum pumps (40 and 25 hp) which followed the air/water separator provided the vacuum to extract the vapors from the ground and move the vapors through the treatment system. The next unit in series was, at different times during the project, either vapor-phase carbon or a catalytic oxidizer. Finally, treated vapors discharged through a 30-ft stack.

Operation began in March 1988. Fourteen of the wells were used at a time to maximize the contaminant loading to the off-gas treatment system based on the VOC concentrations at the wellhead. The total vapor flow, between 1400 and 1600 scfm, was treated using two sets of four vapor-phase carbon units in series, each containing 1000 lb of granular activated carbon. Full-scale operation indicated that the total contaminant mass was roughly 25 times the estimated amount, and that, with a recovery rate over 10 lb per day (lb/d) carbon treatment was not cost effective. As a result the project team replaced the carbon units with a catalytic oxidation unit in January 1990. Operation then continued until November 1990, by which time the system had recovered some 40,000 lb of contaminants.

Between November 1990 and February 1991, the system was shut down to allow for removal of the USTs and reconstruction of the SVE system. The USTs were reportedly left in place during the initial phase of remediation due to health and safety concerns.

The rebuilt SVE system consisted of 20 wells, including 10 existing SVE wells, 8 new SVE wells, and 2 new dual groundwater/SVE wells. The SVE wells were spaced 40 to 60 ft apart. The combined air flow from these wells, approximately 1000 scfm, contained lower levels of VOCs and was treated using activated carbon. This system operated almost continuously from February 1991 to May 1992, when remediation stopped.

To determine whether or not SVE had accomplished the remediation goals, 115 soil samples were collected in a grid pattern. The system met the clean-up levels for all VOCs but PCE (Table 4.3). Of the soil samples, 20 contained PCE at concentrations greater than 0.014 mg/kg. However, the average concentration of PCE was less than that goal.

During 375 d of operation, the SVE system recovered 45,000 lb of VOCs. The recovery rate reached a high of 1000 lb/d during the first 2 weeks of operation. After 250 d of operation, the recovery rate dropped to less than 100 lb/d, and after 400 d, less than 1 lb/d. Costs directly associated with treatment totaled $1,645,281. Thus, remediation cost an estimated $62 per cubic yard of soil treated, or $37 per pound of contaminant removed. Before-treatment activities, such as pre-construction soil sampling and removal of the USTs, added another $535,180 of remediation costs.

Remediation of the source materials at the TSRR site also included recovering over 150 gal (1200 lb) of LNAPL by pumping from an extraction well; groundwater extraction between March 1987 and December 1991 at 300 to 350 gpm (groundwater was pumped to the Verona Well Site for treatment); and, briefly, nitrogen sparging, to enhance recovery of VOCs from groundwater without oxidizing iron.

4.3.1.4 Related Technologies

Several variations of SVE are used in site remediation.

- As noted in Section 3.5.2, SVE is often used to collect the emissions from air sparging volatile contaminants from groundwater.
- A modified version of SVE, *vacuum extraction* or *multiphase extraction*, is used to recover LNAPL. As described in Section 3.6.1.1.3, the vacuum induced by a surface-mounted vacuum pump aspirates LNAPL, air, and water through a suction pipe or slurp tube suspended in a well screened across a LNAPL layer. This configuration can also be used to extract soil vapors and groundwater when NAPL is not present and the yield from a pumping well would be quite low (generally 5 gpm or less).[96]
- Another variation, sometimes called *dual-phase extraction*, combines groundwater extraction with vacuum extraction in a single well or series of wells.[96,97] If the

objective is to treat both unsaturated and saturated soil, the well screen extends through both zones. Alternatively, the well is screened through the saturated zone only in order to treat primarily soil below the water table. A submersible pump is suspended in a sump below the screen. As the pump extracts groundwater or groundwater and LNAPL, it lowers the water table in the vicinity of the well. (The extracted groundwater is pumped through a treatment system, then discharged.) A vacuum is also imposed on the extraction well(s) to withdraw vapors. The system will draw air through the cone of depression, and, to an extent dependent on the position of the well screen, through the vadose zone. A high vacuum blower, which imposes a vacuum of approximately 18 to 26 in. of mercury, is used to treat low-permeability formations (e.g., sandy silts to clays). A low vacuum blower, which imposes a vacuum of approximately 3 to 12 in. of mercury, is used to treat more permeable soils (sands to silty sands).

- In *subslab depressurization*, a SVE system is installed near a building to prevent volatile contaminants in soil or groundwater from entering the building.
- Finally, SVE is used to enhance aerobic bioremediation by increasing the air flow through soil. SVE designed to enhance bioremediation is called *bioventing*. Similar equipment is used for SVE systems and bioventing systems. However, the design goal and operating conditions differ. While SVE systems are designed to maximize the removal of contaminants, bioventing systems are designed to optimize the rate of oxygen transfer to the contaminated soils. Bioventing systems utilize relatively low air flow rates which are high enough to provide sufficient oxygen for biodegradation, but low enough to minimize volatilization and subsequent off-gas treatment.[98] Section 4.4.1.1.3 discusses bioventing further.

4.3.2 Thermal Desorption

Thermal desorption refers to the process of heating soil or sludge to volatilize contaminants and remove them from the soil. The off-gases are collected and treated. Soil can be excavated and treated or treated *in situ*.

Thermal desorption is the BDAT for certain wastes under the RCRA Land Disposal Restrictions [40 CFR 268]. Depending on the operating temperature of the unit, thermal desorption can remove VOCs, such as benzene; petroleum hydrocarbons; creosote or coal tar; PAHs; certain herbicides and pesticides; PCBs; and dioxins.[68,99] Thermal desorption has also been used to destroy complexed cyanides in coal-gasification waste[100] and to remove mercury from soil.[99,101]

4.3.2.1 Treatment of Excavated Soil

Excavated soil can be treated using a mobile unit brought to the site, or taken off-site to a permitted fixed-base unit. As of 1996, over 30 vendors offered full-scale treatment systems.[99]

The basic components of a thermal desorption system include: pretreatment and feed, thermal processor and discharge, and air emissions control. Each of these components is described briefly below. The critical parameters which affect treatment include (1) the type of soil and moisture content, and (2) temperature, residence time, mixing, and the sweep gas flow rate in the thermal processor.

4.3.2.1.1 Pretreatment and Feed — Thermal desorption systems typically have particle size limitations of 0.75 to 4 in., depending on the components in the system.[99] Soil containing oversized particles must be screened to remove large particles before treatment. The oversized particles can be crushed and treated by thermal desorption, or, depending on the level of contamination, backfilled on site or landfilled.

Soil or sludge may also be pretreated to remove or absorb water before treatment. Several vendors suggest that soil or sludge should contain no more than 20% to 50% water, depending on the system, if it is to be treated by thermal desorption.[99] The U.S. EPA notes that a moisture content of 15% to 25% requires some reduction in throughput (treatment rate), although without a significant effect on the cost; a higher moisture content increases the cost of treatment significantly.[102] A high water content affects treatment in three ways. First, wet soils increase the cost for supplemental fuel to provide the energy needed to heat and evaporate the water from the soil. As a result, the moisture content of the soil is one of the critical variables which determines the cost of treatment. Second, fine-grained soils are difficult to feed to and move through the treatment system. Clay balls are difficult to treat, and plastic soils can smear or cake on equipment. Third, wet soils generate a large quantity of water vapor which must be removed from the off-gases and treated.

Soils are loaded into the heating chamber using several types of feed equipment; in many systems, the soil conveyor passes through an air lock to minimize the oxygen entering the thermal processor. Soils are commonly dumped into a hopper using a front-end loader, then fed to the thermal processor using an auger or screw conveyor. Some systems use belt conveyors.

4.3.2.1.2 Thermal Processor — Thermal desorption systems are classified by the temperature at which they operate and the type of equipment used to heat the soils. In general, *low-temperature thermal desorption* units operate at temperatures between 300 and 800°F. Low-temperature units treat soil containing contaminants with relatively low boiling points, such as VOCs and TPH. *High-temperature thermal desorption* units generally operate between 600 and 1200°F. Such units are used to treat soils containing contaminants with higher boiling points, such as PAHs and PCBs. The distinction between high-temperature and low-temperature units is not precise, and refers to different operating conditions rather than different types of thermal processors. High- and low-temperature thermal desorption systems may be constructed of different materials, as appropriate for the operating temperatures, and include different emission-control technologies, as appropriate for the target contaminants which are typically treated at those operating temperatures.[99]

Several different types of equipment are used to heat soils, a rotary drier or drum being the most common. Figures 4.5 and 4.6 show one such system. A rotary drier is a steel cylinder mounted on a slight incline.[99] The soil is fed to the high end of the drier or drum. As the drier rotates the soil moves downward through the unit. The design can incorporate internal mixers or lifters which lift the soil as the drier rotates, carry the soil to the top of the cylinder, then drop it through the hot gases. The retention time in the drier is generally 20 min or less.

Figure 4.5 Thermal desorption system with an indirectly heated rotary drier. (Photograph reprinted with permission of Maxymillian Technologies, Inc., Boston, MA.)

The soil in the drier is heated indirectly or directly. In an *indirectly fired* unit, combustion occurs outside the drum. In a *directly fired* unit, a propane or fuel-oil burner is located either at the feed or discharge end of the unit. Combustion gas from the burner contacts the waste directly. If the burner is located at the feed end, the combustion gas flows co-currently with the waste; if the burner is located at the discharge end, the combustion gas flows counter-current to the waste. Because of the difference in heat transfer efficiency, a co-current system will heat soils to a higher average solids temperature for a longer period of time than an equivalent counter-current unit; however, it will have a higher gas flow rate, requiring larger vapor treatment equipment.[103]

Other thermal desorption systems use a *hollow-flight auger* to heat the soils.[99] One or more hollow-flight augers or screws rotates within a jacketed trough to move the waste through the unit. A heat-transfer fluid such as oil or molten salt flows through the center of the hollow flights of the screw augers and through the trough jacket to provide the thermal energy needed to desorb contaminants. The heat-transfer fluid circulates through a heater fired by propane, natural gas, or fuel oil.

Finally, a few vendors use infrared heating or fluidized bed reactors. See Section 4.4.3 for descriptions of analogous equipment used to incinerate soil.

Regardless of the type of thermal processor, the goal of thermal desorption is to separate, not to incinerate, the contaminants. Vendors achieve this goal by operating at temperatures lower than incinerators and, often, by using a *sweep gas* in the thermal processor which contains little or no oxygen. This gas stream, which may be pure nitrogen or an air stream with some oxygen removed (commonly, combustion

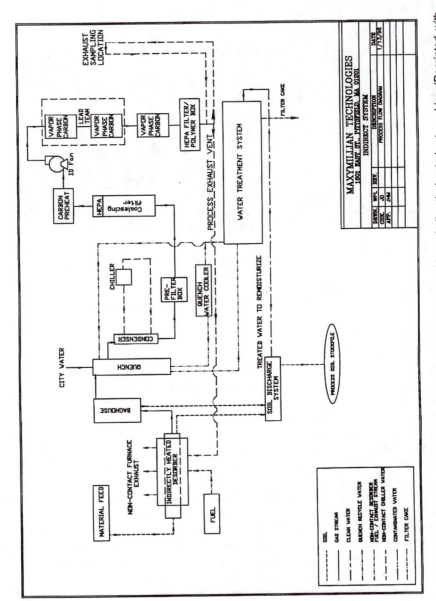

Figure 4.6 Process flow diagram of thermal desorption system with an indirectly heated rotary drier. (Reprinted with permission of Maxymillian Technologies, Inc., Boston, MA.)

gas), sweeps the desorbed contaminants from the thermal processor to the off-gas treatment system. Once treated, an inert gas such as nitrogen is recycled back through the thermal processor.

The rate at which soils can be treated depends primarily on the size and type of the thermal processor, the moisture content of the soils, the type of soils, and the level of contamination. Wetter soils must be processed more slowly than dry soils. All else being equal, sandy soils can be treated at roughly twice the rate of clay soils. Many mobile systems can treat soils at a rate between 20 and 40 tons per hour (tph), with processing rates for some full-scale systems as low as 3 tph or as high as 100 tph. (For efficiency, thermal desorption systems can operate 24 h per day rather than cooling and reheating the system each work day.) A mobile system occupies an area roughly 70 by 80 ft to 200 by 300 ft, depending on the system and not including soil stockpile areas or ancillary facilities.[99]

Treated soil typically exits the thermal processor by a screw conveyor. The clean, hot soils are misted or mixed with water — sometimes obtained from the air pollution control system — to cool the soil and suppress dust. Finally, the soils are stockpiled for testing and eventual backfilling on site.

4.3.2.1.3 Emissions Control

4.3.2.1.3 Emissions Control — The type of emissions control varies with the type of system and the waste stream being treated. Off-gas treatment may include:

- a baghouse or cyclone, to remove particulates;
- a condenser, to cool the gas stream and remove water and/or liquid hydrocarbons from the gas stream; water must be treated, and condensed organics may be recycled or incinerated;
- an afterburner, to destroy organic contaminants; and/or
- activated carbon, to adsorb organic contaminants.

Section 4.5 provides additional information on these air pollution control units.

4.3.2.1.4 Operating Considerations

4.3.2.1.4 Operating Considerations — Operating concerns include the type of soil, the type of contamination, and regulatory and public concerns:[99,102,103]

- Oversized soil particles or debris must be removed before treatment.
- Soils with a high proportion of fines (clays) can cake, and can also generate high levels of particulates in off-gases. Some vendors suggest a limit of 30% fines. Above that level, soil must be fed more slowly, or should be mixed with coarser soils to decrease the level of fines.
- A high water content increases the cost of treatment and can cause materials handling problems.
- In systems which contain some oxygen, operators may limit the acceptable concentration of petroleum hydrocarbons in the feed (ca. 1% to 3%) in order to keep the contaminant concentrations in the vapors below the lower explosive limit. This limit can be met by blending soils.
- The vapors from treating wastes which contain relatively high levels of sulfur or chlorine can corrode system components and may require treatment to remove acid gases before discharge to the atmosphere. Many thermal desorption systems do

not include that type of pollution-control equipment. Similarly, treatment of mercury-bearing wastes requires specialized emissions-control equipment.

- Thermal desorption does not remove most heavy metals. Portland cement or another stabilizing agent can be added to treated soil when it is mixed with water in order to limit the mobility of metals in the treated soil.
- Thermal desorption can result in partial breakdown and reformation of contaminant molecules. As a result, the residuals can (but do not always) contain new contaminants of concern.
- Some systems are noisy and, since they run 24 h per day, the level of noise can cause problems if operated near residential areas or in other sensitive areas.
- Regulatory requirements vary. Thermal desorption systems that vaporize and then burn organic contaminants in hazardous waste are considered incinerators for the purpose of RCRA regulations. Thermal desorption systems which condense the organics so that they can be recycled, rather than incinerated, are considered recycling units which are exempt from RCRA regulations on treatment systems.

4.3.2.2 In Situ Treatment

Heating soil in place volatilizes many contaminants. The vapors are collected and treated in a SVE system (Section 4.3.1). Three techniques have been used to heat soil: electrical resistance, radio frequency (RF) heating, and steam injection.[104] Steam injection differs fundamentally from the six-phase soil heating or RF heating. In the former, steam is a heat-transfer fluid which conveys heat energy to the subsurface. In the latter two processes, the molecules of soil and pore water absorb electromagnetic energy and then dissipate the energy as heat.

As an electrical current passes through soil, the temperature rises due to resistive heating. The increase in temperature increases the vapor pressure of organic contaminants. It can also vaporize water in the soil pores, which has two effects. First, the formation of steam enhances the volatilization of organics (steam stripping). Second, the removal of water from the soil pores as steam increases the air permeability of the soil, enhancing SVE. This technique is particularly effective in clay soils due to the relatively high electrical conductivity of clays.[105]

Six-phase soil heating of unsaturated soil is accomplished using a series of six electrodes installed into contaminated soil in a circular array (e.g., 4.6-m-diameter circle). Multiple arrays are used to treat large areas. The electrodes, constructed of modified well-casing materials, are installed using conventional drilling equipment.

A conventional utility transformer converts three-phase electricity to six phase. Each of the electrodes is provided with a separate current phase. A seventh, electrically neutral electrode is placed in the center of the array. This neutral electrode also functions as a SVE well.

When the current is applied to the electrodes the soil temperature begins to rise. The temperature at any point in the treatment zone depends on the soil type and water content and the location within the array. Increasing the temperature throughout the treatment zone from approximately 20°C to approximately 100°C requires some 7 to 10 d,[106] and roughly 90 kWh hours of energy per cubic meter of clayey soil.[101] As the temperature rises and water vaporizes, water is added to the soil near the electrodes so that the current can continue to flow through the soil.

By 1996, this technology had been demonstrated at pilot scale at two sites and full scale at two sites. The vendor notes two limitations to applying this technology: due to the high voltage, access to the treatment zone must be limited; also, subsurface metal conduits complicate the voltage gradient patterns in the soil.[106]

RF energy has also been used at a limited number of sites to heat soils *in situ* and desorb contaminants. This process, like many others now applied to hazardous waste site remediation, was originally developed for petroleum recovery.[107] A RF heating system has several components:[108-110]

- A RF generator converts AC electricity to the desired frequency radio wave. The frequency is selected from the band designated for industrial, scientific, and medical use by the Federal Communications Commission, based on the size of the site and the soil properties.
- *Exciter electrodes* or *applicators* convey the energy into the soil. The electrodes, which have been made of copper pipe or a combination of components, are installed in boreholes in the vadose zone. A *matching network* regulates the energy flow to the electrodes to adjust for the changes in the characteristics of the soil as it heats.
- *Ground electrodes* made of aluminum piping and installed around the array of exciter electrodes also function as SVE wells.
- One vendor covers the ground surface with a vapor barrier to control air flow and metal shields to contain RF energy.

The soils are heated to a temperature generally between 150 and 200°C to enhance the desorption of organic contaminants, which are captured in a SVE system. Operators monitor temperature, pressure, and electromagnetic field strength.

Finally, soils can be heated in place by injecting steam into wells screened in or below a contaminated zone. Steam can be injected most easily into sandy soils, rather than clayey soils.[100] Section 3.6.2.3 described the injection of steam to mobilize NAPL. A handful of vendors offer steam injection processes.[111]

4.3.3 Soil Washing, Solvent Extraction, and Soil Flushing

Soil washing, solvent extraction, and soil flushing separate organic and inorganic contaminants from soil into an extracting fluid or groundwater. The liquid stream then requires further treatment to remove or destroy the contaminants. These three related treatment technologies each work differently.

- *Soil washing* processes use aqueous solutions to remove contaminants from excavated soil. Soil washing reduces the volume of contaminated material: the fine particles (to which contaminants are adsorbed) are washed out of the soil and disposed of as sludge.
- *Solvent extraction* processes use organic solvents to dissolve contaminants and remove them from the soil.
- *Soil flushing* is an *in situ* process. Aqueous solutions are leached through soil in place to flush contaminants out of the soil and into the groundwater. A soil flushing system must include extraction and treatment or *in situ* treatment of the contaminated groundwater.

Each process is described further below.

4.3.3.1 Soil Washing

Soil washing systems effect treatment by two means: (1) separating the fines, to which contaminants preferentially adsorb, from the coarse soil particles, and (2) dissolving contaminants which have precipitated or adsorbed to soil. Some processes use surfactants or pH adjustment to help to dissolve contaminants. Soil washing generates a stream of contaminated water which requires treatment. The washed sands and gravels can be replaced on site, but the fines must be landfilled or treated further.

Contractors have successfully treated soils containing fuel hydrocarbons, PAHs, PCBs, pentachlorophenol, pesticides, heavy metals, and radioactive wastes. Tars are difficult to wash from soils. Soil washing units have both minimum and maximum particle size limits. For separation and disposal of the fines to be cost effective, the soil or sediment should contain no more than 20% to 30% silt and clay. Different treatment systems have different limits on the maximum particle size (e.g., 2 in., 4 in.). Oversized particles must be screened out of the feed.[112]

Approximately 20 vendors offer full-scale soil washing equipment.[112,113] These vendors use different configurations of unit operations, as well as proprietary additives to enhance treatment. Depending on the vendor and the waste being treated, soil washing systems include different combinations of the following unit operations.[112] Much of this equipment was originally used for ore processing.

- Dry screens separate oversized particles from the feed before treatment, as described in Section 4.1.3.
- A *trommel screen* separates gravel (e.g., particles >0.24 in. in diameter) from the sand and silt in the feed.[114] A trommel screen consists of a cylindrical frame surrounded by wire cloth or a perforated plate and mounted on a slight incline. The size of the openings in the cloth or plate determines the particle size separation. Operators feed the soil or sediment into the upper end. As the cylinder rotates at 15 to 20 revolutions per minute, the finer particles fall through the wire cloth. The oversized particles discharge from the lower end of the cylinder.[115] In some soil washing units, the vendor adds a water stream or high-pressure water spray to a modified trommel screen to enhance treatment and to form the soil slurry.[116,117]
- *Wet screens* are used for two purposes. As an initial step in soil washing, water jets applied to feed soils on a screen help to break up agglomerates, separate out large particles (e.g., >2-in. diameter), and form a soil slurry.[112,113] Later in the soil washing process, shaker screens or vibrating screens can be used to dewater washed soil fractions.
- An *attrition scrubber* agitates a high-solids slurry of coarse sands. As the sand particles rub against each other, fine silt and clay particles and metal precipitates detach from the sand particles.
- A *countercurrent classifier* separates sands from fine soil particles or slimes. A countercurrent classifier is a cylindrical drum containing spiral flights attached to the inside of the drum to form cylindrical troughs. The drum is mounted on a slight incline. As the drum rotates, the feed moves upward through the drum. Wash water introduced into the top of the drum washes fines from the sand. The washed sand exits at the top of the drum. Wash water containing the fines flows from the bottom.[115]

- A *hydrocyclone* separates silts and colloids from sand based on the difference in specific gravity. The top portion of a cyclone is cylindrical and the bottom portion is conical. A soil slurry enters through an opening at the top of the cyclone. The entry pressure and tangential entry direction produce centrifuging action. Coarse sands spiral down the unit and exit the bottom at a solids content on the order of 70%. The water and fines exit the top of the cyclone.[115,116]
- *Froth flotation* separates particles based on their wettability. Separation occurs as a slurry is agitated in an open-top chamber. Fine air bubbles dispersed through the slurry form a froth. As the bubbles rise to the surface, hydrophobic particles rise to the surface. The particles which are readily wetted by water remain in suspension.[115] Froth flotation is used in soil washing to separate fine soil particles, which partition into the froth, from coarser particles such as sands.
- The water used to wash soils is treated to remove soil particles and recycled. Treatment entails conventional wastewater treatment processes such as clarification and dissolved air flotation to remove particulates. The sludge is dewatered by several methods, including gravity settling followed by a belt filter press, and centrifugation.

Vendors combine these unit operations in many ways. Figure 4.7 illustrates one of the simpler equipment configurations used for soil washing.

Mobile soil washing units typically treat some 10 to 50 tons of soil per hour,[112] although larger units (e.g., 300 tph) have been built. A 300-tph unit occupies an area approximately 120 by 100 ft.[116]

4.3.3.2 Solvent Extraction

Solvent extraction processes use organic solvents to dissolve contaminants and remove them from the soil. At least six vendors offer treatment processes.[118,119] These processes differ in the type of equipment used and the type of solvent. Table 4.4 summarizes the processes used by several vendors.

Several concerns must be addressed when using a solvent extraction system:

- solvent handling precautions (e.g., flammability, extreme pH);
- the amount of solvent lost during treatment affects the cost and the emissions to the environment;
- the level of residual solvent remaining in soil after treatment may be of concern due to residual toxicity; and
- treatment may require slurrying soil with water before treatment. Water must be separated from the soil and then treated. Separation of a large proportion of fines from water can be problematic.

4.3.3.3 Soil Flushing

Soil flushing refers to the process of leaching contaminants from the unsaturated zone to the groundwater, where they are treated *in situ* or extracted and treated. Soil flushing is similar to two processes used to treat NAPL and groundwater: water flooding and surfactant flushing (Section 3.6.2). Soil flushing has had limited field application.

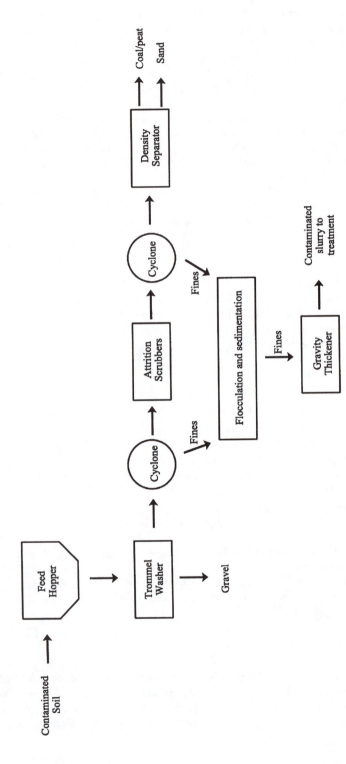

Figure 4.7 Process flow diagram of soil washing system. (Adapted from U.S. EPA, *Technology Demonstration Summary: Toronto Harbour Commissioners (THC) Soil Recycle Treatment Train*, EPA/540/SR-93/517, Center for Environmental Research Information, Cincinnati, OH, September 1993, 6 pp.)

Soil flushing systems can incorporate a variety of techniques to percolate water through soils: surface flooding, surface sprinklers, leach fields, vertical or horizontal injection wells, or basin or trench infiltration systems.[123-125] Surfactants may be added to the water to enhance the solubility of organic compounds, or the pH adjusted to enhance the mobility of metals.

Soil flushing would be most successful at a site where:

- soils are relatively homogenous and permeable, so that the water can be thoroughly distributed through the soil;
- the soil has a relatively low proportion of fines and organic matter that can adsorb contaminants and reduce their mobility;
- the soil has a relatively low cation exchange capacity (CEC), which indicates the affinity of the soil for cationic metals;
- the system is relatively free from natural constituents such as iron which could cause fouling of the water injection or infiltration system; and
- groundwater can be effectively captured and treated.

4.3.4 Electrokinetic Remediation Technologies

When a low-density direct current is applied between electrodes placed in partially or fully saturated soil, ionic contaminants will migrate toward the electrodes based on their charge. This effect can be used to separate contaminants from soil *in situ* or to treat soils *in situ*.

Attempts to use this technology to remove metals from soils date to the 1930s. Electrokinetics has also been used to dewater soils since 1939.[126] Applications to hazardous waste sites began in the late 1970s.[127] By early 1997, the technology had been used for soil remediation at full scale in Europe and at bench and pilot scale in the U.S.[126]

Electrokinetics technologies have been used at bench or pilot scale to treat inorganic soil contaminants such as lead, arsenic, nickel, mercury, copper, zinc, chromium, uranium, thorium, and radium.[126] Electrokinetics processes have been successfully tested on organic soil contaminants such as benzene, toluene, xylene, phenol, and TCE.[128,129] Electrokinetics has also been considered as a means to enhance other remediation technologies, including NAPL removal,[126] soil flushing, and bioremediation.[126,130] Finally, related electrokinetics processes have also been used to treat extracted groundwater.[131]

4.3.4.1 Physical Effects

Electrokinetics remediation processes rely on four physical effects:[126]

- *electromigration*, which is the transport of charged chemical species under an electric gradient;
- *electro-osmosis*, the transport of pore fluid under an electric gradient;
- *electrophoresis*, the movement of charged particles under an electric gradient; and
- *electrolysis*, which are chemical reactions resulting from the electrical field.

Table 4.4 Solvent Extraction Processes

Process (vendor)	Pretreatment required	Process description	Contaminants treated	Ref.
B.E.S.T. Solvent Extraction Process (Ionics/Resources Conservation Company)	Prescreen to remove particles >1 in. diameter	Non-pumpable wastes (soil) treated in batch extraction process; pumpable wastes (sludges, sediments) treated in continuous system. Solvent is secondary or tertiary amine (e.g., triethylamine) whose miscibility with water depends on temperature; at temperatures <20°C, solvent and water are miscible; at temperatures >20°C, solvent and water only partially miscible. Waste and solvent mixed in a mixer/settler; solids are transferred to extractor/dryer; the solvent/oil mixture is decanted, centrifuged to remove small particles, and heated. As the temperature increases, the water separates from the organics/solvent; the remaining mixture of organics/solvent sent to an evaporator for solvent recovery; water is treated.	PAHs, PCBs, pesticides, herbicides; process developed in the mid-1970s for drying wastewater treatment sludge and recovering oils	118, 120, 121
Carver-Greenfield Process® (Dehydro-Tech Corporation)		Soil or sludge mixed with hydrocarbon solvent. Water evaporated, then slurry treated in multistage solvent extract unit where solids contact recycled solvent until contaminants removed. Slurry is then centrifuged to remove solvent from solids; separated solids "desolventized" to evaporate residual solvent; solvent distilled to concentrate contaminants and recover solvent.	PCBs, PAHs, dioxins, municipal wastewater sludge, paper mill sludge, rendering waste, pharmeceutical plant sludge, petroleum-contaminated sludge	118
Liquified gas solvent extraction [LG-SX] technology (CF Systems Corporation)	Yes	Feed mixed with liquified gas under pressure. Slurried soil/sludge treated with propane in a batch extractor/decantor (CO_2 used to treat wastewater in a continuous tray tower). Solvent/contaminant mixture separated from soil. In solvent recovery unit, pressure decreased to vaporize solvent, which is recycled back through process; recovered organics treated. Treated feed discharged from extraction system as an aqueous slurry; slurry is filtered and dewatered.	PCBs, PAHs, pentachlorophenol, dioxins, refinery wastes	119

Low energy extraction process [LEEP®] (ART International, Inc.)	Solids up to 8-in. diameter removed in a gravity settler–floater	Soils washed in a dual-solvent process that uses first a hydrophilic solvent, then a hydrophobic solvent. Contaminants concentrated in sacrificial solvent by liquid–liquid extraction or distillation. Concentrated contaminants taken off-site for treatment.	Coal tar from manufactured gas plants; PCB; refinery wastes	119
Solv-Ex (SRE, Inc.)	Yes	Semicontinuous countercurrent flow extractor; solvent system contains mixture of polar and nonpolar components and is used in food extraction processes; full-scale unit can treat up to 100 ton/d.	Steel mill sludges; VOC; SVOC; oils, grease; coal tar compounds	118
Terra-Kleen Solvent Extraction Unit (Terra-Kleen Response Group Inc.)	Can accept debris up to 3 ft; however, large debris may be removed for ease of handling	Batch treatment using proprietary solvents. Solvents and soil mixed in covered extraction tank, then pumped to sedimentation tank; settled solids are dried to remove residual solvent. Solvent leaving sedimentation tank is filtered, then treated in proprietary solvent purification station to remove and concentrate contaminants. Solvent is recycled through process; concentrated contaminants taken off-site for treatment. Capacity flexible; soil at one site treated at 250 tons/batch in 19 tanks with 16-cy capacity; system occupied 4000 sf.	PAH, PCB, pentachlorophenol, creosote, dioxins, pesticides; does not remove metals	118, 122

Electrokinetic remediation systems create an electrolytic cell in the soil.[128] A current on the order of milliamps per square centimeter of electrode area is applied to the electrodes and creates a voltage gradient on the order of 1 V/cm between electrodes.[132] The electrical energy applied to the electrodes causes nonspontaneous redox reactions to occur.[133]

One electrode, the *anode*, loses electrons. The molecules in the soil, the groundwater, or the other electrode which gain these electrons are reduced. As the anode oxidizes, it gains a positive charge. Negatively charged contaminants and naturally occurring anions in the soil and groundwater migrate toward the anode.

The second electrode, the *cathode*, gains electrons. The molecules which accept these electrons are oxidized. As the cathode is reduced, it gains a negative charge. Positively charged contaminants and naturally occurring cations migrate toward the cathode. Thus, electrokinetic remediation technologies can separate charged contaminants from saturated and partially saturated soils.

The charges at the cathode (negative) and anode (positive) attract cationic contaminants — such as heavy metals — and anionic contaminants, respectively. Contaminants may be immobilized at the electrodes by electroplating onto the electrodes or by precipitation. Alternatively, the remediation system may be designed to extract and treat soil pore water (groundwater) at the cathode.

The movement of water under the influence of the current and changes in the soil pH as a result of the hydrolysis of water also affect the mobility of contaminants.

Water flows toward the cathode under the influence of the current[130] (as noted previously, electrokinetic techniques have been used in dewatering). Freeze and Cherry[134] note that "the mechanism of flow involves an interaction between charged ions in the water and the electrical charge associated with clay minerals in the soil."[134] As the positively charged ions migrate to the cathode, "the water molecules are dragged with them".[128]

The electro-osmotic flow of water depends upon the relative influence of the hydraulic gradient. In soils where the hydraulic gradient is quite high, the gradient imposed by the electric current is negligible relative to the hydraulic gradient.[133] The applied current and the electrical conductivity of the medium, which depends on the water content, cation exchange capacity, and free electrolyte content, also affect the flow of water. Electro-osmosis induces flow most effectively in saturated, low-permeability, fine-grained soil.[133]

Water molecules near the electrodes electrolyze when the current is applied to the electrodes. At the anode, electrolysis produces oxygen gas and hydrogen ions. As a result, the pH of the water drops, potentially to a pH of 2 or less. At the cathode, electrolysis of water produces hydroxyl ions and hydrogen gas. The pore water becomes basic, with a pH of 12 or more. The pH at the anode and the cathode depend on the current applied and the chemical characteristics of the soil and water. The positively charged hydrogen ions migrate toward the negatively charged cathode, creating an *acid front* which acidifies the soils between the electrodes. This change in pH affects the solubility of many contaminants. Hydrogen ions can also exchange with cations adsorbed onto the soil, enabling those cations to migrate toward the cathode.[133]

4.3.4.2 Remediation Processes

Vendors and research institutions have developed different ways to apply electrokinetics to site remediation. Certain processes enhance the removal of contaminants, while others enhance treatment *in situ* without removal. Some of the commercial processes under development are described below.

Electrokinetics, Inc. holds the patent/license to an *in situ* soil processing technique.[129] Electrodes are inserted into the soil through boreholes. In unsaturated soils where the pore water does not suffice to sustain the current, "pore fluids" or "processing fluid" may be added to the soils through these boreholes. Contaminants are electroplated on the electrodes or separated from the processing fluide in a post-treatment unit.

Environmental & Technology Services used electrokinetics to remediate a site contaminated by a gasoline spill from an underground tank. To remediate the site, 56 electrodes were installed into the upper clay layer of the 2400-square-ft site. The current applied to the electrodes drew the contaminants and water down 15 ft into dense cemented conglomerate sandstone. A bioventing system removed the contaminants from the sandstone. Hydroxyl ions produced from the hydrolysis of water may have promoted the oxidation of the contaminants. Initial gasoline levels of 100 to 2200 mg/kg were reportedly reduced to well below the target level of 100 mg/kg after about 90 d.[135]

A consortium of parties which includes Monsanto, E. I. DuPont de Nemours and Company, General Electric, in cooperation with the U.S. EPA's Office of Research and Development and the DOE, is developing the *Lasagna*™ *process*.[126,129] This process combines electro-osmosis with layers of various other materials used to treat contaminants *in situ*. These layers may be installed either vertically or horizontally. In general, the vertical configuration would be used for relatively shallow contamination (<50 ft deep). In the horizontal configuration, where the layers are installed by hydraulic fracturing (Section 4.3.1.2.2), deeper soils could be treated.

Electrodes comprise the outer layers. Steel rods are used for vertical electrodes, or granular graphite is injected into horizontal soil fractures to create horizontal electrodes. The current applied to the electrodes drives the movement of contaminants through the middle treatment layers. These middle layers can include zero-valent iron, for reductive dehalogenation, and/or granular activated carbon, to provide a medium for biodegradation by methanotrophic microorganisms. Other treatment media have been considered, e.g., oxidants or buffers. The direction of contaminant transport through the treatment media can be reversed, if necessary, by reversing the charge on the electrodes.

As of late 1996, the consortium had tested this technology primarily on TCE in low-permeability soils.[129]

Example 4.5: The U.S. DOE tested the Lasagna™ process at the Cylinder Drop Test Area of the Paducah Gaseous Diffusion Plant (PGDP) in Paducah, KY. The following summary is based on three reports of those tests.[136-138]

The PGDP, a U.S. DOE facility, enriches uranium for use in commercial reactors. At times during the 1960s and 1970s, the facility performed drop tests to demonstrate the structural integrity of steel cylinders used to store and transport uranium hexafluoride. In these tests, a crane lifted the cylinders to a specified height and then dropped the cylinders onto a concrete and steel pad. In tests performed in February 1979, testers submerged one cylinder in a concrete in-ground pit containing TCE and dry ice in order to chill the cylinder. Leakage from that pit apparently contaminated shallow soil and groundwater in an area approximately $105 \times 60 \times 45$ ft deep.

In 1988, TCE and the radionuclide technetium-99 were detected in an off-site drinking water well. As a result, the DOE entered into an Administrative Consent Order with the U.S. EPA and the Kentucky Department of Environmental Protection (KDEP) to define the sources and extent of pollution at PGDP. The Cylinder Drop Test Area was one of the source areas studied. The field investigation occurred in four phases between April 1993 and March 1996.

Data from the field investigation indicated that a gravel road base, 3 to 4 ft thick, lies under most of the Cylinder Drop Test Area. A layer of clay and silty clay lies under the road base and extends to a depth of 20 ft. The clay layer is underlain by a 3- to 5-ft-thick discontinuous layer of sandy gravel, and then a layer of clay, interbedded clay, and sand which is approximately 36 ft thick. The uppermost aquifer for the area is in a thick interval of sand and gravel beneath that clay.

Analysis of soil samples indicated that the uppermost portion of the clay was contaminated with TCE. The concentrations of TCE in the clay ranged from 1 ng/kg to 1760 mg/kg. Groundwater samples from a monitoring well completed in water perched in the shallow sand and gravel layer contained up to 160 mg/L TCE; subsequent samples of groundwater from that well contained levels around 20 mg/L. The groundwater also contained other organic compounds. Site investigators concluded from the soil and groundwater data that DNAPL could be present.

Investigators selected an area approximately 15×15 ft within the Cylinder Drop Test Area to test the Lasagna™ process (Phase I). TCE concentrations in this test area ranged from 1 to 500 mg/kg, with an average concentration of 83 mg/kg.

Steel sheet piling comprised the electrodes. Piling was driven to a depth of 16 ft on two parallel sides of the 15-ft-square test area. The outside face of each pile was electrically insulated using geomembrane panels. Workers installed carbon-filled wick drains (18 in. wide $\times$ 2 in. thick $\times$ 15 ft deep) between the electrodes. These so-called treatment zones were spaced 21 in. apart.

The initial voltage applied to the electrodes was 138 V at approximately 41 A. Operators maintained the current at approximately 40 A. As the soil temperature increased, the electrical conductivity of the soil increased, and, as a result, the voltage dropped. After 1 month of operation, the voltage stabilized at 105 V. The voltage gradient ranged from 0.45 to 0.35 V/cm throughout the test.

The current induced the flow of water through the clay at an elecro-osmotic flow rate between 4 and 5 L/h. The current also heated the soil, from an initial temperature of 15°C to an average temperature of 25 to 30°C and a maximum temperature of 45.2°C.

As the groundwater migrated under the influence of the current, it flowed through the activated carbon in the treatment zones. Contaminants adsorbed onto the carbon.

After 4 months of treatment, an estimated three pore volumes of water had moved between adjacent treatment zones. Post-treatment soil samples indicated that the concentration of

TCE in the soil had decreased by an average of 98%, to an average concentration of 1.2 ppm. The effectiveness of treatment varied thoughout the treatment area. Analysis of the activated carbon indicated TCE concentrations ranging from several thousand to greater than 10,000 mg/kg. A mass balance on the TCE accounted for about 50% of the TCE. The difference was attributed to physical mechanisms which could remove or degrade TCE, analytical limitations, and the limitations of data obtained from a nonuniform system. Researchers concluded that electrokinetics could move the TCE out of the soil and into the treatment zones, and that the activated carbon treatment zones had effectively trapped the contamination.

In 1996, the project team began Phase IIa tests of the Lasagna™ process. The goal of these tests was to reduce the soil contamination to an average of 5.6 mg/kg or less, using iron filings to degrade the TCE by reductive dehalogenation.

An area 20 ft long × 30 ft wide was selected for the Phase IIa tests. Data from pre-test soil borings indicated that the contamination tapered off to below the detection limit at a depth of 30 to 35 ft. As a result, the sheet-pile electrodes were installed to a depth of 45 ft. Soil data indicated that the concentration of TCE ranged between 10 and 500 mg/kg throughout most of the test area.

Bench-scale tests were performed using groundwater from the site to develop the design of treatment zones. These tests indicated half lives for TCE, 1,2-DCE, and vinyl chloride of 2.2, 5.2, and 3.3 h, respectively (at room temperature). At a target voltage gradient of 0.25 V/cm, the velocity of groundwater through the clay at the site would be approximately 0.32 cm/d. Using this estimate of the flow rate and the half-life data from the laboratory tests, the design team calculated that a 2-in.-thick treatment zone would theoretically more than suffice to treat the groundwater. As a result, the design team decided to mix the iron filings with clay for ease of construction. The initial design was based on a mixture of clay and iron containing 8% iron filings by volume.

Three treatment zones were installed between the electrodes, perpendicular to the direction of groundwater flow. The first zone was installed 7 ft from the anode; the second zone 5 ft from the first zone; and the third zone 2 ft from the second. Each treatment zone was installed by pouring a slurry of wet kaolin mixed with iron filings into a series of hollow mandrels. These mandrels were 3 in. thick, 2 ft wide, and 45 ft deep. During installation, the design mix was changed to improve the ease of construction: the water content of the slurry was increased, and the iron content was increased to 26 vol%.

The test was initially designed to last for 3 months. However, the initial operating results suggested that free-phase DNAPL was present in the area, potentially in the sand layer approximately 20 ft deep. As a result, the test period extended to nearly a year.

During this operating period, an electro-osmotic flow equivelent to 2.5 pore volumes occurred between the second and third treatment zones. The average soil temperature in the treatment area increased from 18°C to a maximum of 60°C after 5 to 6 months, then decreased to 50°C for the remainder of the test. Flux boxes placed over the test area indicated that a negligible amount of TCE was volatilizing.

Researchers monitored the concentrations of gaseous daughter products ethane, ethylene, and acetylene (HCCH) in samples from monitoring wells. The nature of the daughter products depended on whether or not dissolved-phase TCE or TCE DNAPL was moving through the treatment zone. Degradation of dissolved-phase TCE produced primarily ethane and ethylene in equal amounts, and little acetylene. Degradation of TCE DNAPL produced primarily acetylene, followed by ethylene, and little ethane. (These field results were duplicated in laboratory tests using dissolved-phase and separate-phase TCE.)

Researchers also monitored the levels of TCE in groundwater, soil, and activated carbon cells installed into the test zone. Analysis of soil samples indicated that the soils had, on average, approached, but not reached the target concentration of 5.6 mg/kg. The project team concluded that the site could be remediated within 2 years using this technology.

4.4 DESTRUCTION TECHNOLOGIES

Technologies such as enhanced biodegradation, dechlorination, and incineration can destroy organic contaminants and certain inorganic contaminants. As elements, metals cannot be destroyed.

4.4.1 Biological Processes

Bioremediation refers to a group of treatment methods or processes designed to enhance the natural microbial or fungal degradation of organic contaminants. *Phytoremediation* is a biological treatment process which uses plants to concentrate or degrade contaminants.

4.4.1.1 *Bioremediation*

4.4.1.1.1 Overview — Biodegradation can occur *aerobically* or *anaerobically*. *Aerobic* microorganisms use oxygen as the electron acceptor. *Anaerobic* microorganisms use other electron acceptors such as nitrate, sulfate, iron, manganese, or certain organic compounds. *Facultative anaerobic* organisms can use oxygen, if it is present, or other electron acceptors if it is not. Most soil bioremediation processes are aerobic. A few are anaerobic or designed to be capable of operating in an anaerobic mode.[139,140]

Aerobic biodegradation involves a series of oxidation reactions where oxygen serves as the electron acceptor. Microorganisms convert organic compounds to intermediate breakdown products, then ultimately to carbon dioxide, water, inorganic compounds, and biomass. Degradation to these final end products is called *mineralization*. As described in Section 3.4.3.1, biological degradation generally follows first-order or pseudo first-order kinetics.

In general, hydrocarbons — alkanes and aromatics — can be bioremediated. Bioremediation was historically used to treat waste sludges from petroleum refining. Aliphatic hydrocarbons with eight carbon atoms or less degrade readily. Longer-chain hydrocarbons (C12 and longer) degrade more slowly.[141] The U.S. EPA's records indicate that at Superfund sites, bioremediation is commonly used to treat soils containing (in decreasing order of frequency of selection) PAHs, VOCs other than benzene, toluene, ethyl benzene, and xylenes (BTEX), and BTEX. Bioremediation was selected for nearly half the former wood-preserving sites in the Superfund program.[142]

Highly chlorinated PCBs and some of the higher-molecular-weight PAHs degrade slowly or not at all. The solubility of high-molecular-weight compounds can be a limiting factor, as contaminants adsorbed to soil must dissolve into the water in soil pores to be available to microorganisms. High-molecular-weight PAHs (e.g., 5- and 6-ring molecules) are very difficult to treat effectively by bioremediation.[141,142] Because of their potential carcinogenicity, the clean-up levels for some of the higher-molecular-weight are quite low. As a result of the difficulty in degrading these compounds, bioremediation is not always feasible for PAH-contaminated soil.

As described in Section 3.5, chlorinated solvents generally biodegrade anaerobicly. However, PCE and TCE can be degraded aerobically to carbon dioxide, water, and chloride in the presence of cometabolites. Cometabolites include methane (natural gas) and aromatic compounds (e.g., toluene).

Bioremediation processes are designed to optimize the conditions for microbial growth and degradation of contaminants. Treatment can be performed in a variety of ways. At Superfund sites, excavation and land treatment was the most commonly selected form of bioremediation, followed by slurry-phase treatment. *In situ* treatment, such as bioventing, is also common.[142] Process options for treatment are discussed further below.

Microbial growth and metabolism of contaminants requires:[143]

- the macronutrients phosphorus and nitrogen, generally at the C to N to P ratio (by weight) of 120:10:1;
- micronutrients;
- acceptable pH, generally between 5.5 and 8.5;
- an adequate supply of electron acceptors; aerobic microorganisms require greater than 0.2 mg/L dissolved oxygen and a minimum air-filled pore space of 10%. Anaerobic microorganisms require less than 0.2 mg/L dissolved oxygen and an oxygen concentration less than 1% air-filled pore space;
- moisture content of unsaturated soils generally between 25 and 85% of field capacity (the *field capacity* is the amount of water which the soil can hold against the force of gravity); optimum biodegradation rates typically occur at a moisture content between 60 and 80% of field capacity; and
- acceptable redox level: for aerobes and facultative anaerobes, greater than 50 mV, and for anaerobes, less than 50 mV.

The rate of biodegradation may also be limited by:

- temperature;
- the toxicity of concentrated contaminants; or
- mass transfer limitations.

4.4.1.1.2 Testing — Bench- and/or pilot-scale testing is usually required to evaluate natural conditions and optimize the treatment process. These tests may examine one or more of the following factors, depending on the scale and complexity of remediation:

- Microbial characterization — tests such as plate counts determine the number and type of microorganisms naturally present in the soil.
- Soil characterization — soil samples are analyzed for nitrogen, phosphorus, pH, and moisture content; the project team considers (or analyzes for) the availability of electron acceptors and temperature; bench-scale tests may be performed, varying these parameters, to determine the conditions for optimum treatment.
- Presence of inhibitors — samples may be analyzed for chemical content or tested using a bioassay to evaluate whether or not concentrations of target compounds, other hazardous substances such as metals, or soil characteristics such as the salt content could inhibit biological growth. Chemical pretreatment (e.g., pH adjustment) or dilution may be necessary to alleviate inhibition.
- Ability to degrade target compounds — microbial activity may be measured indirectly, by measuring the levels of oxygen and carbon dioxide, or directly, by measuring the change in the concentration of target compounds.
- Kinetics — bioremediation specialists evaluate the kinetics of degradation under controlled conditions to determine (roughly) the length of treatment time required to achieve clean-up goals. The results obtained from controled laboratory tests are usually best-case projections.
- Byproducts of biodegradation — large organic molecules do not degrade immediately to carbon dioxide and water: biodegradation proceeds though a series of intermediate compounds; the intermediate products of biodegradation can be more or less soluble, volatile, or toxic than the parent compound. One objective of treatability testing can be to identify the intermediate products of biodegradation and evaluate the degradability of those intermediates.
- Volatilization — testing may also need to evaluate the mass of contamination lost through volatilization rather than biodegradation, as full-scale bioremediation of soil containing volatile compounds can require collection and treatment of air emissions.

4.4.1.1.3 Process Options — Bioremediation can be accomplished in several ways. The nomenclature for these variations is not precise. Just as for solidification/stabilization/fixation, people often refer to the same form of treatment by different names.

The oldest version of bioremediation for sludges, formerly used to treat byproducts of petroleum refining, is *land farming*. Sludges were spread over the ground and tilled into the earth. Current regulations severely restrict land farming.

In the simplest and potentially least costly version of *in situ* bioremediation, *land treatment,* shallow contaminated soils can be watered, supplemented with nutrients, and tilled to mix the nutrients and aerate the soil.

Five processes can occur in this form of bioremediation:[141]

- biodegradation, which typically accounts for most of the contaminant transformation;
- at the soil surface, certain compounds such as some PAH, degrade by ultraviolet light;
- lower-molecular weight compounds can volatilize;

- certain compounds, including some pesticides, degrade by hydrolysis; and
- *humification*, which is a polymerization reaction whereby some molecules — including some PAHs — are added to the humic materials in soil.

Two aspects of land treatment are particularly important: tilling and moisture control. Tilling mixes nutrients into soil, aerates the soil, and helps to ensure contact between microorganisms and contaminants. Most tractor-mounted tillers cultivate the soil to a depth of 1 ft; specialized equipment can reach a depth of 3 ft or more. Tilling has some negative effects as well as positive effects. Tilling very wet or saturated soil tends to destroy the soil structure, reduce the intake of oxygen and water, and reduce microbial activity. Excessive tilling can compact soil below the tilling zone.[141]

As noted above, soil moisture is best maintained at some 60 to 80% of field capacity. If the soil dries out, microbial activity stops. If the soil is saturated for over an hour, oxygen transfer is limited and microbial action slows. Maintaining soil moisture within the desired range is a difficult task.[141]

In order for land treatment to be acceptable, emissions to the air, leaching to groundwater, and surface water runoff must be minimal or acceptable to regulatory agencies. In order for land treatment to be effective, the ambient temperature and rainfall (supplemented by irrigation) must allow for treatment to occur at a reasonable rate. If these conditions cannot be met, but bioremediation is a viable option, the soil can be excavated and treated by one of the other methods described below.

Contaminated soil which is deeper than the limits of tilling equipment can be treated *in situ* by bioventing. *Bioventing* refers to the process of injecting and/or extracting air into the vadose zone to supply the oxygen needed for aerobic biodegradation. While the equipment and materials used for bioventing are similar to those used for air sparging (Section 3.5.2) and SVE (Section 4.3.1), bioventing has a different objective: to supply oxygen for bioremediation, not to strip volatile contaminants. As a result, bioventing systems operate at lower air flow rates than air sparging or SVE systems.

The fate of contaminants in a bioventing system depends on their volatility. Highly volatile compounds — particularly those with a vapor pressure above approximately 760 mmHg — volatilize too rapidly to biovent. Compounds with vapor pressures below 1 mm Hg will not substantially volatilize, but may degrade in a bioventing system. Finally, compounds with vapor pressures between those bounds may be amenable to either volatilization or biodegradation.[144]

The effectiveness of bioventing depends on the ability of microorganisms to degrade the contamination, as described above, and the ability to distribute sufficient oxygen through the subsurface. The air permeability depends on the soil structure and particle size as well as the soil moisture content. While uniform, permeable soils are most easily treated, soils with a high proportion of silts and clays have been successfully biovented. A high moisture content limits the air permeability and thus the potential for bioventing. Field experience suggests that soil with an air permeability above 10^{-9} cm^2 can be readily treated. At an air permeability below approximately 10^{-10} cm^2, soil gas flows through fractures or lenses or more permeable material, and this preferential flow limits the potential for bioventing.[145]

Bioventing can be accomplished by air injection or SVE. Critical design variables in either case are the air flow rate and the spacing of injection/extraction wells.

An air injection system consists of a blower or compressor, distribution piping, and injection wells. Because an air injection system does not include vapor-phase treatment, it is less expensive to construct and operate than a bioventing system which utilizes SVE.

The objective for air injection is to supply enough oxygen to stimulate biodegradation without causing emissions to the atmosphere. That objective can be met most easily for relatively low-volatility contaminants. Injection of air into the vadose zone can have several physical effects which may enhance remediation. Air injection can depress the water table slightly, allowing for treatment of soil in the capillary fringe and increasing the air permeability of the soil. Injection of air can also expand the volume of soil available for effective use as a bioreactor: volatile compounds will migrate in the gas phase to surrounding (uncontaminated) soil, where they can biodegrade.[144]

Air injection is not practicable at every site. If air injection would cause unacceptable contaminant levels in the air in a nearby building, for example, air extraction would be used to biovent the contamination rather than air injection.

Bioventing by air extraction is similar to SVE: a blower or vacuum pump induces a vacuum on one or more wells, drawing air through the soil pores to the wells. Extracted vapors are treated to remove water vapor and contaminants, then discharged.

Bioventing by air extraction has several disadvantages relative to air injection, in addition to cost. Air extraction causes the water table and capillary fringe to rise or upwell near the extraction points. This effect can saturate soil in the smear zone (the soil just above and below the water table where LNAPL may have smeared over the soil with the natural rise and fall of the water table). As a result, the soil in the smear zone is not effectively treated. Upwelling can also increase the soil moisture in the capillary fringe, decreasing the air permeability and thus the ROI of the extraction well.[144]

A design engineer determines the air flow rate based on the amount of oxygen needed to sustain biodegradation. An *in situ* respiration test provides the data needed to estimate that amount. In that test, a field team places narrowly screened soil gas monitoring points into the soil which is to be treated, measures the levels of carbon dioxide and oxygen in the soil gas, and then injects air containing an inert tracer gas such as helium. After approximately 24 h, the team shuts off the gas flow. The levels of carbon dioxide, oxygen, and the tracer gas are measured periodically. An increase in carbon dioxide levels indicates that aerobic biodegradation has occurred. The decrease in oxygen levels over time indicates the oxygen utilization rate.[146] This oxygen utilization rate is then used to calculate the required air flow rate based on the volume of soil and its porosity, bulk density, and moisture content.[144]

The spacing of injection/extraction wells is based on the ROI, which is "the maximum distance from the air extraction or injection well where a sufficient supply of oxygen for microbial respiration can be delivered".[144] The oxygen ROI depends on the soil properties, configuration of the air injection/extraction well and air flow rate, and the rate of microbial activity. The pressure ROI, i.e., the distance over which an injection or extraction well can affect the soil gas pressure by 0.1 in. of water, is a conservative estimate of the oxygen ROI.

Bioventing can be a relatively simple, cost-effective remediation alternative. However, several conditions may preclude *in situ* treatment of biodegradable contaminants:

- soils are highly heterogeneous (difficult to distribute oxygen due to short circuiting through more permeable zones), or
- soils have very low permeability, or
- highly concentrated contamination/waste is present (e.g., coal tar deposits) that could be toxic to microorganisms or so concentrated that mass transfer limitations could slow biodegradation.

In such cases, excavation of the soil and treatment in a land-treatment unit, composting, or a slurry-phase reactor might be more appropriate ways to treat the soil.

When excavated soil will be treated in a *land treatment unit*, a lined treatment cell, sometimes called a *prepared bed*, is typically constructed. The impermeable liner contains and controls leaching from the material undergoing treatment. The treatment cell is also bermed to control run on/run off of precipitation. Water is collected in a sump at the lowest point on the treatment pad, then used to irrigate the soil, or it is treated and discharged.

Construction begins with clearing and grading the area where the treatment unit will be built. The impermeable layer may be constructed of clay or a geomembrane. A layer of clean sandy soil or a geonet is placed over the impermeable layer to allow for water drainage.

The contaminated soil is spread on the treatment pad in *lifts* or layers approximately 1 ft thick. Clayey soils are applied in shallower lifts than sandy soils due to oxygen transfer limitations. Carbonaceous materials such as wood chips, sawdust, straw, or animal manure may be mixed into the soil, typically at 3 to 4% by weight of the soil. Adding such material can serve several purposes, depending on the soil and the contaminants: to supply carbon, increase the water-holding capacity of sandy soil, improve the tilth (workability) of clayey soil, and/or increase the sorptive properties of the soil and thereby decrease contaminant mobility.[141]

Nutrients, water, and chemicals to control the pH (if necessary) are added. The soil is tilled periodically to mix the soil and nutrients and provide oxygen. As a lift meets the remediation goal, another lift of contaminated soil is added to the treatment unit.

Use of a land-treatment unit may not be practical when:

- a very large area may be required to treat a very large volume of soil, but space on site is limited, or
- air emissions from VOCs are of concern, or
- highly concentrated contamination/waste is present (e.g., coal tar deposits), or
- the ambient temperatures during much of the year are too low for effective bioremediation.

In such cases, a biopile, composting, or a slurry-phase reactor might be more appropriate ways to treat the soil.

A *biopile* or *static pile* represents the next level of engineering complexity and process control. A treatment pad is prepared similar to that used for a land treatment unit. Perforated piping is laid on the pad, then excavated soil is piled on the unit to a height of up to 3 to 10 ft. As with a land treatment unit, the soil may be mixed with carbonaceous material, and provisions are made to irrigate the soil and supply nutrients. A blower or fan either forces air through the piping in the biopile under positive pressure, or pulls air through the pile under negative pressure. In the latter case, the system collects volatile emissions for treatment. In either case, the ventilation system enhances the supply of oxygen needed for aerobic metabolism.[147] The biopile may be covered or placed in a building to control air emissions and temperature. While building and operating a biopile costs more than a simple land treatment unit, a biopile occupies less space and allows for better control of volatile emissions.

In *composting*, soil or sludge is mixed with an inert bulking agent such as wood chips or sawdust. The bulking agent provides pore spaces for air and water to penetrate through the waste. Composting entails adding a much larger volume of a bulking agent than the level of carbonaceous material added to soil in a land treatment unit. For example, one series of tests with PAH-contaminated soils used bulking agents at some 30 to 50% of the soil by weight.[147]

Soil or sludge can be composted in a biopile. Alternatively, the mixture of soil and the bulking agent is placed in a *windrow*, a long pile some 6 to 8 ft wide which is mechanically turned to mix and aerate the soil. Finally, some smaller-scale composting has occurred in vessels which allow for mixing, nutrient and water addition, and aeration.

The aerobic compost process occurs in four stages. Each stage is characterized by a different temperature range and diversity and type of microbial population. In the mesophillic stage, the temperature ranges from 35 to 55°C. The mesophillic stage has the greatest microbial diversity. The second phase is the thermophillic phase, characterized by temperatures between 55 to 75°C. Spore-forming bacteria and thermophillic fungi have been observed during the thermophilic stage. During the third phase, cooling, microbial recolonization brings the appearance of mesophilic fungi whose spores withstood the temperatures of the previous phase. The final phase is maturation.[147]

Composting improves the mass transfer of oxygen and nutrients, and allows for treatment of highly contaminated waste. By conserving the heat generated by microbial activity, compost piles enable bioremediation to continue through colder months. In addition, the elevated heat can kill pathogenic bacteria (more of a concern for sewage sludge than for typical hazardous waste).

A slurry-phase bioreactor provides the highest level of process control of all of the bioremediation processes. It can also be the most mechanically complicated and expensive process option.

Before treatment in a slurry-phase reactor, soil or sludge is screened to remove oversized particles (typcially over ¼ in. diameter) and slurried with water to form a mixture which is between 10 and 40% solids by weight. Nutrients and other amendments are added, and the slurry is mixed and aerated. When treatment of a batch of soil is completed, the soil and water are separated using equipment such as clarifiers and filter presses. Slurry bioreactors have been used to treat soil containing creosote, fuel hydrocarbons (PAHs), and other contaminants.[148-150]

Slurry-phase bioreactors can have several advantages over other forms of bioremediation: less space is required on site for treatment; optimum mass transfer, and as a result, more highly contaminated waste and/or less soluble waste can be treated; good control of temperature, oxygen levels, and nutrient levels; and control of air emissions. Slurry-phase bioreactors have several disadvantages compared to other forms of bioremediation: relatively high capital and operating costs; more complicated equipment required; limits on the maximum particle size; and potential materials handling problems with silty or clayey soil, similar to soil washing.

4.4.1.1.4 Operating Issues and Variations — All of the process options for bioremediation of soil share these operating issues:

- maintenance of optimum conditions;
- air and water emissions;
- waste variability in contaminant concentrations and soil types at a single site, and the resulting need for an adaptable treatment process; and
- kinetics of degradation and ability to meet remediation goals in a timely way.

Several process variations have been tested to overcome some of the limitations of bioremediation. These include augmenting soil with nonindigenous microorganisms or fungi, and pretreatment by chemical oxidation.

Many vendors offer cultured microorganisms which can be added to contaminated soil to stimulate biodegradation.[151,152] However, adding microbial cultures may not significantly improve treatment. First, microbes which are not indigenous to the soil must compete with those which are. Nonindigenous microbes rarely compete well enough with naturally occurring microbes to develop and sustain a useful population. Second, the soil at most sites which are not newly contaminated already contains indigenous microorganisms which can degrade the contaminants under the proper conditions. However, few controlled studies have been performed.[141]

A second enhancement is the addition of white rot fungus to soil. White rot fungi can degrade lignin, a polymeric component of wood which otherwise resists biodegradation. Lignin is structurally similar to certain contaminants. As a result, white rot fungi can degrade organic contaminants such as PAHs, phenolic compounds such as pentachlorophenol, and certain herbicides and pesticides.[151,153] By mid-1996, fungal bioremediation had only been applied in the field to wood treating wastes (pentachlorophenol and creosote).[154]

White rot fungi were initially characterized in 1974. Commercial applications in the pulp and paper industry were first considered in 1981. Environmental applications were developed by the U.S. EPA in conjunction with the USDA Forest Service. Other early researchers also hold patents on the environmental application. Several vendors are licensed to apply the technology.[151]

Treatment is effected by inoculating contaminated soil with cultured white rot fungi. In one field trial, the fungi were added using spore-innoculated/infested wood chips.[154] The soil is treated *in situ,* in a land treatment unit, or in a biopile. Results to date for wood-preserving wastes, albeit from a limited number of trials, suggest that white rot fungi may effect treatment more quickly than traditional solid-phase bioremediation.[154,155]

Bioremediation can also be enhanced chemically. Fenton's Reagent has been used to pretreat soils containing recalcitrant contaminants such as high-molecular-weight PAHs. Chemical oxidation breaks these compounds down into smaller fragments which can then be biodegraded.[148,156]

4.4.1.2 Phytoremediation

The term *phytoremediation* refers to processes which treat soil or sediment in place using plants to concentrate or degrade contaminants. Remediation occurs in the *rhizosphere*, or soil that surrounds and is influenced by the roots of plants.[157] While plants have been used to treat wastewater for over 300 years and plant-based treatment of soils and sediments has been proposed since the mid-1970s, phytoremediation is still a developing technology.[158]

Phytoremediation has be used to treat many types of organic and inorganic contamination, primarily in laboratory or field trials: BTEX; chlorinated solvents; nitrotoluene ammunition wastes; PAHs; chlorinated pesticides and organophosphate insecticides; excess nutrients such as nitrate, ammonium, and phosphate; and metals such as Cd, Cr (VI), Co, Cu, Pb, Ni, Se, and Zn.[159] Effective treatment of lead requires that the soil be amended with chelating agents to make lead more bioavailable.[158,160]

4.4.1.2.1 Treatment Mechanisms — Plants can effect treatment of soil and shallow groundwater by several mechanisms: phytoextraction, phytostabilization, or enhancement of biological degradation. The operative mechanism depends on the type of plant, the type of contaminant, and the soil composition and chemistry. Each mechanism is described briefly below.

In *phytoextraction*, plants remove contaminants from soils and concentrate those contaminants in the plant tissues. Ideally, the plants used for phytoextraction can accumulate and tolerate high concentrations of contaminants in harvestable tissue; grow rapidly; and produce a large amount of biomass which can contain contaminants.[159]

Certain plants, which typically grow in areas with naturally high levels of metals in soil, have an unusual ability to accumulate metals. The leaf tissues from certain hyperaccumulative plants can contain percentage levels of metals on a dry weight basis.[158] For example, the herb *Thlaspi caerulescens* can accumulate up to 25,000 mg zinc per kilogram of dry matter of shoots and 1000 mg cadmium per kilogram. In field trials in amended soil, indian mustard (*Brassica juncea*) has accumulated up to 1.5% (dry weight) lead in the plant shoots.[160]

The shoots and leaves of metal-laden plants grown in contaminated soil are harvested and landfilled, or incinerated to further concentrate the metals in the ash. A third option in the future, depending on the economics, may be to treat the metals as an ore and recover the metals content. Phytoextraction can concentrate the metals by a factor of up to 20.[158]

Phytoextraction has also been used to remove certain organic contaminants, such as BTEX and chlorinated solvents, from soil. Highly hydrophobic contaminants (i.e., those with log K_{ow} greater than 3.0) bind so strongly to plant roots that they cannot be phytoextracted. Hydrophilic contaminants (i.e., those with log K_{ow} less than 0.5) are too water soluble to adsorb to the roots or to be transported through the plant membranes. Plants can metabolize, volatilize, or store an organic contaminant or degradation products within plant tissue. The roots of plants can also release enzymes into the soil which degrade certain contaminants.[161] Tests using cottonwood trees to remediate shallow soil and groundwater containing TCE began in 1996.[162] Poplar trees have also been field tested in several applications at sites with organic contaminants.[161]

In *phytostabilization*, plants limit the mobility and bioavailability of contaminants in soils. Phytostabilization involves both physical and chemical mechanisms. The root systems of the plants hold soils in place and can absorb contaminants. Transpiration, or the movement of water up through a plant from the roots to the leaves, affects the percolation of water through soil. Plant roots can produce chemical compounds which adsorb or complex contaminants, immobilizing them at the interface between the roots and the soil. Plant roots can also release chemical compounds which increase the soil pH, thereby immobilizing metals. As a result of these physical and chemical mechanisms, phytostabilization can limit the migration of contaminants in wind-blown dust, surface water runoff, or leaching to groundwater. Phytostabilization contains, but does not remove contaminants. As a result, the plants must be cultivated indefinitely or until more active remediation occurs.[158,161]

Unlike phytoextraction, plants which accumulate contaminants in their shoots are not used for phytostabilization.[159] Poplar trees and certain grasses have been used to phytostabilize metals.[158]

Finally, phytoremediation can enhance the growth of microbes and fungi which can degrade contaminants. Rhizosphere soils contain 10 to 100 times more metabolically active microorganisms than unplanted soils.[163] Bacterial colonies can cover up to 4 to 10% of the surface area of plant roots. Mycorrhizae fungi can grow symbiotically with the roots of plants. Plant roots can release oxygen and provide a supply of electron acceptors to aerobic microbes. Roots can also exude sugars, alcohols, and acids, which supply organic carbon to the rhizosphere. The decay of fine-root biomass also provides organic carbon which can enhance microbial growth.[161]

4.4.1.2.2   Limitations — Phytoremediation has some limitations:[158,159,161]

- Site-specific applications require careful selection of the plants, testing, and knowlege of agricultural techniques as well as more typical waste-site remediation skills.
- Plant roots can only reach relatively shallow soils and groundwater, typically less than 3 ft deep and 10 ft deep, respectively.
- Plant tissues containing concentrated contaminants still require treatment or disposal, potentially at substantial cost. Wind-blown leaves from trees used for phytoremediation may spread contamination.
- Toxic levels of contaminants prevent plants from growing.

- Contaminants must desorb from the soil into pore water before plant uptake can occur. Desorption can be a rate-limiting step.
- Remediation may require several growing seasons. The rate of remediation depends strongly on site-specific conditions and cannot be predicted for a specific site without testing. One estimate for metal-contaminated soils illustrates the potential rate of reduction: an annual decrease in contaminant concentrations of 2.5 to 100 mg/kg contaminant.[158]
- Like any agricultural process, phytoremediation depends on the weather and is subject to plant disease and pests.

4.4.2 Dechlorination

Chlorinated aromatic compounds can be dechlorinated using chemical reagents or by electrochemical reactions. Soils and sludges containing PCBs, dioxins and furans, and chlorobenzenes have been treated by dechlorination.

Several variations on the chemical dechlorination process have been patented, including glycolite dechlorination and base-catalyzed dechlorination. These variations rely on different chemical reactions to effect treatment. Dechlorination processes have also been used in conjunction with thermal desorption to treat soils.

4.4.2.1 Glycolite Dehalogenation

Glycolite dehalogenation was the first dechlorination technology developed for soil. Technology development began in the late 1970s.[164,165] The treatment process is based on the reaction between polyethylene glycol and a chlorinated compound, such as a PCB, under alkaline conditions. (The treatment process is sometimes abbreviated APEG, for alkaline–polyethylene glycol; KPEG, for potassium (K) hydroxide–polyethylene glycol; or NaPEG, for sodium (Na) polyethylene glycol, depending on the base used to achieve alkaline conditions.) In this nucleophilic aromatic substitution reaction, a portion of the polyethylene glycol molecule (abbreviated ROH) is substituted for a chlorine atom on the aromatic ring to form an aromatic ether.

In the initial applications, contaminated soil or sludge was excavated, screened to remove oversized particles, and slurried with water and reagents for treatment in a batch reactor. The mixture was heated to ca. 150°C and agitated for several hours to allow the reaction to occur. Treated soil was separated from the reagent mixture, washed with water, neutralized with acid, and dewatered.

Slurry-phase glycolite dechlorination was selected as the treatment method for soils at a handful of Superfund sites during the 1980s and early 1990s. Based on preliminary results which showed materials handling problems due to clayey soils or the presence of debris, or concerns about effectiveness or cost, none of these projects were completed at full scale by slurry-phase glycolite dehalogenation. Instead, site remediation was completed by a combination of thermal desorption and dehalogenation, incineration, capping, or other remedial options.[166,167]

KPEG is now used in conjunction with thermal desorption to treat contaminated soils. After soil or sludge is fed to the desorption unit, KPEG is added. Heating the mixture initiates the dechlorination reaction.[168]

4.4.2.2 Base-Catalyzed Dechlorination

The U.S. EPA's National Risk Management Research Laboratory developed the base-catalyzed decomposition process. Under a license agreement, ETG Environmental, Inc. and Separation and Recovery Systems, Inc. (SRS) have developed the SAREX® and K/S THERM-O-DETOX® systems to treat soil using the process.[169,170]

Figure 4.8 shows a block flow diagram of the SAREX® and K/S THERM-O-DETOX® systems, which each comprise a medium-temperature thermal desorption unit, a vapor recovery system, and a dechlorination system to treat liquids recovered from the vapor recovery system.[169,170] Soils and sludges must be pretreated by screening to remove particles over 0.5-in. diameter. Sodium bicarbonate is mixed with the soil, which is then fed to the thermal desorption unit. Batch and continuous-flow reactors have been used. In the thermal desorption unit, the soil is heated to 600 to 950°F under an inert gas for approximately 1 h. Some dechlorination can reportedly occur in the thermal desorption unit. Water and organic compounds which desorb from the soil discharge to the vapor recovery system. Treated soils pass to a cooling system, and then are discharged for replacement on site or off-site disposal.

A series of units treat the off-gases from the thermal desorption system. Sequential scrubbers use oil and water to capture and condense the desorbed organic compounds, particulates, and water vapor. The off-gases then pass through a demister and vapor-phase carbon units before discharge.

The condensate from the vapor-recovery system is treated in the liquid tank reactor (LTR) using the base-catalyzed decomposition process. The condensate is mixed with a high-boiling-point hydrocarbon, sodium hydroxide, and a proprietary catalyst. Dechlorination occurs as the solution is heated to 600 to 650°F and mixed within the reactor for 3 to 6 h. The treated residuals may be recycled or incinerated.

The EPA tested this system on soils from a former wood-preserving site at pilot scale.[169,170] The soil used in the test initially contained 1600 to 8100 mg/kg pentachlorophenol and up to 17 mg/kg total dioxins. Treatment in the thermal desorption unit removed 99.97% or better of pentachlorophenol and 99.56% or better of total dioxins and total furans from soils. In the LTR, pentachlorophenol concentrations were reduced by 96.89% or better and total dioxin and total furan concentrations by 99.97% or better.

4.4.3 Incineration

Incineration destroys organic materials by thermal oxidation. Organic compounds oxidize to form carbon dioxide, water, sulfur dioxide, nitric oxide, ash, and/or *products of incomplete combustion* (PIC). Incineration does not destroy metals. Metals concentrate in the ash or enter the off-gas treatment system.

Virtually any organic contaminant can be incinerated. Because of the relatively high cost to handle soil and heat soil to incineration temperatures, other technologies — such as SVE or thermal desorption — are sometimes used to separate organic contaminants from soil so that just the contaminants can be incinerated in vapor or liquid phase. However, concentrated liquid or solid hazardous wastes are often incinerated directly. Incineration is the BDAT for certain organic wastes under the RCRA Land

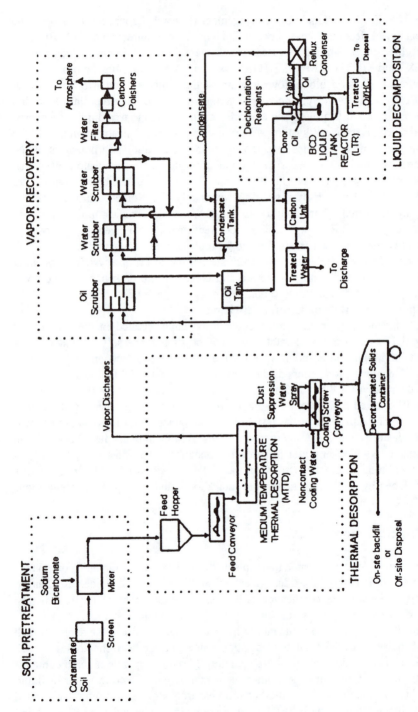

Figure 4.8 Base-catalyzed dechlorination process. (From U.S. EPA, *Superfund Innovative Technology Evaluation Program: Technology Profiles*, 9th ed., EPA/540/R-97/502, Office of Research and Development, Washington, D.C., December 1996, 108–109.)

Disposal Restrictions [40 CFR 268]. TSCA regulations require that concentrated PCB wastes (>500 ppm) must be incinerated [40 CFR 761].

In permitting jargon, a target contaminant is a *principle organic hazardous constituent* (POHC) of the waste. The *destruction and removal efficiency*, or *DRE*, characterizes the extent of oxidation of a POHC:

$$DRE = \frac{W_i - W_o}{W_i} \cdot 100 \tag{4.4}$$

where W_i is the mass feed rate of the POHC into the incinerator and W_o is the mass rate at which the POHC exits the system in the exhaust gas stream. The DRE for organic compounds is typically 99.99% or more.

4.4.3.1 Overview

An incineration system comprises three main systems, as shown in Figure 4.9: feed preparation and handling, primary combustion unit, and off-gas treatment. Feed preparation usually includes screening to remove oversized particles that could clog or jam the combustion unit. Oversized waste may be shredded so that it can be incinerated, or treated or landfilled separately. Sludge is usually dewatered to make the material easier to feed to the incinerator, and to limit the fuel cost associated with heating, volatilizing, and treating the water in the incinerator.

The combustion unit is designed to provide the temperature, turbulence, and time necessary to oxidize contaminants. Several types of combustion units are used to treat soils or sludges containing hazardous waste. Rotary kilns are among the most common units; fluidized bed incinerators and other designs, such as infrared incineration systems, are also used. Section 4.4.3.2 describes these process options.

Ash leaving the combustion unit must be backfilled into the soil excavation or landfilled off-site. If the ash contains high levels of metals, it may require treatment before disposal. The quantity of ash (as a percentage of the feed) varies, but depends on the level of combustible organics in the original waste. For example, incineration of a sludge with a high organic content will produce less ash than incineration of relatively sandy soil.

The off-gases exiting the combustion unit enter a series of air pollution control units. The number and type of air pollution control units varies. Off-gas treatment commonly includes a *secondary combustion chamber* or *afterburner*. This chamber provides sufficient oxygen and temperature to destroy organic contaminants in the off-gases which desorbed from the soil in the combustion unit without oxidizing. Typically, an afterburner provides some 2 sec of retention time with 10% to 60% excess air at a temperature of approximately 2200°F.[171] Air pollution control units also commonly include units designed to remove particulates, such as baghouses, venturi scrubbers, and cyclones; and units designed to remove acids (e.g., sulfuric, hydrochloric), such as wet scrubbers and packed column scrubbers. Section 4.5 provides additional information on such air pollution control units.

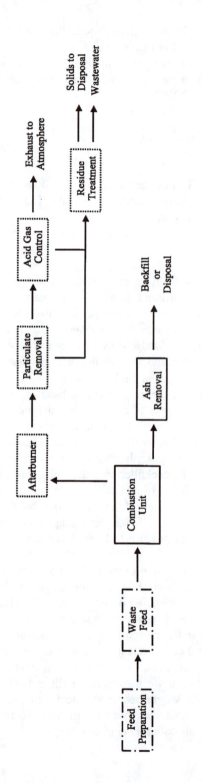

Figure 4.9 Process flow diagram of typical incineration system. (Adapted from U.S. EPA, *Superfund Engineering Issue: Issues Affecting the Applicability and Success of Remedial/Removal Incineration Projects*, EPA/540/2-91/004, Office of Solid Waste and Emergency Response, Washington, D.C., February 1991, 6.)

Incinerators are commonly and successfully used to treat a wide variety of wastes. However, incineration is subject to some limitations. In general,

- Wastes with a high water content can be expensive and difficult to burn, as noted above.
- High levels of halogens such as chloride cause acid emissions. While the emissions-control system is designed to remove some acids, removal of high concentrations can be difficult and expensive.
- Volatile metals (such as lead, mercury, and arsenic) or their oxides will volatilize. These metals cannot be easily removed from off-gases using conventional air pollution control equipment.
- Cr(III) can be oxidized to the more toxic and mobile form, Cr(VI), in an incinerator.
- High metal concentrations in the ash can render the ash a hazardous waste by characteristic, requiring treatment of the ash before disposal.
- A high proportion of fine particles such as clays and silts in the soil fed to an incinerator can result in high particulate loading in off-gases and resulting clogging.

Waste can be incinerated on a hazardous waste site or at an off-site commercial incinerator permitted to handle such wastes. The choice between on-site and off-site incineration depends on logistics, safety issues, and cost. Public reaction to on-site incineration in a populated area often makes that option infeasible. For relatively small quantities of waste, off-site incineration is more cost effective than on-site incineration.

Physical/chemical parameters used to characterize waste for incineration include:

- contaminants and levels, in particular of PCBs, metals, cyanide, and pentachlorophenol;
- waste classification under RCRA and TSCA;
- *proximate analysis*, which includes the moisture content, ash content, volatile matter, fixed carbon, and heating value (BTU/lb) of the waste;
- *ultimate analysis*, which provides the levels of the elements C, H, N, S, and Cl as a percentage of the waste material; and
- viscosity and density, for sludges, tars, or similar materials.

Commercial facilities may require other information.

4.4.3.2 Process Options

The combustion unit in a rotary kiln incinerator is a refractory-lined cylinder mounted on a slight incline.[172] Wastes and auxiliary fuel injected into the high end of the kiln move downward through the combustion zone as the kiln slowly rotates. This rotation creates turbulence, which promotes thorough combustion. The retention time in the kiln can vary from several minutes to an hour or more, but is typically 30 to 60 min.[171,172] At the operating temperature of some 1200 to 1800°F, organic materials oxidize to gaseous compounds and ash. The ash is removed from the lower end of the kiln. It may be sprayed with water to cool the hot ash and to control dust. The off-gases contain water vapor, some desorbed contaminants, volatile metals, particulates, and the products of combustion. The off-gases pass through a secondary combustion chamber, then through air pollution control units for particulate and acid gas removal.

Rotary kilns have several operating limitations, including:

- Debris over a maximum particle size of approximately 6 in. can jam the equipment.
- Unless the incinerator is being operated deliberately in the slagging mode (Section 4.2.5), high levels of certain compounds may cause slagging which clogs the incinerator. Sodium and potassium have low slagging temperatures and can be a problem particularly when the waste contains a large amount of plastics, glasses, and wood. Alkali metal salts, particularly as sodium and potassium sulfates, can also attack the refractory lining of the kiln. Iron, calcium, silicon, and phosphorus have moderate slagging temperatures and can also be of concern.[172,173]
- Highly variable waste can cause operating problems. Solids with high BTU value (e.g., plastics, wood) mixed with soil burn rapidly, spiking the temperature up and decreasing oxygen levels. The control system may stop the feed system in response.[173]

The combustion unit in a *fluidized bed incinerator* consists of a refractory-lined vessel containing a bed of inert, granular, sand-like material. Solids, sludges, and liquids can be injected directly into the bed or at its surface; if soil is being treated, the soil itself can act as the bed. Auxiliary fuels are added as necessary to sustain combustion at ca. 1450 to 1600°F, depending on the heating value of waste. As combustion air is forced upward through the bed, the air flow fluidizes the bed. The turbulent bed enhances heat transfer and mixing of the waste. As a result, fluidized bed incinerators can operate at a lower temperature (and resulting lower cost for auxiliary fuel) than rotary kilns. Fluidized bed incinerators are also relatively resistant to fluctuations in temperature and retention time due to variations in the moisture, ash, or BTU content of the waste.[172]

Off-gases pass through a secondary combustion chamber, then through air pollution control units. In a circulating bed combustor, one variation of a fluidized bed incinerator, limestone is added to the solid feed to neutralize acid gases and minimize the need for off-gas treatment units.[174]

Fluidized bed incinerators have several limitations, including:[172]

- The maximum particle size is 1 to 3 in., depending on the unit.
- High concentrations of low-melting point (i.e., <1600°F) constituents, particularly alkali metal salts and halogens, can cause problems. Defluidization of the bed may occur when particles begin to melt and become sticky. Alkali metal salts >5% by weight and halogen >8% by weight contribute to refractory attack, defluidization of the bed, and slagging problems.
- Waste with high ash content can foul the bed.

An *infrared thermal unit* uses silicon carbide elements to generate thermal radiation beyond the red end of the visible spectrum. Materials to be treated are exposed to the radiation while they pass through the combustion unit on a moving belt. From the primary combustion chamber, combustion gases pass into a secondary combustion chamber, which can be either a combination gas-fired/infrared unit or a conventional afterburner. Gases exiting the secondary combustion chamber pass through air pollution control units for particulate and acid gas removal.[175] As of 1996, this technology is no longer available through vendors in the U.S.[176]

4.4.3.3 Regulatory Requirements

Incinerators treating RCRA hazardous wastes must comply with regulations at 40 CFR 264.340-351. These regulations require, among other things,

- a *trial burn* to demonstrate performance and set permit conditions [40 CFR 270.62];
- 99.99% DRE for *POHC*, in general, and 99.9999% DRE for dioxins;
- limits on HCl and particulate emissions; and
- compliance with operating conditions specified in permit (e.g., combustion temperature, feed rate, CO levels in stack gases).

Incineration of certain PCB-contaminated materials is regulated under TSCA [40 CFR 761.70]. FIFRA regulates the incineration of certain pesticides [40 CFR 165].

4.5 CONTROL OF AIR EMISSIONS FROM SOIL TREATMENT UNITS

Many of the air pollution control units used in soil remediation are also used in groundwater treatment systems. Section 3.7 describes those technologies, which include demisters, knock-out tanks, activated carbon, oxidizers, and biofilters. Off-gases from some soil remediation systems require other types of treatment to remove particulates and other contaminants. Air pollution control units include condensers, to remove water or liquid organics; units designed to remove particulates, such as cyclones, baghouses, and wet scrubbers; and units designed to remove acid gases (e.g., sulfuric, hydrochloric), such as packed column scrubbers. Each of these units is briefly described below.

A *condenser* simply cools the gas temperature to the point where an organic vapor or water vapor condenses into a liquid.[177] Condensed hydrocarbons are incinerated or recycled. Condensed water often requires treatment before it can be discharged.

A *cyclone* removes particulates from a gas based on the difference between the specific gravity of the gas and of the particulate. Separation is most effective for particulates between 5 and 200 μm in size. A cyclone consists simply of a cylindrical section atop a conical section; cyclones have no moving parts. The gas enters the cyclone tangentially at one or more points. While the gas stream spins around the cyclone, the inertia of the particles resists the change in direction. As a result, the particles are propelled into the wall of the cyclone. Particulates fall down the chamber and exit at the point of the cone. The clean gas exits through the top of the cyclone.[178,179] (Cyclones are also used to separate soil from water as described in Section 4.3.3.1).

A *baghouse* comprises a housing containing a number of "bags" made of filter fabric such as woven cloth, felt, or porous membrane. Once the filter fabric is precoated with an initial layer of dust or particulates, fabric filters can collect even very fine particles. The various fabrics used in filters have different temperature limitations and chemical compatibility.[178,180]

Wet scrubbers can remove particulates from gas streams. A wet scrubber consists of two functional parts. In the contactor stage, water is sprayed into the gas stream to collect particulates in the water droplets. In the entrainment separator stage, the particulates/water are removed from the gas stream for further treatment and/or discharge. These two stages can occur in a single unit, or in different units. In the latter case, a cyclone is sometimes used as a separator.[178] A *venturi scrubber* is a type of wet scrubber. A venturi scrubber contains a narrow throat or orifice across the flow path to increase the velocity of the gas and atomize the scrubbing fluid.[178,181]

Wet scrubbers are also used for treating acid gases. When an alkaline aqueous solution is sprayed through a gas containing acids (e.g., SO_2), the bases and acids react to form neutral salts and water. The salts are collected as precipitates.

Packed column scrubbers or *absorption towers*, which can also remove acids from off-gases, are analogous to air strippers in design. In a countercurrent system, a scrubbing liquid flows downward through packing or perforated trays, while the gas to be treated flows upward through the scrubber. The packing or trays are designed to allow extensive contact between the liquid and the gas so that the liquid can absorb the acidic compounds in the gas stream. Like an air stripper, the packing in a scrubber can foul with particulates. As a result, particulates are usually removed before a gas stream enters a packed bed scrubber.[178,182]

Problems

4.1   Develop a plan to excavate contaminated soils at the ASR site based on the clean-up goal that you derived for PCE for industrial use of the site (Problem 2.9) and clean-up levels for TCE and PCBs in unsaturated soils of 520 and 41 mg/kg, respectively. Include the following, noting all assumptions:

[a] Sketch of area to be excavated on a site plan.
[b] Estimate of total volume of soil to be excavated, in bcy and lcy.
[c] Estimated time and equipment cost to excavate the soil and stockpile it on site. Assume that the soil is excavated using a backhoe with a 1½-cy bucket, a production rate of 70 bcy/h in uncontaminated soils, and a unit cost, for equipment plus crew, of $1123/d.[183] Adjust your estimate of equipment productivity for work in Level C health and safety protection. (See Section 2.7.4.)
[d] Optional: repeat [a] and [b], using clean-up levels developed for residential site use: 12 mg/kg PCE, 58 mg/kg TCE, and 1.6 mg/kg PCBs. By what percentage does this change in clean-up goals increase the volume estimated in part [b]?
[e] Optional: repeat [a] and [b], using clean-up levels developed for protection of groundwater: 0.04 mg/kg PCE, 0.02 mg/kg TCE, and 2 mg/kg PCBs. By what percentage does this change in clean-up goals increase the volume estimated in part [b]?

4.2 In the first ROD for the Abex Corporation Superfund Site (Example 4.1), the EPA estimated that soil remediation would cost $28,891,243. The EPA later selected a different remedy costing an estimated $31,507,670. Based on the information provided in the example, which remedy was more cost effective? Explain your reasoning.

4.3 Consider the site described in Problem 3.4. The State regulatory agency requires that the impoundments be closed. The impoundments, their contents, and underlying contaminated soil must be remediated.

[a] The data indicate that the sludges in impoundments A, D, and E contain similar levels of total chromium. However, the TCLP levels were much higher for sludge from impoundment A than for sludges in impoundments D and E. Considering the original use of each impoundment, what characteristic of chromium contamination might cause that difference?

[b] Briefly describe a plausible closure program and explain why that program would be appropriate.

[c] Indicate any site-specific testing that would be necessary to design your closure program.

4.4 A natural wetland historically received surface water and wastewater discharges from a chemical plant. The wetland is the home to a type of turtle which is a potentially endangered species.

During the remedial investigation, the field team collected shallow (3-ft) samples from the sediments at 38 locations in the wetland. These samples were analyzed for the compounds predominantly used and manufactured at the chemical plant. Fish were also collected and the tissue analyzed.

All but one of the sediment samples analyzed for mercury contained detectable mercury concentrations, ranging from 0.42 to 690 mg/kg. Figure 4.10 summarizes the data distribution. Clean-up standards have not been promulgated for mercury in sediment; the ERL for inorganic mercury is 0.15 mg/kg.[184] Mercury was also found in fish tissue at levels from 0.59 mg/kg fish to 3.1 mg/kg. These levels exceeded the Food and Drug Administration's criterion of 0.5 mg/kg. Mercury has not been detected in the groundwater.

[a] Briefly describe a remediation program for this wetland. If additional site-specific testing or data are necessary, indicate the type of tests or data in your answer.

[b] Indicate at least two logistic or implementation problems associated with remediation of the wetlands.

4.5 Develop a conceptual design for a SVE system for the ASR site, including:

- preliminary layout of system shown on site plan;
- cross-section showing typical subsurface components;
- block flow diagram or preliminary piping and instrumentation diagram showing vapor treatment system;
- list of major components and, to the extent possible, their sizes/quantities to be included in estimates of capital and operation and maintenance costs;
- assumptions used to develop conceptual design; and
- description of any tests that you would perform to complete the design.

Base the design on clean-up levels intended to protect groundwater.

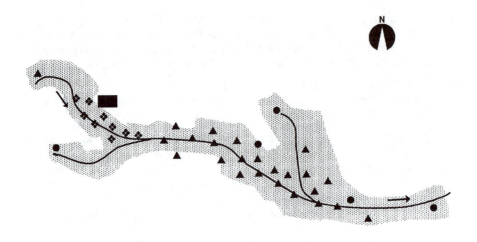

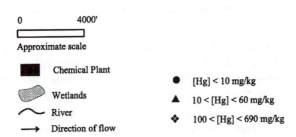

Figure 4.10   Problem 4.4: distribution of mercury in wetland sediments.

4.6   A SVE system has been in operation at a site contaminated as the result of a solvent spill. The soil comprises silty sand. A total of 2900 cy of unsaturated soil has been contaminated at levels above the clean-up level. The average concentration of the solvent in that volume of soil is 214 mg/kg, computed on a volume-weighted basis. The clean-up level is a concentration of 5 mg/kg, on average.

The average air flow rate is 250 cfm. The concentration of the solvent in the off-gas stream has been monitored at a point before the emissions control system. Table 4.5 summarizes the data.

[a]   Can the system be shut off? Explain why or why not.

[b]   The concentrations of volatile compounds in the off-gas from a SVE system typically decrease rapidly during the initial period of operation, then level off. Soil vapor concentrations may rebound after a SVE system is shut down. Why?

4.7   A third field investigation at the HypoChem site provided data on the levels of contamination in soil. Since the first two field investigations (see Problems 2.6 and 3.8), the above-ground storage tanks have been removed. During the third field program, soil borings were installed in the vicinity of the former tanks and soil samples collected for analysis. Figure 4.11 shows the locations of the soil borings, and Table 4.6 summarizes the data. One additional well (MW-6) was installed at the

Table 4.5   Problem 4.6 Operating Results for
             SVE System

Time (days)	Off-gas concentration (mg/m³)
1	2980
5	2375
10	2000
15	1500
20	1375
25	1005
30	1000
40	450
50	500
60	250
70	125
80	150
90	110
100	85
110	130
120	120

Table 4.6   Problem 4.7
             Soil Data from the HypoChem Facility

Location	Depth (ft)	Toluene (mg/kg)
SB-1	0 to 2	37
	4 to 6	9,600
SB-2	2 to 4	602
	6 to 8	8,700
SB-3	2 to 4	11,200
	4 to 6	57,700
SB-4	0 to 2	83
	4 to 6	9,800
SB-5	2 to 4	24,900
	6 to 8	52,200
SB-6/MW-6	0 to 2	102

site, as shown on Figure 4.11. The well was screened in the sand layer above the clay. Samplers measured 0.3 ft of LNAPL in that well.

State regulations mandate a clean-up level for toluene of 5 mg/kg for protection of groundwater. Develop a conceptual design for remediation of the soils near the former tanks. Include in your conceptual design the six elements listed in Problem 4.5.

4.8   Remediation Is Us, Inc. (RIU) has submitted an unsolicited proposal to you for bioremediation of VOC-contaminated soil at the ASR site. RIU proposes to remediate the soil, from excavation to backfill, for $50/ton. RIU emphasized in the proposal that they would work quickly — in fact, the proposal states that RIU will mobilize to the site for full-scale remediation within 3 weeks of contract signature.

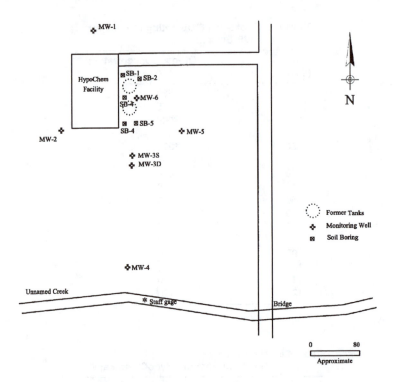

Figure 4.11   Problem 4.7: location of soil borings, HypoChem facility.

In brief, RIU has proposed to:

- Construct a lined treatment unit on the southern portion of the facility. The treatment area will be located to the south of the former disposal pit. The area will be cleared and graded for construction of a treatment pad 150 by 200 ft in size. A 6-in.-thick base of sand will be emplaced, then covered with 40-mil HDPE, and a 1-ft-thick layer of sand. The treatment pad will be sloped to a sump where water will be collected. The treatment area will be bermed to control run on/run off of surface water.
- Excavate soil and stockpile for treatment. Stockpiled soil will be placed on the treatment pad and treated in batches.
- Treat soil in 1-ft lifts. Soil from the stockpile will be spread onto the treatment pad in 1-ft lifts. Fertilizer containing nitrogen and phosphorus will be added to each lift, then water will be sprayed over the soil. The soil will then be tilled to thoroughly mix the contaminated soil with fertilizers and water. As water accumulates in the sump, it will be sprayed back over the soil undergoing treatment.
- Backfill excavation. As each lift achieves the remediation goals, the soil will be removed from the treatment pad and used to backfill the excavation. As each lift is removed, another lift will be placed on the treatment pad until all of the soil is treated.

Is this a sensible remediation approach? Why or why not? Should you hire RIU to remediate this site?

4.9   Each group of remediation technologies below includes technologies which use similar equipment to effect treatment. For each group, indicate (i) the common type of equipment, and the differences between the technologies in terms of (ii) the primary objective of treatment, and (iii) typical operating conditions.

[a] Thermal desorption, incineration, and vitrification;
[b] Air sparging, bioventing; and
[c] Six-phase soil heating, electrokinetics, *in situ* vitrification.

4.10 Consider the application of each of the following five remediation technologies to the VOC- and PCB-contaminated soils at the ASR site.

[a] Thermal desorption;
[b] Solvent extraction;
[c] Asphalt batching;
[d] Containment using a multi layer (RCRA) cap; and
[e] *In situ* vitrification.

For each technology, indicate briefly:

• Could the technology be used?
• Why or why not?
• One site-specific concern regarding application of the technology.

REFERENCES

1. U.S. EPA, *Innovative Treatment Technologies: Annual Status Report (Eighth Edition)*, EPA-542-R-96-010, Number 8, Office of Solid Waste and Emergency Response, Washington, D.C., November 1996, 6–11, A-1.
2. 29 CFR 1926.652 (Requirements for Protective Systems), Appendix A to Subpart P — Soil Classification, and Appendix B to Subpart P — Sloping and Benching.
3. Church, H. K., *Excavation Handbook*, McGraw-Hill, New York, 1981, Appendix A.
4. Church, H. K., *Excavation Handbook*, McGraw-Hill, New York, 1981, chap. 17.
5. U.S. EPA, *Handbook: Remedial Action at Waste Disposal Sites (Revised)*, Office of Emergency and Remedial Response, Washington, D.C., October 1985, 7:2–10.
6. Caterpillar, Inc., *Caterpillar Performance Handbook*, Edition 24, Caterpillar, Inc., Peoria, IL, October 1993, chap. 1.
7. Rast, R. R., *Environmental Remediation Estimating Methods*, R. S. Means Company, Kingston, MA, 1997, 179–186.
8. Church, H. K., *Excavation Handbook*, McGraw-Hill, New York, 1981, 12:12–25.
9. Caterpillar, Inc., *Caterpillar Performance Handbook*, Edition 24, Caterpillar, Inc., Peoria, IL, October 1993, chap. 4.
10. U.S. EPA, *Engineering Bulletin: Control of Air Emissions From Materials Handling During Remediation*, EPA/540/2-91/023, Office of Emergency and Remedial Response, Washington, D.C., 7 pp.
11. U.S. EPA, *Demonstration of a Trial Excavation at the McColl Superfund Site, Application Analysis Report*, EPA/540/AR-92/015, Office of Research and Development, Washington, D.C., October 1992, 9–12.
12. Jones, L. W., Cullinane, M. J., and Hathaway, A. W., Stabilization/Solidification of Hazardous Waste, U.S. EPA Technical Handbook, Presentation to the Office of Waste Programs Enforcement, Washington, D.C., 1985.

13. Goldberger, W. M., Robbins, L. A., Fiedler, R. A., Jepson, T. L. B., Knoll, F. S., Malony, J. O., Mitchell, D. W., Parekh, B. K., Sorenson, T. C., Stavenger, P. L., Thelen, R. L., Treybal, R. E., and Wechsler, I., Solid-Solid and Liquid-Liquid Systems, in *Perry's Chemical Engineer's Handbook*, 6th ed., Perry, R. H., Green, D. W., and Malony, J. O., Eds., McGraw-Hill, New York, 1984, 21:14.

14. Hersh, R., Probst, K., Wernstedt, K., and Mazurek, J., *Linking Land Use and Superfund Cleanups: Uncharted Territory*, prepared for the U.S. EPA by Resources for the Future, Washington, D.C., Internet Edition (http://www.rff.org), 1997, 36–40.

15. U.S. EPA, *Abex Corporation*, Web page last revised August 7, 1996, EPA National Priority List-0302667y.htm, 3 pp.

16. U.S. EPA, *Abex Corporation Superfund Site, Portsmouth, Virginia*, Fact Sheet January 1997, 3 pp.

17. U.S. EPA, *Abex Corporation Superfund Site*, Update Fact Sheet, May 23, 1997, 2 pp.

18. U.S. EPA, Record of Decision Abstracts, Superfund Sites — Virginia, downloaded from http://www.epa.gov August 20, 1997, 12 pp.

19. U.S. EPA, *Record of Decision — Abex Corporation Site*, EPA/ROD/R03-92/159, September 29, 1992.

20. U.S. EPA, *Presumptive Remedy Guidance for CERCLA Municipal Landfill Sites*, EPA/540-F-93-035, Office of Solid Waste and Emergency Response, Washington, D.C., September 1993, 14 pp.

21. U.S. EPA, *Guide to Technical Resources for the Design of Land Disposal Facilities*, EPA/625/6-88/018, Risk Reduction Engineering Laboratory, Cincinnati, OH, December 1988, 47–52.

22. U.S. EPA, *Design and Construction of RCRA/CERCLA Final Covers*, EPA/625/4-91/025, Office of Research and Development, Washington, D.C., May 1991, 1–4.

23. Koerner, R. M., *Designing with Geosynthetics*, 2nd ed., Prentice-Hall, Englewood Cliffs, NJ, 1990, 40.

24. Koerner, R. M., *Designing with Geosynthetics*, 2nd ed., Prentice-Hall, Englewood Cliffs, NJ, 1990, 364–402.

25. Schroeder, P. R., Lloyd, C. M., Zappi, P. A., and Aziz, N. M., *The Hydrologic Evaluation of Landfill Performance (HELP) Model, Users Guide for Version 3*, EPA/600/R-94/168a, U.S. EPA Office of Research and Development, Cincinnati, OH, September 1994.

26. Koerner, R. M., *Designing with Geosynthetics*, 2nd ed., Prentice-Hall, Englewood Cliffs, NJ, 1990, 334, 340–345.

27. Koerner, R. M., *Designing with Geosynthetics*, 2nd ed., Prentice-Hall, Englewood Cliffs, NJ, 1990, 3.

28. U.S. EPA, *Design and Construction of RCRA/CERCLA Final Covers*, EPA/625/4-91/025, Office of Research and Development, Washington, D.C., May 1991, chap. 12.

29. Koerner, R. M., Halse, Y. H., and Lord, A. E., Long-Term Durability and Aging of Geomembranes, in *Waste Containment Systems: Construction, Regulation, and Performance*, Geotechnical Special Publication No. 26, Bonaparte, R., Ed., ASCE, New York, 1990, 106–134.

30. Massachusetts Department of Environmental Quality Engineering, The On-Site Treatment of Petroleum Contaminated Soils by Mobile Asphalt Emulsion Fixation Technology — Standard Operating Practices, Draft, March 6, 1990, 13 pp.

31. Czarnecki, R. C., Making Use of Contaminated Soil, *Civil Eng.*, December 1988, 72–72.

32. Troxler, W. L., Cudahy, J. J., Zink, R. P., Yezzi, J. J., and Rosenthal, S. I., Treatment of Nonhazardous Petroleum-Contaminated Soils by Thermal Desorption Technologies, *Air Waste,* 43, 1512–1525, 1993.

33. U.S. EPA, *Stabilization/Solidification of CERCLA and RCRA Wastes: Physical Tests, Chemical Testing Procedures, Technology Screening and Field Activities,* EPA/625/6-89/022, Office of Research and Development, Washington, D.C., May 1989, 1.

34. Marks, P. J., Wujcik, W. J., and Loncar, A. F., *Remediation Technologies Screening Matrix and Reference Guide,* 2nd ed., PB95-104782, U.S. Army Environmental Center, Aberdeen Proving Ground, MD, October 1994, 4:77–78.

35. U.S. EPA, *Handbook — Remedial Action at Waste Disposal Sites (Revised),* EPA/625/6-85/006, Office of Emergency and Remedial Response, Washington, D.C., October 1985, 9: 51, 58.

36. PRC Environmental Management, *Recent Developments for In Situ Treatment of Metal Contaminated Soils,* prepared for: U.S. EPA Office of Solid Waste and Emergency Response, Washington, D.C., March 1997, 42–45.

37. U.S. EPA, *Technology Demonstration Summary — International Waste Technologies In Situ Stabilization/Solidification Hialeah FL,* EPA/540/S5-89/004, RREL, Cincinnati, OH, June 1989, 5 pp.

38. U.S. EPA, *Handbook — Remedial Action at Waste Disposal Sites (Revised),* EPA/625/6-85/006, Office of Emergency and Remedial Response, Washington, D.C., October 1985, 10:105–117.

39. U.S. EPA, *Superfund Innovative Technology Evaluation Program Technology Profiles,* 9th ed., EPA/540/R-97/502, Office of Research and Development, Washington, D.C., December 1996, 46.

40. Arniella, E. F. and Blythe, L. J., Solidifying traps, *Chem. Eng.,* 92–102, February 1990.

41. U.S. EPA, *Handbook for Stabilization/Solidification of Hazardous Wastes,* EPA/540/2-86/001, Office of Solid Waste and Emergency Response, Washington, D.C., June 1986.

42. U.S. EPA, *Technology Demonstration Summary, Chemfix Solidification/Stabilization Process, Clackamas, Oregon,* EPA/540/S5-89/011, Center for Environmental Research Information, Cincinnati OH, December 1990, 5 pp.

43. U.S. EPA, *Stabilization/Solidification of CERCLA and RCRA Wastes: Physical Tests, Chemical Testing Procedures, Technology Screening, and Field Activities,* EPA/625/6-89/022, Office of Research and Development, Washington, D.C., May 1989, 5:4–12.

44. U.S. EPA, *Superfund Innovative Technology Evaluation Program Technology Profiles,* 9th ed., EPA/540/R-97/502, Office of Research and Development, Washington, D.C., December 1996, 46–47, 76–77, 151.

45. U.S. EPA, *Superfund Innovative Technology Evaluation Program, Technology Profiles,* 9th ed., EPA/540/R-97/502, Office of Research and Development, Washington, D.C., December 1996, 78–79.

46. U.S. EPA, *Geosafe Corporation, In Situ Vitrification, Innovative Technology Evaluation Report,* EPA/540/R-94/520, Office of Research and Development, Washington, D.C., March 1995, 115, 120.

47. U.S. EPA, *Vendor Information System for Innovative Treatment Technologies, Version 5.0, Technology Type: Vitrification,* Office of Solid Waste and Emergency Response, Technology Innovation Office, Washington, D.C., August 1996.

48. Queneau, P. B., Cregar, D. E., and Karwaski, L. J., Slag Control in Rotary Kilns, *Pollut. Eng.*, 26–32, January 15, 1992.
49. Vesilind, P. A. and Rimer, A. E., *Unit Operations in Resource Recovery Engineering*, Prentice-Hall, Englewood Cliffs, NJ, 328–329.
50. Marks, P. J., Wujcik, W. J., and Loncar, A. F., *Remediation Technologies Screening Matrix and Reference Guide,* 2nd ed., PB95-104782, U.S. Army Environmental Center, Aberdeen Proving Ground, MD, October 1994, 4:109.
51. U.S. EPA, *Superfund Engineering Issue: Issues Affecting the Applicability and Success of Remedial/Removal Incineration Projects*, EPA/540/2-91, Office of Solid Waste and Emergency Response, Washington, D.C., February 1991, 6.
52. U.S. EPA, *Technology Demonstration Summary — Technology Evaluation Report; SITE Program Demonstration Test; Retech, Inc. Plasma Centrifugal Furnace; Butte, MT*, EPA/540/S5-91/007, Center for Environmental Research Information, Cincinnati, OH, August, 1992, 6 pp.
53. U.S. EPA, *Babcock & Wilcox Cyclone Furnace Vitrification Technology, Application Analysis Report*, EPA/540/AR-92/017, Center for Environmental Research Information, Cincinnati, OH, August 1992, 35 pp.
54. Williams, I., EnVitCo, Inc., personal communication, January 17, 1992.
55. FitzPatrick, V. F., Timmerman, C. L., and Buelt, J. L., In situ vitrification — a new process for waste remediation, presented at the Second International Conference on New Frontiers for Hazardous Waste Management, Pittsburgh, PA, September 27–30, 1987.
56. U.S. EPA, *Geosafe Corporation, In Situ Vitrification, Innovative Technology Evaluation Report*, EPA/540/R-94/520, Office of Research and Development, Washington, D.C., March 1995, 17–18.
57. U.S. EPA, *Geosafe Corporation, In Situ Vitrification, Innovative Technology Evaluation Report*, EPA/540/R-94/520, Office of Research and Development, Washington, D.C., March 1995, 62, 80.
58. U.S. EPA, *Geosafe Corporation, In Situ Vitrification, Innovative Technology Evaluation Report*, EPA/540/R-94/520, Office of Research and Development, Washington, D.C., March 1995, 118–119.
59. U.S. EPA, *Innovative Treatment Technologies: Annual Status Report,* 8th ed., EPA-542-R-96-010, Number 8, Office of Solid Waste and Emergency Response, Washington, D.C., November 1996, Appendices B, E.
60. U.S. EPA, *Vendor Information System for Innovative Treatment Technologies, Version 5.0, Profile — Geosafe Corporation*, Office of Solid Waste and Emergency Response, Technology Innovation Office, Washington, D.C., August 1996.
61. PRC Environmental Management, *Recent Developments for in Situ Treatment of Metal Contaminated Soils*, prepared for U.S. EPA, Office of Solid Waste and Emergency Response, Washington, D.C., March 5, 1997, 46.
62. U.S. EPA, *Geosafe Corporation, In Situ Vitrification, Innovative Technology Evaluation Report*, EPA/540/R-94/520, Office of Research and Development, Washington, D.C., March 1995, 93, 122.
63. U.S. Army Corps of Engineers, *Engineering and Design, Soil Vapor Extraction and Bioventing*, Engineer Manual (EM) 1110-1-4001, U.S. Army Corps of Engineers, Washington, D.C., 30 November 1995, 3:1.
64. *Webster's Ninth New Collegiate Dictionary,* Merriam-Webster, Springfield, MA, 1990, 59.
65. U.S. EPA, *Guide for Conducting Treatability Studies Under CERCLA: Soil Vapor Extraction (Interim Guidance)*, EPA/540/2-91/019A, September 1991, 7.

66. U.S. Army Corps of Engineers, *Engineering and Design, Soil Vapor Extraction and Bioventing*, Engineer Manual (EM) 1110-1-4001, U.S. Army Corps of Engineers, Washington, D.C., 30 November 1995, 2:1.

67. U.S. EPA, *Soil Vapor Extraction Technology Reference Handbook*, EPA/540/2-91/003, February 1991, Glossary.

68. U.S. EPA, *Innovative Treatment Technologies: Annual Status Report,* 8th ed., EPA-542-R-96-010 Number 8, Office of Solid Waste and Emergency Response, Washington, D.C., November 1996, 15.

69. U.S. EPA, *How to Evaluate Alternative Cleanup Technologies for Underground Storage Tank Sites*, EPA/510-B-94-003, September 1996, II: 12–28.

70. U.S. EPA, *Soil Vapor Extraction Technology Reference Handbook*, EPA/540/2-91/003, February 1991, 17–18.

71. U.S. EPA, *Accutech Pneumatic Fracturing Extraction and Hot Gas Injection, Phase I, Application Analysis Report*, EPA/540/AR-93/509, Risk Reduction Engineering Laboratory, Cincinnati, OH, July 1993.

72. U.S. EPA, *Engineering Bulletin: In Situ Steam Extraction Treatment*, EPA/540/2-91/005, Office of Emergency and Remedial Response, Washington, D.C., May 1991, 7 pp.

73. U.S. EPA, *In Situ Steam Enhanced Recovery Process, Hughes Environmental Systems, Inc., Innovative Technology Evaluation Report*, EPA/540/R-94/510, Office of Research and Development, Washington, D.C., July 1995.

74. U.S. Army Corps of Engineers, *Engineering and Design, Soil Vapor Extraction and Bioventing*, Engineer Manual (EM) 1110-1-4001, U.S. Army Corps of Engineers, Washington, D.C., 30 November 1995, 2:8–9; 4:4.

75. U.S. Army Corps of Engineers, *Engineering and Design, Soil Vapor Extraction and Bioventing*, Engineer Manual (EM) 1110-1-4001, U.S. Army Corps of Engineers, Washington, D.C., 30 November 1995, 5:27–30.

76. U.S. Army Corps of Engineers, *Engineering and Design, Soil Vapor Extraction and Bioventing*, Engineer Manual (EM) 1110-1-4001, U.S. Army Corps of Engineers, Washington, D.C., 30 November 1995, chap. 2, Appendix C.

77. Jordan, D. L., Mercer, J. W., and Cohen, R. M., *Review of Mathematical Modelling for Evaluating Soil Vapor Extraction Systems*, EPA/540/R-95/513, U.S. EPA National Risk Management Research Laboratory, Cincinnati, OH, July 1995.

78. U.S. Army Corps of Engineers, *Engineering and Design, Soil Vapor Extraction and Bioventing*, Engineer Manual (EM) 1110-1-4001, U.S. Army Corps of Engineers, Washington, D.C., 30 November 1995, 2:11; 5:1; 5:16–17.

79. Johnson, P. C. and Ettinger, R. A., Considerations for the design of in situ vapor extraction systems: radius of influence vs. zone of remediation, *Groundwater Monit. Rev.,* 123–128, Summer 1994.

80. U.S. Army Corps of Engineers, *Engineering and Design, Soil Vapor Extraction and Bioventing*, Engineer Manual (EM) 1110-1-4001, U.S. Army Corps of Engineers, Washington, D.C., 30 November 1995, 5:3.

81. Shan, C., Falta, R. W., and Javandel, I., Analytical solutions for steady state gas flow to a soil vapor extraction well, *J. Water Resour. Res.,* 28(4), 1105-1120, 1992.

82. U.S. Army Corps of Engineers, *Engineering and Design, Soil Vapor Extraction and Bioventing*, Engineer Manual (EM) 1110-1-4001, U.S. Army Corps of Engineers, Washington, D.C., 30 November 1995, 5:18–26.

83. U.S. Army Corps of Engineers, *Engineering and Design, Soil Vapor Extraction and Bioventing*, Engineer Manual (EM) 1110-1-4001, U.S. Army Corps of Engineers, Washington, D.C., 30 November 1995, 5:31–34.

84. U.S. Army Corps of Engineers, *Engineering and Design, Soil Vapor Extraction and Bioventing*, Engineer Manual (EM) 1110-1-4001, U.S. Army Corps of Engineers, Washington, D.C., 30 November 1995, 3:5–8; 5:52–54.
85. U.S. Army Corps of Engineers, *Engineering and Design, Soil Vapor Extraction and Bioventing*, Engineer Manual (EM) 1110-1-4001, U.S. Army Corps of Engineers, Washington, D.C., 30 November 1995, 5:17–18.
86. U.S. EPA, *Hydraulic Fracturing Technology — Technology Evaluation Report*, EPA/540/R-93/505, Risk Reduction Engineering Laboratory, Cincinnati, OH, September 1993, 4.
87. Murdoch, L. C., Losonsky, G., Cluxton, P., Patterson, B., Klich, I., and Braswell, B., *Project Summary: Feasibility of Hydraulic Fracturing to Improve Remedial Actions*, EPA/600/S2-91/012, U.S. EPA Risk Reduction Engineering Laboratory, Cincinnati, OH, August 1991, 8 pp.
88. U.S. EPA, *Hydraulic Fracturing Technology — Application Analysis Report*, EPA/540/R-93/505, Risk Reduction Engineering Laboratory, Cincinnati, OH, September 1993, 1:4–8.
89. U.S. EPA, *Accutech Pneumatic Fracturing Extraction and Hot Gas Injection, Application Analysis Report*, EPA/540/AR-93/509, Risk Reduction Engineering Laboratory, Cincinnati, OH, July 1993, 23.
90. U.S. Army Corps of Engineers, *Engineering and Design, Soil Vapor Extraction and Bioventing*, Engineer Manual (EM) 1110-1-4001, U.S. Army Corps of Engineers, Washington, D.C., 30 November 1995, 5:7–15, 5:38–41.
91. Johnson, P. C., Stanley, C. C., Kemblowski, M. W., Byers, D. L., and Colthart, J. D., A practical approach to the design, operation, and monitoring of in situ soil venting systems, *Groundwater Monit. Rev.*, 159–178, Spring 1990.
92. Leeson, A. and Hinchee, R., *Manual: Bioventing Principles and Practice, Volume II: Bioventing Design*, EPA/625/XXX/001, EPA Office of Research and Development, Washington, D.C., September 1995, 30–32.
93. U.S. EPA, *Engineering Forum Issue Paper: Soil Vapor Extraction Implementation Experiences*, EPA 540/F-95/030, Office of Solid Waste and Emergency Response, Washington, D.C., April 1996, 11 pp.
94. U.S. Army Corps of Engineers, *Engineering and Design, Soil Vapor Extraction and Bioventing*, Engineer Manual (EM) 1110-1-4001, U.S. Army Corps of Engineers, Washington, D.C., 30 November 1995, 8:4–6.
95. U.S. EPA, *Cost and Performance Report: Soil Vapor Extraction at the Verona Well Field Superfund Site, Thomas Solvent Raymond Road (OU-1) Battle Creek, Michigan*, Office of Solid Waste and Emergency Response, http://clu-in.com/Verona.htm#SITEINFO, March 1995, 30 pp.
96. U.S. EPA, *Analysis of Selected Enhancements for Soil Vapor Extraction*, EPA-542-R-97-007, Office of Solid Waste and Emergency Response, Washington, D.C., September 1997, 4:1–28.
97. U.S. EPA, *Presumptive Remedy: Supplemental Bulletin Multi-Phase Extraction (MPE) Technology for VOCs in Soil and Groundwater*, Office of Solid Waste and Emergency Response, Washington, D.C., April 1997, 11 pp.
98. U.S. Army Corps of Engineers, *Engineering and Design, Soil Vapor Extraction and Bioventing*, Engineer Manual (EM) 1110-1-4001, U.S. Army Corps of Engineers, Washington, D.C., 30 November 1995, 3:2–3.
99. U.S. EPA, *Vendor Information System for Innovative Treatment Technologies, Version 5.0, Technology Type: Thermal Desorption, Full Scale*, Office of Solid Waste and Emergency Response, Technology Innovation Office, Washington, D.C., August 1996.

100. U.S. EPA, *Demonstration Bulletin: Thermal Desorption System, Clean Berkshires, Inc.*, EPA/540/MR-94/507, Center for Environmental Research Information, Cincinnati, OH, April 1994.

101. Perry, R. A., Mercury from process sludges, *Chem. Eng. Prog.*, 70(3), 73–80, 1974.

102. Blanchard, J. and Stamnes, R., *Engineering Issue Paper: Thermal Desorption Implementation Issues*, EPA 540/F-95/831, U.S. EPA Office of Emergency and Remedial Response, Washington, D.C., January 1997, 8 pp.

103. Troxler, W. L., Cudahy, J. J., Zinc, R. P., Yezzi, J. J., and Rosenthal, S. I., *Thermal Desorption of Petroleum Contaminanted Soils*, presented at Sixth Annual Conference Hydrocarbon Contaminated Soils, University of Massachusetts, Amherst, September 23–26, 1991.

104. U.S. EPA, *In Situ Remediation Technology Status Report: Thermal Enhancements*, EPA/542-K-94-009, Office of Solid Waste and Emergency Response, Technology Innovation Office, Washington, D.C., April 1995, 1.

105. Bergsman, T. M., Gauglitz, P. A., Roberts, J. A., and Schlender, M. H., Soil-heating technology shown to accelerate the removal of volatile organic compounds from clay soils, *Fed. Fac. Environ. J.*, 69–79, 1995/1996.

106. U.S. EPA, *Vendor Information System for Innovative Treatment Technologies, Version 5.0, Profile — Six Phase Soil Heating, Terra Vac/Battelle PNL*, Office of Solid Waste and Emergency Response, Technology Innovation Office, Washington, D.C., August 1996.

107. Dev, H., Sresty, G. C., and Downey, D., Field Test of the Radio Frequency In Situ Soil Decontamination Process, in *Proceedings of Superfund '88*, Washington, D.C., November 1988, Hazardous Materials Control Research Institute, 498–502.

108. U.S. EPA, *Vendor Information System for Innovative Treatment Technologies, Version 5.0, Profiles for IIT Research Technology*, Office of Solid Waste and Emergency Response, Technology Innovation Office, Washington, D.C., August 1996.

109. U.S. EPA, *IITRI Radio Frequency Heating Technology Innovative Technology Evaluation Report*, EPA/540/R-94/527, Office of Research and Development Washington, D.C., June 1995, 141 pp.

110. U.S. EPA, *Radio Frequency Heating, KAI Technologies, Inc., Innovative Technology Evaluation Report*, EPA/540/R-94/528, Office of Reseach and Development, Washington, D.C., April 1995, 151.

111. U.S. EPA, *Vendor Information System for Innovative Treatment Technologies, Version 5.0, Profiles for Steam Injection and Vacuum Extraction*, SIVE Services; CROW, Western Research Institute; Praxis Environmental Technologies, Inc., Office of Solid Waste and Emergency Response, Technology Innovation Office, Washington, D.C., August 1996.

112. U.S. EPA, *Vendor Information System for Innovative Treatment Technologies, Version 5.0, Technology Type: Soil Washing/Full Scale*, Office of Solid Waste and Emergency Response, Technology Innovation Office, Washington, D.C., August 1996.

113. U.S. EPA, *Superfund Innovative Technology Evaluation Program, Technology Profiles*, 9th ed., EPA/540/R-97/502, Office of Research and Development, Washington D.C., December 1996. 26–27, 32–33, 38–41, 110–111, 114–115, 158–159, 248–249.

114. U.S. EPA, *Technology Demonstration Summary: Toronto Harbour Commissioners (THC) Soil Recycle Treatment Train*, EPA/540/SR-93/517, Center for Environmental Research Information, Cincinnati, OH, September 1993, 6 pp.

115. Goldberger, W. M., Robbins, L. A., Fiedler, R. A., Jepson, T. L. B., Knoll, F. S., Malony, J. O., Mitchell, D. W., Parekh, B. K., Sorenson, T. C., Stavenger, P. L., Thelen, R. L., Treybal, R. E., and Wechsler, I., Solid-solid and liquid-liquid systems, in *Perry's Chemical Engineer's Handbook*, 6th ed., Perry, R. H., Green, D. W., and Malony, J. O., Eds., McGraw-Hill, New York, 1984, 21:24.

116. U.S. EPA, *Bergmann U.S.A. Soil Sediment Washing Technology, Application Analysis Report*, EPA/540/AR-92/075, Office of Research and Development, Washington, D.C., September 1995, 33.

117. U.S. EPA, *BESCORP Soil Washing System for Lead Battery Site Treatment, Applications Analysis Report*, EPA/540/AR-93/503, Office of Research and Development, Washington, D.C., January 1995, 10.

118. U.S. EPA, *Vendor Information System for Innovative Treatment Technologies, Version 5.0, Technology Type: Solvent Extraction/Full Scale*, Office of Solid Waste and Emergency Response, Technology Innovation Office, Washington, D.C., August 1996.

119. U.S. EPA, *Superfund Innovative Technology Evaluation Program, Technology Profiles*, 9th ed., EPA/540/R-97/502, Office of Research and Development, Washington, D.C., December 1996, 44–45, 52–53, 96–97, 152–153, 252–253.

120. Culp, G. L., *Handbook of Sludge-Handling Processes, Cost and Performance*, Garland STPM Press, New York, 1979, 162–164.

121. U.S. EPA, *Resources Conservation Company B.E.S.T.® Solvent Extraction Technology, Application Analysis Report*, EPA/540/AR-92/079, Office of Research and Development, Washington, D.C., June 1993.

122. U.S. EPA, *SITE Technology Capsule: Terra-Kleen Solvent Extraction Technology*, EPA 540/R-94/521a, Office of Research and Development, Cincinnati, OH, February 1995, 9 pp.

123. U.S. EPA, *Vendor Information System for Innovative Treatment Technologies, Version 5.0, Technology Type: Soil Flushing-In Situ, Vendor Name: Horizontal Technologies Inc.*, Office of Solid Waste and Emergency Response, Technology Innovation Office, Washington, D.C., August 1996.

124. PRC Environmental Management, *Recent Developments for In Situ Treatment of Metal Contaminated Soils*, prepared for U.S. EPA, Office of Solid Waste and Emergency Response, Washington, D.C., March 1997, 33–40.

125. Roote, D. S., *In Situ Flushing*, TO-97-02, Groundwater Remediation Technology Analysis Center, Pittsburgh, PA, June 1997, 19 pp.

126. PRC Environmental Management, *Recent Developments for In Situ Treatment of Metal Contaminated Soils*, prepared for U.S. EPA Office of Solid Waste and Emergency Response, Washington, D.C., March 5, 1997, 8–20.

127. Horng, J. J., Banerjee, S., and Herrmann, J. G., Evaluating electro-kinetics as a remedial action technique, in *Proceedings of the Second International Conference on New Frontiers for Hazardous Waste Management*, EPA/600/9-87/018F, U.S. EPA, Washington, D.C., August 1987, 65–77.

128. Bruell, C. J., Segall, B. A., and Walsh, M. T., Electroosmotic removal of gasoline hydrocarbons and TCE from clay, *J. Environ. Eng.*, 118(1), 68–83, 1992.

129. U.S. EPA, *Superfund Innovative Technology Evaluation Program, Technology Profiles*, 9th ed., EPA/540/R-97/502, Office of Research and Development, Washington, D.C., December 1996, 194–195, 198–199.

130. Segall, B. A. and Bruell, C. J., Electroosmotic Contaminant-Removal Processes, *J. Environ. Eng.*, 118(1), 84–100, 1992.

131. U.S. EPA, *Superfund Innovative Technology Evaluation Program, Technology Profiles*, 9th ed., EPA/540/R-97/502, Office of Research and Development, Washington, D.C., December 1996, 74–75 .

132. Acar, Y. B. and Hamed, J., *Electrokinetic Soil Processing in Waste Remediation/Treatment*, Presented at the session on "Remediation of Contaminated Soil", Transportation Research Board, 70th Annual Meeting, Washington, D.C., January 1991.

133. Masterton, W. L. and Slowinski, E. J., *Chemical Principles,* 4th ed., W. B. Saunders, Philadelphia, 1977, 541.

134. Freeze, R. A. and Cherry, J. A., *Groundwater,* Prentice-Hall, Englewood Cliffs, NJ, 1979, 25.

135. U.S. EPA, *In Situ Remediation Technology Status Report: Electrokinetics*, EPA 542-K-94-007, Office of Solid Waste and Emergency Response, Washington, D.C., April 1995, 7.

136. U.S. Department of Energy, *Lasagna™ Soil Remediation — Innovative Technology Summary Report*, U.S. DOE Office of Environmental Management and Office of Science and Technology, Washington, D.C., April 1996, 14 pp.

137. Clausen, J. L., Johnstone, E. F., Zutman, J. L., Pickering, D. A., and Smuin, D. R., *Preliminary Site Characterization/Baseline Risk Assessment/Lasagna™ Technology Demonstration at Solid Waste Management Unit 91 of the Paducah Gaseous Diffusion Plant, Paducah, Kentucky,* Document Number KY/EM-128, Lockheed Martin Energy Systems, Oak Ridge, TN, May 1996, 223 pp.

138. Monsanto Company, *Rapid Commercialization Initiative (RCI) Report for an Integrated in situ Remediation Technology (Lasagna™),* submitted to U.S. Department of Energy, Morgantown Energy Technology Center, Morgantown, WV, 1998.

139. U.S. EPA, *J. R. Simplot Ex-Situ Bioremediation Technology for Treatment of TNT-Contaminated Soils, Innovative Technology Evaluation Report,* EPA/540/R-95/529, Office of Research and Development, Washington, D.C., September 1995.

140. U.S. EPA, *Demonstration Bulletin: New York State Multi-Vendor Bioremediation, ENSR Consulting and Engineering/Larsen Engineers Ex Situ Biovault,* EPA/540/MR-95/524, National Risk Management Research Laboratory, Cincinnati, OH, August 1995.

141. Pope, D., *Land Treatment,* Seminars on Bioremediation of Hazardous Waste Sites: Practical Approaches to Implementation, EPA/625/K-96/001, U.S. EPA, Office of Research and Development, Washington, D.C., May 1996, 6:1–16.

142. U.S. EPA, *Innovative Treatment Technologies: Annual Status Report,* 8th ed., EPA-542-R-96-010 Number 8, Office of Solid Waste and Emergency Response, Washington, D.C., November 1996, 16–17 and Appendix E.

143. Sims, R. C., *Background Information for Bioremediation Applications,* Seminars on Bioremediation of Hazardous Waste Sites: Practical Approaches to Implementation, EPA/625/K-96/001, U.S. EPA, Office of Research and Development, Washington, D.C., May 1996, 1:1–16.

144. Leeson, A. and Hinchee, R., *Manual — Bioventing Principles and Practice, Volume I: Bioventing Principles,* EPA/540/R-95/534a, U.S. EPA, Office of Research and Development, Washington, D.C., September 1995, 18–28.

145. Leeson, A. and Hinchee, R., *Manual — Bioventing Principles and Practice, Volume I: Bioventing Principles,* EPA/540/R-95/534a, U.S. EPA, Office of Research and Development, Washington, D.C., September 1995, 11–12.

146. Leeson, A. and Hinchee, R., *Manual — Bioventing Principles and Practice, Volume II: Bioventing Design,* EPA/625/XXX/001, U.S. EPA, Office of Research and Development, Washington, D.C., September 1995, 11–16, Appendix C.

147. Potter, C. L., *Development and Application of Composting Techniques for Treatment of Soils Contaminated with Hazardous Waste,* Seminars on Bioremediation of Hazardous Waste Sites: Practical Approaches to Implementation, EPA/625/K-96/001, U.S. EPA, Office of Research and Development, Washington, D.C., May 1996, 9:1–8.

148. U.S. EPA, *Emerging Technology Summary: Innovative Methods for Bioslurry Treatment,* EPA/540/SR-96/505, Center for Environmental Research Information, Cincinnati, OH, August 1997.

149. U.S. EPA, *Pilot-Scale Demonstration of a Slurry-Phase Biological Reactor for Creosote-Contaminated Soil, Application Analysis Report*, EPA/540/A5-91/009, Office of Research and Development, Washington, D.C., January 1993.

150. McCauley, P. and Glaser, J., *Slurry Bioreactors for Treatment of Contaminated Soils, Sludges, and Sediments*, Seminars on Bioremediation of Hazardous Waste Sites: Practical Approaches to Implementation, EPA/625/K-96/001, U.S. EPA, Office of Research and Development, Washington, D.C., May 1996, 12:1–9.

151. U.S. EPA, *Vendor Information System for Innovative Treatment Technologies, Version 5.0, Technology Type: Bioremediation, Solid Phase*, Office of Solid Waste and Emergency Response, Technology Innovation Office, Washington, D.C., August 1996.

152. U.S. EPA, *Demonstration Bulletin: Augmented In Situ Subsurface Bioremediation Process™, BIO-REM, Inc.*, EPA/540/MR-93/527, Center for Environmental Research Information, Cincinnati, OH, November 1993.

153. U.S. EPA, *Demonstration Bulletin — Fungal Treatment Bulletin*, U.S. EPA-RREL/USDA-FPL, EPA/540/MR-93/514, Center for Environmental Research Information, Cincinnati, OH, June 1993.

154. Glaser, J. A., *Effective Treatment of Hazardous Waste Constituents in Soil by Lignin-Degrading Fungi*, Seminars on Bioremediation of Hazardous Waste Sites: Practical Approaches to Implementation, EPA/625/K-96/001, U.S. EPA, Office of Research and Development, Washington, D.C., May 1996, 11:1–8.

155. U.S. EPA, *Superfund Innovative Technology Evaluation Program, Technology Profiles*, 9th ed., EPA/R-97/502, Office of Research and Development, Washington, D.C., December 1996, 274–275.

156. U.S. EPA, *Emerging Technology Bulletin: Institute of Gas Technology (Chemical and Biological Treatment)*, EPA/540/F-94/504, Center for Environmental Research Information, Cincinnati, OH, May 1994.

157. *Webster's Ninth New Collegiate Dictionary*, Merriam-Webster, Springfield, MA, 1990, 1012.

158. U.S. EPA, *Recent Developments for In Situ Treatment of Metal Contaminated Soils*, Office of Solid Waste and Emergency Response, Washington, D.C., March 1997, 24–30.

159. Miller, R. R., *Technology Overview Report: Phytoremediation*, TO-96-03, Ground Water Remediation Technologies Analysis Center, Pittsburgh, PA, October 1996, 2–6.

160. Watanabe, M. E., Phytoremediation on the brink of commercialization, *Environ. Sci. Technol.*, 31(4), 182A–186A, 1997.

161. Schnoor, J. L., Light, L. A., McCutcheon, S. C., Wolfe, N. L., and Carreira, L. H., Phytoremediation of organic and nutrient contaminants, *Environ. Sci. Technol.*, 29(7), 318A–323A, 1995.

162. U.S. EPA, *Superfund Innovative Technology Evaluation Program, Technology Profiles*, 9th ed., EPA/540/R-97/502, Office of Research and Development, Washington, D.C., December 1996, 230–231.

163. U.S. EPA, *Superfund Innovative Technology Evaluation Program, Technology Profiles*, 9th ed., EPA/540/R-97/502, Office of Research and Development, Washington, D.C., December 1996, 360.

164. GRC Environmental, sales brochure, November 1991.

165. U.S. EPA, *Project Summary: Destruction of PCBs — Environmental Applications of Alkali Metal Polyethylene Glycolate Complexes*, EPA/600/S2-85/108, Hazardous Waste Engineering Research Laboratory, Cincinnati, OH, December 1985, 3 pp.

166. U.S. EPA, *Innovative Treatment Technologies: Semi-Annual Status Report*, 3rd ed., EPA/540/2-91/001, Number 3, Washington, D.C., April 1992.

167. U.S. EPA, *Innovative Treatment Technologies: Annual Status Report*, 8th ed., EPA-542-R-96-010, Number 8, Washington, D.C., November 1996, Appendix E.

168. U.S. EPA, *Vendor Information System for Innovative Treatment Technologies, Version 5.0, Profile — SDTX Technologies Inc.*, Office of Solid Waste and Emergency Response, Technology Innovation Office, Washington, D.C., August 1996.

169. U.S. EPA, *Superfund Innovative Technology Evaluation Program: Technology Profiles*, 9th ed., EPA/540/R-97/502, Office of Research and Development, Washington, D.C., December 1996, 108–109.

170. PRC Environmental Management, *Base Catalysed Decomposition (BCD) Process/SAREX® and K/S THERM-O-DETOX® System Technology SITE Demonstration at Koppers Company, Inc. (Morrisville Plant) Site, Preliminary Data Summary Report for the Remedial Design/Remedial Action Study*, prepared for U.S. EPA Risk Reduction Engineering Laboratory, Cincinnati, Ohio, April, 1994.

171. U.S. EPA, *Superfund Engineering Issue — Issues Affecting the Applicability and Success of Remedial/Removal Incineration Projects*, EPA/540/2-91/004, Office of Solid Waste and Emergency Response, Washington, D.C., February 1991, 6.

172. U.S. EPA, *Technology Screening Guide for Treatment of CERCLA Soils and Sludges*, EPA/540/2-88/004, Office of Solid Waste and Emergency Response, Washington, D.C., September 1988, 35–42.

173. Stumbar, J. P., Sawyer, R. H., Gupta, G. D., Perdek, J. M., and Freestone, F. J., Effect of feed characteristics on the performance of EPA's mobile incineration system, in *Remedial Action, Treatment, and Disposal of Hazardous Waste*, Proceedings of the Fifteenth Annual Research Symposium, EPA/600/9-90/006, Risk Reduction Engineering Laboratory, Cincinnati, OH, February 1990, 480–498.

174. U.S. EPA, *Superfund Innovative Technology Evaluation Program, Technical Profiles*, 9th ed., EPA/540/R-97/502, Office of Solid Waste and Emergency Response, Washington, D.C., December 1996, 72–73.

175. U.S. EPA, *Shirco Infrared Incineration System Application Analysis Report*, EPA/540/A5-89/010, Office of Research and Development, Washington, D.C., June 1989.

176. U.S. EPA, *Superfund Innovative Technology Evaluation Program, Technical Profiles*, 9th ed., EPA/540/R-97/502, Office of Solid Waste and Emergency Response, Washington, D.C., December 1996, 84–85.

177. Buonicore, A. J., Theodore, L., McKinney, R. E., and Tchobanoglous, G., Waste management, in *Perry's Chemical Engineers' Handbook*, 6th ed., Green, D. W. and Maloney, J. O., Eds., McGraw-Hill, New York, 1984, 30–31.

178. Porter, H. F., Schurr, G. A., Wells, D. F., and Semrau, K. T., Solids drying and gas-solid systems, in *Perry's Chemical Engineers' Handbook*, 6th ed., Green, D. W. and Maloney, J. O., Eds., McGraw-Hill, New York, 1984, 82–121.

179. Ashbee, E. And Davis, W. T., Control of particulates: cyclones and inertial separators, in *Air Pollution Engineering Manual*, Buonicore, A. J. and Davis, W. T., Eds., Air & Waste Management Association, Van Nostrand Reinhold, New York, 1992, 71–78.

180. McKenna, J. D. and Furlong, D. A., Fabric filters, in *Air Pollution Engineering Manual*, Buonicore, A. J. and Davis, W. T., Eds., Air & Waste Management Association, Van Nostrand Reinhold, New York, 1992, 114–131.

181. Schifftner, K. and Hesketh, H., Wet scrubbers, in *Air Pollution Engineering Manual,* Buonicore, A. J. and Davis, W. T., Eds., Air & Waste Management Association, Van Nostrand Reinhold, New York, 1992, 78–88.

182. Donnelly, J. R., Refuse, in *Air Pollution Engineering Manual,* Buonicore, A. J. and Davis, W. T., Eds., Air & Waste Management Association, Van Nostrand Reinhold, New York, 1992, 273–274, 280.

183. R. S. Means Company, *Site Work and Landscape Cost Data 1998,* 17th Annual Edition, Kingston, MA, 1997, 430–431.

184. U.S. EPA, *ECO Update: Ecotox Thresholds*, EPA 540/F-95/038, Office of Solid Waste and Emergency Response, Washington, D.C., January 1996, 10.

CHAPTER 5

Solutions

5.1 CHAPTER ONE

The answers to the problems in this chapter are very subjective. There are no "right" answers.

5.2 CHAPTER TWO

Problem 2.2

[a] Use Henry's Law to estimate the concentration (Equation 2.10). Assume that the system is at 20°C and is at equilibrium.

$$k_{H,styrene} = \frac{5 \ mmHg \ styrene}{300 \ mg/L \ styrene} = \frac{P_{tank,styrene}}{42 \ mg/L \ styrene}$$

$$P_{tank,styrene} = 0.7 \ mmHg$$

Converting units,

$$\frac{0.7 \ mmHg \ styrene}{760 \ mmHg \ total} \cdot \frac{1 \ kg \ mole}{22.41 \ m^3} \cdot \frac{104.15 \ kg \ styrene}{1 \ kg \ mole} \cdot \frac{10^6 \ mg}{kg} = 4400 \ mg/m^3$$

[b] Use Raoult's Law (Equation 2.13), assuming the system is at 20°C and at equilibrium, and behaves as an ideal solution.

$$P_{tank,styrene} = (5 \ mmHg) \cdot (0.15) = 0.75 \ mmHg$$

Converting units as in part [a], the concentration of styrene is 4600 mg/m³.

[c] The headspace concentration of styrene is the partial pressure, $p_{tank, \ styrene} = 5 \ mm \ Hg$.

Converting the units as in part [a], the concentration of styrene is 31,000 mg/m³.

Problem 2.3

Several different answers are possible for each part to this problem. For example,

[a] trichloroethylene, tetrachloroethylene

[b] benzene, toluene

[c] bezene, phenol

Problem 2.4

[a] biological oxidation (highly water soluble, so air stripping and carbon adsorption are limited)

[b] air stripping, carbon adsorption, biological oxidation

[c] carbon adsorption (low volatility, so air stripping infeasible; difficult to oxidize)

Problem 2.5

LNAPL compounds are not present in the soil or groundwater. The release history and type of contamination suggest that DNAPL residuals may be present.

Problem 2.6

[a] Based on the information provided, the lower aquifer is not likely to be contaminated. The physical forces which would drive contamination to the lower aquifer are not in effect:

  - toluene is less dense than water, so free product (if any) would float in the upper aquifer rather than sink to the lower aquifer; and
  - water level measurements in MW-3S and MW-3D indicate an upward gradient.

[b] This is a mass balance problem. Calculate the mass of toluene entering and leaving the reach of the stream near the site to estimate the concentration in the stream. Using C to represent the concentration of toluene, and Q to represent the flow of water, and assuming that no reaction occurs,

$$C_{upstream} \cdot Q_{upstream} + C_{groundwater} \cdot Q_{groundwater} = C_{downstream} Q_{downstream} + mass\ volatilized$$

Assume that there are no upstream sources of toluene, i.e., $C_{upstream} = 0$. To be conservative, assume that no mass is volatilized. Then the mass balance on toluene reduces to

$$C_{groundwater} \cdot Q_{groundwater} = C_{downstream} \cdot Q_{downstream}$$

Estimate the discharge from the groundwater to the stream using Darcy's Law (Equation 2.16). Begin by estimating the hydraulic gradient between MW-4 and the creek (Equation 2.15):

$$\frac{\Delta H}{\Delta L} = \frac{201.32\ ft - 200.10\ ft}{20\ ft} = 0.061$$

The width of the plume is unknown. The data indicate that the plume extends to the west to MW-2; assume that the plume is symmetric around the tanks, so that it extends to an equal distance to the east. The estimated plume width is then approximately 120 ft. The plume discharges to the creek through a saturated thickness of approximately 3 ft. Using Darcy's Law,

$$Q_{groundwater} = 1\times10^{-3} \ cm/sec \cdot 0.061 \cdot (120 \ ft \cdot 3 \ ft) \cdot 1.969 \frac{ft/min}{cm/sec} = 0.043 \ ft^3/min$$

The worst-case concentration of toluene in this discharge would be the concentration at MW-4, 18.2 mg/L. Alternatively, the calculation could be performed assuming an average across the width of the plume 9 (4.1 mg/L).

Estimate the flow in the creek. Assuming a trapezoidal cross section, 5 ft wide at the top and with the sides sloped at 45°, the cross-sectional area is 6 ft². Assuming a rectangular cross section, the cross-sectional area is 9 ft². If an object at the surface floated at a rate of 5 ft/min, the average velocity (over depth) would be slower. Using the maximum velocity, the flow rate in the stream is 30 ft³/min. (This estimate is not precisely $Q_{downstream}$, as the depth measurement was taken in the middle of the site and it is not clear from the site profile where the flow rate was "measured". However, given that this estimate of stream flow is much greater than the contribution from groundwater, the change in the flow rate across the site is probably not significant.)

Returning to the mass balance, estimate the worst-case concentration in the creek:

$$C_{downstream} = 18.2 \ mg/L \cdot \frac{0.0431 \ ft^3/min}{28.8 \ ft^3/min} = 0.02724 \ mg/L \ or \ 30 \ \mu g/L$$

This concentration is much less than the criterion of 130 μg/L.

Problem 2.7

Calculated clean-up levels depend upon the assumptions used. See Table 5.1 for EPA-calculated values based on conservative exposure assumptions.

Table 5.1   Solution to Problems 2.7 and 2.9
EPA-Derived Values
Soil Clean-Up Levels

Compound	Residential, ingestion (mg/kg)	Industrial, ingestion (mg/kg)	Inhalation (mg/kg)	Protection of groundwater (mg/kg)
B(a)P	0.088	0.78	11	4
TCDD	0.000004	0.000037	—	—
PCE	12	110	11	0.04

From U.S. EPA, Region III Risk-Based Concentration Table, June 1996.

Problem 2.8

[a] From Equations 2.33 and 2.34, and beginning by converting units,

$$K = (2.3 \times 10^{-5} \, cm/sec)(10^2 \, m/cm)(86400 \, sec/day)(365 \, days/yr) = 72532.8 \, m/yr$$

$$i = (20.2 \, in/yr)(0.0254 \, m/in) = 0.51206 \, m/yr$$

$$L = (100 \, ft)(1 \, m/3.281 \, ft) = 30.48 \, m$$

$$d_a = (35 \, ft)(1 \, m/3.281 \, ft) = 10.67 \, m$$

$$d = \left(0.0112(30.48 \, m)^2\right)^{0.5} + (10.67 \, m)\left(1 - e^{-\frac{(30.48 \, m)(0.51206 \, m/yr)}{(72532.8 \, m/yr)(0.025)(10.67 \, m)}}\right) = 3.23 \, m$$

$$DAF = 1 + \frac{(72532.8 \, m/yr)(0.025)(3.23 \, m)}{(0.51206 \, m/yr)(30.48 \, m)} = 376$$

Calculate the target concentration in water:

$$C_w = DAF \cdot MCL = (376)(0.10 \, mg/L) = 37.6 \, mg/L$$

Then from Equation 2.32,

$$Clean\text{-}up \, Level = (37.6 \, mg/L)\left((219 \ L/kg)(0.002) + \frac{0.30 + (0.40 - 0.30)(0.152)}{1.5 \ g/cm^3}\right)$$

$$= 24 \ mg/kg$$

[b] This solution uses the following variables (in addition to those previously defined),

$C_{gw,up}$ = concentration of contaminant immediately upgradient of the area of contamination;

$Q_{gw,up}$ = flow of groundwater entering the area of contamination;

Q_{infilt} = flow of water infiltrating through the ground;

$C_{gw,down}$ = concentration of contaminant at downgradient boundary of the area of contamination;

$Q_{gw, down}$ = flow of groundwater at downgradient boundary of the area of contamination; and

W = width of source area (dimension perpendicular to groundwater flow direction).

The mass balance on the contaminant, assuming no attenuation and no NAPL in the source area, is

$$C_{gw,up} \cdot Q_{gw,up} + C_w \cdot Q_{infilt} = C_{gw,down} \cdot Q_{gw,down}$$

If there are no upgradient sources of contamination,

$$C_w \cdot Q_{infilt} = C_{gw,down} \cdot Q_{gw,down}$$

If infiltration occurs over an area of size $W \times L$, then

$$Q_{infilt} = iWL$$

According to Darcy's Law, the groundwater flow through a cross-sectional area of width W and depth through the mixing zone d is

$$Q_{gw,down} = K I (Wd)$$

Then

$$DAF = \frac{C_w}{C_{gw,down}} = \frac{iWL + KIWd}{iWL} = 1 + \frac{KId}{iL}$$

[c] Accounting for biodegradation between the source area and the receptor,

$$MCL = \frac{C_w}{DAF} \cdot e^{-kt}$$

Calculate the travel time t between the downgradient edge of the contaminated area and the receptor from the average linear velocity:

$$v = KI/n = (2.3 \times 10^{-5} \text{ cm/sec})(0.025)(1 \text{ in}/2.54 \text{ cm})(1 \text{ ft}/12 \text{ in})$$
$$(60 \text{ sec/min})(1440 \text{ min/day}) \, 0.40 = 0.004075 \text{ ft/day}$$

$$t = \frac{1000 \text{ ft}}{0.004075 \text{ ft/day}} = 24,541 \text{ days}$$

(Note that this simplified estimate does not account for retardation as a result of adsorption. Retardation would increase the travel time and thus the time available for biodegradation.)

Then, using the minimum value for the rate constant k from Table 5.2,

Table 5.2 Solution to Problem 2.10[a]

Compound	$t_{1/2}$ (days)	k (days^{-1})	$t_{(100 \to 1)}$ (days)
Benzene	7	0.0990210258	50
Phenol	0.5	1.38629436112	3
	7	0.0990210258	50
Chlorobenzene	136	0.00509667	904
	300	0.002310491	2000

$$C_w = \frac{37.6 \ mg/L}{e^{-(0.0023 \ day^{-1})(24541 \ days)}} = 1.23 \times 10^{26} \ mg/L$$

$$Clean\text{-}up \ Level = 3 \times 10^{27} \ mg/kg$$

This clean-up level is greater than one million milligrams per kilogram. The calculation implies that a clean-up level cannot be derived based on source control when biodegradation is account for. Where the contaminant is degradable and the source relatively distant from the receptor, accounting for attenuation significantly changes a clean-up level which is based on source control.

Problem 2.9

Calculated clean-up levels depend upon the assumptions used. See Table 5.1 for EPA-calculated values using conservative exposure assumptions.

Problem 2.10

[a] Use Equation 2.44 to calculate the rate constant k using the values of $t_{\frac{1}{2}}$ given in the problem, then Equation 2.43 to calculate the range of time $t_{(100\rightarrow1)}$ required for degradation of each compound. See Table 5.2.

[b] The half-life data and calculated degradation times indicate that, in general:

- An alcohol or phenol (i.e., compound containing the hydroxyl group, –OH) is more biodegradable than the corresponding hydrocarbon; and
- A chlorinated compound is less biodegradable than the corresponding hydrocarbon.

5.3 CHAPTER THREE
Problem 3.1

Initial projections for the duration of remediation at the Verona Well Field site did not account for the continuing release of material from NAPL residuals.

Problem 3.2

[a] Estimate the radius of influence of the pumping well using Equation 3.1.

$$T = Kb = 75 \ gal/day/ft^2 \cdot 10 \ ft = 750 \ gal/day/ft$$

$$R_0 = 0.25 + \sqrt{\frac{750 \ gal/ft/day \cdot 960 \ min}{4790 \cdot 0.3}} = 23 \ ft$$

[b] Use Equation 3.5 to estimate the flow rate corresponding to the radius of influence. Equation 3.5 models the flow to a well which fully penetrates an unconfined aquifer. Assuming a drawdown of 67% (or 6.7 ft, so that $h_w = 3.3$ ft),

$$Q = \frac{\pi \cdot 75 \ gal/day/ft^2 \cdot ((10 \ ft)^2 - (3.3 \ ft)^2)}{\ln((23 \ ft)/(0.25 \ ft))} = 4700 \ gal/day \ or \ 3.3 \ gal/min$$

Problem 3.3

Based on the data provided, the remediation system is not likely to achieve the remediation goal of 5 µg/L within 10 years. The concentrations measured during the first 5 months of operation show a significant and rapid reduction in concentration. After about 5 months, the concentrations of benzene continue to decrease, but much more slowly (see Figure 5.1). Concentrations appear to be steadily decreasing from about month 6 until the last measurement in month 19. Assuming that this decreasing trend will continue, this trend in the data can be used to predict future concentrations of benzene.

Linear regression techniques can be used to determine whether or not the apparent relationship between time and benzene concentration can be described by a linear expression (i.e., whether or not concentrations decrease at a constant rate over time). Using the data from months 6 through 19 (where the slope of the concentrations vs. time plot in Figure 5.1 appears linear), the following best-fit linear relationship is observed between concentration and time (t, in months):

$$Concentration\ of\ benzene\ (\mu g/L) = 306.92 - 1.3t$$

Assuming concentrations decrease at a constant rate (i.e., the slope remains constant), this relationship can be used to predict future concentrations of benzene (see Table 5.3). The predicted data indicate that after 10 years of operation, the concentration of benzene in the aquifer will be 151 µg/L, considerably higher than the remediation goal of 5 µg/L. Therefore, using the measured data to predict future concentrations, it is not likely that the remediation goal of 5 µg/L will be achieved after 10 years.

It is noteworthy that the overall regression is not significant ($p = 0.29$). While concentrations tend to decrease over time (the best-fit estimate of the slope is negative), the trend is not statistically significant. It is possible that aquifer conditions may not be stable enough after only 19 months to reliably predict aquifer conditions 10 years into the future. (This situation is not unusual: management often requires predictions of the future duration and cost of remediation regardless of statistical accuracy.) Alternatively, it is possible that asymptotic conditions have been achieved in the aquifer.

The 95% confidence interval of the slope of the line is –3.89 to 1.29. Since the confidence interval includes the value zero, it cannot be concluded that the slope of the line is significantly different from zero. This would suggest that an asymptote has been reached and substantial additional reduction in concentration may not be possible, given the current configuration of the remediation system. If this is the case, several options may be considered to manage the problem. First, additional wells could be installed and/or the pumping rate may be adjusted, e.g., by pulsed pumping. Second, an alternative approach to remediation, such as *in situ* treatment, could be considered. Alternatively, the remediation goal may be re-evaluated to determine whether or not the asymptotic concentration is acceptable.

Based on the linear expression described above, the mass of benzene that can be removed from the aquifer can be estimated, assuming a certain pumping rate and a duration of operation. Assuming a pumping rate of 50 gpm is maintained during the entire operation, an estimated mass of 164 kg benzene may be removed from the system during the first 5 years of operation (see Figure 5.2). During the first year, an estimated 58 kg were removed from the system (note that this estimate is based on measured concentrations during this

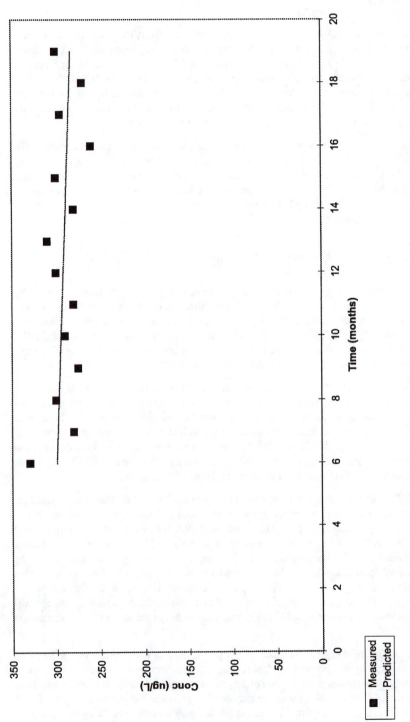

Figure 5.1 Solution to Problem 3.3: concentration vs. time, measured data (6 to 19 months).

Table 5.3   Solution to Problem 3.3
             Benzene Recovery from Pump-and-Treat System

Type of data	Time (month)	Concentration benzene (µg/L)	Mass benzene (during period) (kg)	Cumulative mass benzene (kg)
Actual data	0	1600	0	0
	0.5	1620	6.78810132	6.79
	0.75	1379	3.161104947	9.95
	1	1273	2.795348556	12.75
	2	1174	10.317070764	23.07
	3	700	7.901181288	30.97
	5	420	9.44431488	40.41
	6	330	3.162159	43.57
	7	280	2.57188932	46.14
	8	300	2.44540296	48.59
	9	275	2.4243219	51.01
	10	290	2.38215978	53.39
	11	280	2.40324084	55.79
	12	300	2.44540296	58.24
	13	310	2.57188932	60.81
	14	280	2.48756508	63.3
	15	300	2.44540296	65.75
	16	260	2.36107872	68.11
	17	295	2.33999766	70.45
	18	270	2.38215978	72.83
	19	300	2.40324084	75.23
Estimated data	*24*	*275.802197802*	*12.13852068*	*87.37*
	30	*268.021978022*	*13.757268096*	*101.13*
	36	*260.241758242*	*13.363631424*	*114.49*
	42	*252.461538462*	*12.969994752*	*127.46*
	48	*244.681318681*	*12.57635808*	*140.04*
	54	*236.901098901*	*12.182721408*	*152.22*
	60	*229.120879121*	*11.789084736*	*164.01*

Note: Mass benzene/period = (50 gal/min)(1440 min/day)(30.42 day/month)•
       (__month)(__ µg/L benzene)(10^{-9} kg/µg)(3.85 L/gal) [=] __ kg.

time). This quantity represents 58/164, or about 35% of the estimated 5-year total mass removal. During years two through five, an additional mass of 106 kg is estimated to be removed. This corresponds to about 65% of the estimated 5-year total mass removal. Clearly, the efficiency of the system declines after the first year of operation. It appears that even after 5 years of operation, a significant mass of benzene may be removed from the aquifer. However, consider that to achieve the remediation goal (assuming the linear expression above remains valid over time), an estimated 334 kg benzene will be removed from the aquifer. About half this amount is removed during the first 5 years; the remaining half requires an additional 15 years to remove!

Problem 3.4

Several alternative designs could solve the problem. Rather than presenting a complete conceptual design for a single solution, the following discussion identifies critical points which should be incorporated into the many possible solutions.

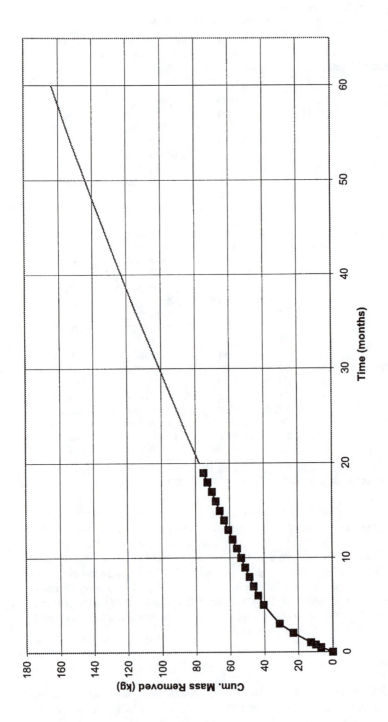

Figure 5.2   Solution to Problem 3.3: cumulative mass removed vs. time (0 to 60 months).

A pump-and-treat system would entail a minimum of two pumping wells. Two wells, spaced slightly less than 400 ft apart at the downgradient edge of the plume and pumping at a total of ~200 gpm containing ~0.1 mg/L Cr, could potentially prevent the plume from migrating further. However, that minimal design may not effectively prevent the plume from continuing to migrate, nor would it remove much of the mass of chromium in the plume. A more effective design would include a series of wells. The most extensive system would place wells throughout the plume. Alternatively, one or more additional wells would be placed at the downgradient edge of the plume — to ensure containment — and a series of wells would be placed in the most concentrated portion of the plume.

The degree of treatment required would depend upon whether treated groundwater was discharged to surface water or reinjected to the aquifer. Less chromium removal would be required to reinject the water than to discharge to surface water. A reinjection system would have to be modeled to — at a minimum — ensure that it did not compromise the groundwater extraction/containment system or, ideally, to demonstrate that it enhanced the containment of the plume.

Conventional treatment would entail reducing $Cr(VI)$ to $Cr(III)$, and then precipitating $Cr(III)$ as the hydroxide, in a process similar to that formerly used in the original wastewater treatment impoundments. However, this treatment process might not reach the proposed discharge limits if the treated water was discharged to surface water: the theoretical solubility limit of $Cr(OH)_3$ is approximately 0.5 mg/L. Additional treatment would be required to meet the discharge limits.

EPA's Risk Reduction Engineering Laboratory data base on treatment effectiveness indicates that several treatment techniques have reportedly achieved levels close to or below the discharge limits for this site,[2] e.g.,

- chemical precipitation, ~6 µg/L;
- chemical precipitation and filtration, 3 to 100 µg/L;
- sedimentation, 4 to 30 µg/L;
- granular activated carbon, 4 to 18 µg/L;
- ion exchange, 10 µg/L;
- reverse osmosis, 0.5 to 1.7 µg/L.

Additional research, site-specific treatability tests of chromium removal techniques, and cost estimates of feasible techniques would be necessary to select the treatment techniques and complete the design of a treatment system. Additional data on other groundwater constituents (e.g., iron, manganese, hardness, pH) would also be needed to design a groundwater treatment system, determine the quantity of byproducts generated, and, potentially, to design a reinjection system.

The wastes generated from the treatment system would include, depending on the unit processes: hydroxide sludge; spent granular activated carbon; spent ion-exchange resin and/or regenerant solution; or a concentrated aqueous waste from a reverse osmosis unit.

Problem 3.5

[a] Approximately 59,000 lb of liquid-phase carbon per year would be needed to treat the groundwater.

Table 5.4 Solution to Problem 3.5[a]

Compound	C (µg/L)	K_p (µg/g)(L/µg)$^{1/n}$	1/n	x/m (µg/g)	x (µg/year)	m (kg/yr)
Benzene	1,000	1,260	0.533	50,046.1352	3.9792e+11	7,951.08845
Toluene	900	5,010	0.429	92,727.3316	3.5813e+11	3,862.17437
Xylene	850	12,600	0.418	211,284.509	3.3823e+11	1,600.84174
Sum					1.0943e+12	13,414.1046
Estimated carbon usage, lb/year (double sum and convert units)						59,100

Note: x = C * (200 gal/min)(1440 min/day)(365 days/year)(3.7854 L/gal), assuming no down time.

Table 5.4 summarizes the calculation, which is based upon Equation 3.11. Several factors affect the accuracy of the estimate:

- Isotherm data tabulated in the text were derived for deionized distilled water. Natural groundwater characteristics can affect carbon adsorption. The isotherm data were also collected under test conditions (e.g., carbon type, contact time) which may not represent field conditions.
- The concentrations in this problem are outside the concentration ranges for the isotherm data given in the text.
- Estimating the carbon usage for groundwater containing several compounds using isotherms for individual compounds provides only a rough estimate of the actual usage.
- The estimate is based on a "snapshot" of the flow rate and contaminant concentrations. As concentrations decrease over time, carbon usage will decrease.

Site-specific testing would be necessary to refine this crude estimate.

[b] Approximately 24,000 lb of vapor-phase carbon per year would be needed to treat the air-stripper emissions. This estimate is based on the sum of the values of *x* calculated in Table 5.4 and the rule of thumb regarding vapor-phase carbon use in Section 3.7.2.

[c] In general, vapor-phase treatment is more efficient than liquid-phase treatment.

Problem 3.7

Water flooding or injected surfactants could mobilize contamination into fractured rock.

Problem 3.8

Several alternative designs could solve the problem. Rather than presenting a complete conceptual design for a single solution, the following discussion identifies critical points which should be incorporated into the solution.

Options for remediation include groundwater pump-and-treat, air sparging/soil vapor extraction, and *in situ* bioremediation. The most important data necessary to determine a comprehensive approach to site remediation is the definition of the source of the toluene in groundwater. The source should be defined, and the need to remove a source determined, before proceeding with groundwater remediation (see Problem 4.7). Additional data on the width of the plume near the unnamed stream would be helpful.

For a pump-and-treat system, estimate the groundwater extraction rate and the number and spacing of extraction wells.

- In Problem 2.6, the flow of groundwater across the downgradient site boundary was estimated to be 0.0431 cfm. Adjust this estimate for revised hydraulic conductivity:

$$0.0431 \frac{ft^3}{min} \cdot 7.48 \frac{gal}{ft^3} \cdot \frac{1.9 \times 10^{-3}}{1 \times 10^{-3}} = 0.61 \frac{gal}{min}$$

  This estimate provides a reality check for estimates of pumping from extraction wells.
- Estimate the radius of influence of a pumping well from Equation 3.1. Assume a 6-in. pumping well (r_w) reaches steady state in 1 d (t). S ranges from 0.1 to 0.3; assume a value of 0.2. Estimate the transmissivity:

$$T = Kb = 1.9 \times 10^{-3} \frac{cm}{sec} \cdot 6\,ft \cdot 2.12 \times 10^4 \frac{gal/day/ft^2}{cm/s} = 242\ gal/day/ft$$

$$R_0 = 0.25 + \sqrt{\frac{242\ gal/ft/day \cdot 1440\ min}{4790 \cdot 0.2}} = 19\ ft$$

- Equation 3.5 models the flow to a well which fully penetrates an unconfined aquifer. For a well with a radius of influence of 19 ft, and assuming a drawdown of 67%,

$$Q = \frac{\pi \cdot 1.9 \times 10^{-3}\ cm/sec \cdot \left((6\,ft)^2 - (2\,ft)^2\right) \cdot \dfrac{1\ in}{2.54\ cm} \cdot \dfrac{1\,ft}{12\ in} \cdot \dfrac{60\ sec}{1\ min} \cdot \dfrac{7.48\ gal}{1\,ft^3}}{\ln(19\,ft/0.25\,ft)}$$

$$= 0.6\ gal/min$$

- Three extraction wells, spaced approximately 30 ft apart and located on a line perpendicular to the direction of groundwater flow and near the leading edge of the plume, would capture approximately 1.8 gpm. The wells should be located far enough from the unnamed stream so that pumping does not dewater the stream. The difference between the estimate of the groundwater flow through the cross-sectional area (0.6 gpm) and the estimate of the total pumping rate (1.8 gpm from three wells) is not unusual given the number of assumptions in the latter calculation. The pumping rate based on Equation 3.5 is an overestimate of the probable pumping rate.
- Wells would be placed throughout the plume to contain the plume and restore the aquifer, ultimately up to eight to ten total. While the calculation of Q based on Equation 3.5 would suggest that ten wells could pump 60 gpm, the estimate of the groundwater flux through the site suggests that the extraction system would likely pump much less water, i.e., less than 5 gpm.

- Extracted groundwater could be treated by several conventional techniques. The choice would depend on cost and on ease of operation. While toluene could be treated in a bioreactor, biological treatment probably would not be used on such a small remediation system due to the capital cost and the level of operation and maintenance required. Most likely, the groundwater would be treated with a small package system comprising an air stripper/carbon.
- The data in Table 3.9 indicate that the total concentration of iron and manganese is 10.9 to 41 mg/L, and the hardness is between 40 and 65 mg/L as $CaCO_3$. The total suspended solids data sugggest that some of the metals may be particulate or colloidal. However, aerobic biodegradation may have increased the levels of iron and manganese in portions of the plume. These data suggest that hardness should not cause a fouling problem in a groundwater treatment system. The concentrations of iron and manganese, however, could cause a fouling problem with activated carbon or a packed-tower air stripper.
- Groundwater could be treated in a tray stripper with air emissions control, filtered, and polished with carbon to meet the treatment limits. The tray stripper would require periodic cleaning to remove iron precipitates.
- Treatment would generate spent activated carbon, used filters, and, occasionally, sludge from cleaning the stripper. The quantities of activated carbon and sludge could be estimated using a mass balance on the system.

Alternatively, the plume could be treated *in situ* by air sparging. Coupled with soil vapor extraction and multiphase extraction (Problem 4.7), an air sparging system could be part of a comprehensive remediation program which would also address the soils in the source area. The design would probably be based on a field test. See Section 3.5.2 for the components and approximate sizing of an air sparging system.

Problem 3.9

Measure the levels of perchloroethylene and its daughter products (including Cl and CO_2/alkalinity). In addition, see Table 5.5.

Table 5.5 Parameters to Monitor During Natural Attenuation Solution to Problem 3.9

Analysis	Range	Interpretation
Redox potential	<50 mV against Ag/AgCl	Reductive pathway possible
Sulfate	<20 mg/L	Competes at higher concentrations with reductive pathway
Nitrate	<1 mg/L	Competes at higher concentrations with reductive pathway
Oxygen	<0.5 mg/L	Tolerated; toxic to reductive pathway at higher concentrations
	>1 mg/L	Vinyl chloride oxidized
Iron (II)	>1 mg/L	Reductive pathway possible
Sulfide	>1 mg/L	Reductive pathway possible
H	>1nM	Reductive pathway possible; vinyl chloride may accumulate
H	<1 nM	Vinyl chloride oxidized
pH	5 <pH <9	Tolerated range

From Wilson, B. H., Wilson, J. T., and Luce, D., in *Proceedings of the Symposium on Natural Attenuation of Chlorinated Organics in Ground Water*, Dallas, TX, September 11–13, 1996, EPA/540/R-96/509, Washington, D.C., September 1996, pp. 21–28.

Problem 3.10

See Table 5.6.

5.4 CHAPTER FOUR

Problem 4.1

[b] Approximately 4500 bcy of soil would be excavated. Volume estimates will vary depending on where the limits of excavation are defined based on interpolation between existing data points.

[c] Time required to excavate the soil:

$$\frac{4500\ bcy}{(70\ bcy/hour)(0.75\ productivity)} = 86\ hours,\ or\ about\ 11\ days$$

Cost to excavate the soil: $(11\ days)(\$1123/day) = \$12,000$

(Note that this cost is simply the cost to dig up the soil; it does not reflect associated work such as health and safety monitoring, decontamination, oversight, post-excavation sampling, stockpiling, etc.)

Problem 4.2

At the level of accuracy typical of a pre-design estimate, the difference between the two estimates is not significant. The estimator reported far too many significant figures.

Problem 4.3

[a] Impoundment A received wastewater containing primarily hexavalent chromium, whereas impoundments D and E contain trivalent chromium hydroxide sludge. Hexavalent chromium is more soluble than trivalent chromium. As a result, the sludge in impoundment A leaches higher levels of chromium than the sludge in impoundments D and E and would require additional treatment before disposal.

[b] Closure could entail (but the options are not limited to):

- pumping water out of the impoundments, potentially to the groundwater treatment plant; and
- either: (i) removal of the sludge and underlying soil contaminated above the clean-up level, for solidification/stabilization and off-site disposal; (ii) similar to (i), except the sludge would be removed for recycling as raw material, if the economics were feasible; (iii) remove the sludge for disposal or resource recovery, and treat the underlying soil by soil washing; or (iv) stabilizing the sludge, backfilling with clean material to create the necessary grade, and capping the area.

Table 5.6 Solution to Problem 3.10

Remediation Option	BTEX	PAH	PCB	Metals	Chlorinated Ethenes	TPH
				Contaminant		
Air stripping	Yes	No, except limited removal naphthalene	No	No, except that Fe(II) can be oxidized and removed upon aeration	Yes	Lighter hydrocarbons
Carbon adsorption	Yes	Yes	Yes	Yes; not regenerable	Yes	Yes
Bioremediation	Yes	Yes, limited for heavier PAH	No	No, except in very rare applications	Yes, under certain conditions	Yes
Precipitation	No	No	No	Yes	No	No
Chemical oxidation	Yes	Yes	Yes	No; may reduce certain metals such as Cr(VI)	Yes	No, except for rare use of Fenton's reagent
Ion exchange	No	No	No	Yes	No	No

[c] Site-specific testing could include:

- tests of water treatment techniques, if a groundwater treatment plant was not in place;
- characterization for waste classification;
- feasibility of resource recovery;
- treatability tests for soil washing with an acidic solution; and/or
- treatability tests for solidification/stabilization.

Problem 4.4

Several alternative designs could potentially solve the problem. Rather than presenting a complete conceptual design for a single solution, the following discussion identifies critical points which should be incorporated into the solution:

- Remediation of mercury-contaminated sediments would destroy wetlands, altering or eliminating wildlife habitat. While wetlands can be restored, the process is as much art as science and can take many years.
- Remediation goals must be determined. While sediments exceed the Effects Range Low, that concentration is only a screening level and does not reflect site-specific conditions such as the speciation and bioavailability of mercury in sediments and the types of organisms. Additional work must be done — potentially based on toxicity testing and further evaluation of the existing data — to develop remediation goals for the site. Ideal remediation goals must be balanced against the damage to the wetlands which would result from extensive remediation.
- Depending on the remediation goals/clean-up level, additional sampling may be necessary to define the extent of contamination. The contaminant distribution shown in Figure 4.10 suggests that mercury was not transported solely through stream flow; airborne contamination, sediment disturbance by flood waters, and/or other sources may have contributed to the extent of the problem.
- Remediation options for mercury in sediment depend on the form of the mercury; the physical characteristics of the sediments; and the hydrology of the wetlands. Options potentially include dredging, dewatering, and landfilling sediments off-site; subaqueous capping; and enhanced sedimentation, to cover contaminated sediments with clean silt. A "hot spot" approach could be warranted: highly contaminated sediments or areas with a high impact could be removed, and remaining sediments covered or no action taken.
- Additional testing (toxicity testing) would be useful in developing a clean-up level or approach. Additional testing would be necessary to design a remediation system: for dredging and disposal of sediments, waste characterization, testing of solidification/stabilization methods if necessary, and potentially tests of dredging and dewatering techniques; for capping or enhanced sedimentation, potentially a test of proposed materials/configuration in a relatively small area to determine effective design parameters.
- Logistics or implementation problems include: wetlands would be destroyed to remediate the area, potentially destroying the habitat of the endangered species; dredging can mobilize sediment particles, so measures would have to be taken to limit the spread of contamination; data were obtained from the top 3 ft of sediments, but the contamination may well be concentrated in the top few inches of sediments; moving heavy equipment through wetlands can be difficult, requiring construction of temporary roads; remediation would require permits for work in wetlands, and, likely, in the flood plain.

Table 5.7   Solution to Problem 4.6
Estimated Mass Removal

Time (days)	Off-gas concentration (mg/m³)	Estimated mass removed during time period (kg)
1	2980	30.37812
5	2375	109.17774
10	2000	111.496875
15	1500	89.1975
20	1375	73.269375
25	1005	60.6543
30	1000	51.097425
40	450	73.9065
50	500	48.4215
60	250	38.2275
70	125	19.11375
80	150	14.01675
90	110	13.2522
100	85	9.93915
110	130	10.95855
120	120	12.7425
	Total:	735.471615

Problem 4.6

Estimate the mass removed by (1) estimating the mass removed during each time period, by multiplying the off-gas concentration by the air flow rate and converting units; (2) plotting removal vs. time and summing the area under the curve. See Table 5.7. The total mass removed is approximately 740 kg.

The total mass that must be removed to meet the remediation goal is approximately:

$$(2900 \ cy)(1.5 \ ton/cy)(2000 \ lb/ton)(1 \ kg/2.2046 \ lb)(214 - 2 \ mg/kg)(1 \ kg/10^6 \ mg)$$

$$= 840 \ kg$$

[a] This calculation suggests that remediation is nearly complete.

[b] Concentrations of contaminants in soil gas sometimes rebound for the same reason that dissolved concentrations in groundwater typically rebound after an extraction system is shut down. If the fluid (gas or water) velocities are too high during operation to allow dissolution/vaporization or desorption to reach equilibrium, the fluid will flow through the pore spaces without removing an appreciable amount of the contaminant. When operation of the system stops, the fluid in the pore spaces can reach equilibrium with NAPL residual and/or sorbed contaminants, and the vapor concentration will increase or rebound.

Problem 4.8

RIU's proposal has several critical flaws. The primary problem is that the proposed treatment system will volatilize, not bioremediate, the soil.

Problem 4.9

See Table 5.8.

Problem 4.10

See Table 5.9.

Table 5.8   Solution to Problem 4.9

Part	Technology	(I) Equipment type	(ii) Treatment objective	(iii) Typical operating conditions
[a]	Thermal desorption	Rotary kiln or other equipment designed to mix and heat soils; materials of construction and emissions-control equipment vary with type of treatment, waste	Volatilize contaminants	Temperature 300 to 1200°F; retention time 20 min or less
	Incineration		Oxidize contaminants	Temperature 1200 to 1800°F, retention time from several minutes to an hour
	Vitrification		Contain contaminants within glassy matrix	Temperature 2200 to 2600°F, retention time 60 to 100 min
[b]	Air sparging	System comprising series of injection wells, blower or compressor, used to inject air into the subsurface	Strip volatile contaminants from groundwater	Air flow rate (typ. 3 to 25 scfm) designed to enhance volatilization; often coupled with soil vapor extraction system to capture air emissions
	Bioventing		Provide oxygen for aerobic biodegradation	Air flow rate at the minimum necessary to supply oxygen for bioremediation based on field test; ideally air flow at a low enough rate to minimize volatilization and eliminate need for emissions control
[c]	Six-phase soil heating	Series of electrodes inserted into the ground to convey electric current	Heat soil in order to volatilize contaminants	Moderate temperature increase (e.g., to 212°F); water volatilizes; cost limits application in saturated soil
	Electrokinetics		Induce flow of groundwater and/or enhance mobility of contaminants	Relatively low temperature increase (e.g., to 60°F or less in the PGDP tests); voltage gradient on the order of 1 V/cm induces groundwater flow
	In situ vitrification		Contain contaminants within glassy matrix	Severe temperature increase (2900 to 3600°F); water volatilizes; cost limits application in saturated soil

Table 5.9 Solution to Problem 4.10

Part	Technology	Could the technology be used?	Why or why not?	Site-specific concern regarding application
[a]	Thermal desorption	Yes	VOC and PCB volatilize at the treatment temperature; has been used to treat similar contaminants	Particle size limits on equipment; extent of excavation near/beneath building
[b]	Solvent extraction	Yes	Has been used to treat similar contaminants	Particle size limits on equipment; extent of excavation near/beneath building
[c]	Asphalt batching	No	Asphalt batching is appropriate for petroleum hydrocarbons, not the chlorinated contaminants at the ASR site	
[d]	Multilayer cap	Yes	Could prevent contact with and minimize leaching from the soils in the vadose zone	Change in grade would interfere with use of site, existing building
[e]	In situ vitrification	Yes	Volatile compounds would desorb; less volatile PCBs would be immobilized within matrix	Potential effect on building; potential production of dioxin; limited availability of technology; cost

REFERENCES

1. U.S. EPA, *Region III Risk-Based Concentration Table*, http://www.epa.gov/reg3hwmd/riskmenu.htm?=Risk+Guidance, as of June 1996.
2. U.S. EPA, *Treatability Database, Version 5.0,* Risk Reduction Engineering Laboratory, Cincinnati, OH, April 1994.
3. Wilson, B. H., Wilson, J. T., and Luce, D., Design and Interpretation of Microcosm Studies for Chlorinated Compounds, in *Proceedings of the Symposium on Natural Attenuation of Chlorinated Organics in Ground Water,* Dallas TX, September 11–13, 1996, EPA/540/R-96/509, September 1996, pp. 21–28 (Table 1).

Index